Membrane Practices For Water Treatment

Membrane Practices For Water Treatment

English Language
First Edition

Steven J. Duranceau, Technical Editor

American Water Works Association
Dedicated to Safe Drinking Water

Acquisitions Editor: Colin Murcray
Copy Editor: Lori Kranz
Production Editor and Cover Design: Carol Magin
Production Services: Claro Systems

Library of Congress Cataloging-in-Publication Data has been applied for.

Disclaimer

Much of the material in this book comes from papers submitted to AWWA conferences. Information regarding the employment of the various authors in this book is meant to reflect the status of the authors at the time their paper was submitted. These papers were selected based on their relevance and value to those working with, or considering membrane technology. Because these papers come from various parts of the world, some use inch/pound units, and some use metric. In addition, though these papers were selected due to there valuable content, many deal with case studies, and are meant to inform readers about the experiences of the authors. Therefore, the authors, editors and publisher cannot assume responsibility for the validity of the content, or any consequences of their use. In no event will AWWA be liable for direct, indirect, special, incidental, or consequential damages arising out of the use of information presented in this book. In particular AWWA will not be responsible for any costs, including, but not limited to, those incurred as a result of lost revenue. In no event shall AWWA's liability exceed the amount paid for the purchase of this book. The views expressed in this publication represents the opinions of the authors, and do not constitute an endorsement by AWWA.

American Water Works Association
6666 West Quincy Avenue
Denver CO 80235

ISBN: 1-58321-147-0

Table of Contents

Part 2 Fouling

Part 3 Membrane Monitoring

Part 5 RO and NF Applications and Operations

Part 6 Integrated Membrane Systems

Preface

Many communities today face a variety of environmental infrastructure challenges, and often find it difficult to comply with the multitude of environmental regulations. This most certainly holds true for the Safe Drinking Water Act (SDWA) as many water systems face multicontaminant compliance problems. Many water purveyors have to often deal with a mixture of contaminants while attempting to maintain control of lead and copper levels at the consumer tap. Membrane processes allow drinking water communities a distinct advantage over other available processes for multicontaminant removal requirements. Membrane processes have been shown to be reliable, cost-competitive, and easily operated and maintained; they can be remotely monitored and produce a superior finished water quality. The finished water quality does not vary significantly from day to day with fluctuations in raw source water quality, unlike conventional processes that are dynamic and tend to be operator-dependent and that can be affected by raw source water quality.

The utilization of membrane processes for the production of potable water has become a competitive alternative to conventional treatment processes. Trends indicate that membrane filtration and ultrafiltration will play a more dominant role in the US surface water treatment market in future years. The interest and use in membrane technology have been spurred by increasingly stringent water quality regulations, diminishing fresh water supplies in the US, and advances in membrane materials and construction. Continual use of membrane processes for water processing will increase because of improved membrane performance and lower costs due to technological advances. The development of membrane processes for widespread drinking water treatment application is limited by source water fouling and because of the fact that the disposal of concentrates from these processes can pose regulatory burdens for water purveyors relative to compliance with the Clean Water Act and NPDES requirements.

Historically, the primary use of membrane technology in the United States was limited to the removal of dissolved solids from brackish water supplies. More recently, membrane applications have been implemented for the removal of disinfection by-product (DBP) precursors, pathogens (disinfection), particulate matter, hardness, and turbidity removal (clarification). Membrane processes have also become useful in controlling arsenic, nitrate, sulfate, radionuclides, and synthetic organic chemicals in drinking water supplies. In addition, membrane processes are now being used for the treatment of many different types of wastewater processes, primarily for reclamation and reuse of those waters.

This book is a compilation of the more significant published work relative to membrane processes. Most of the articles presented in this document were selected from several of the American Water Works Association's (AWWA's) Membrane Technology Conference Proceedings. Other articles were taken from publications provided at the Water Quality Technology Conference and recent Annual Conferences. This book was edited by Dr. Steven J. Duranceau, P.E., who serves as director of Water Quality and Treatment at Boyle Engineering Corporation in Orlando, Florida. Dr. Duranceau has been intimately involved in researching, piloting, designing, and operating membrane processes for over 15 years. With the high level of interest in the subject of membrane processes, the chapters in this book are expected to serve as excellent sources of reference material on this dynamic technology.

Dr. Steven J. Duranceau, P.E.
Director of Water Quality and Treatment
Boyle Engineering Corporation
320 East South Street
Orlando, FL 32801 USA

Part 1
Cost

The Cost of Membrane Filtration for Municipal Water Supplies

Cost Model for Low-Pressure Membrane Processes

Comparative Life-Cycle Costs for Operation of Full-Scale Conventional Pretreatment/RO and MF/RO Systems

CHAPTER · 1

The Cost of Membrane Filtration for Municipal Water Supplies

Joseph R. Elarde, Process Engineer
CH2M HILL, Gainesville, FL

Robert A. Bergman, Principal Technologist
CH2M HILL, Gainesville, FL

INTRODUCTION

Membrane filtration processes, microfiltration and ultrafiltration, are now mainstream filtration processes competing with conventional technologies. Water providers are selecting membrane because of their lower-cost trends and superior treatment performance of a wide variety of source waters. Many new large-capacity municipal projects are now under construction in North America, and even more are being planned. Water utility decision makers require meaningful and accurate construction and operation and maintenance (O&M) cost data to determine the feasibility of membrane technologies.

DRIVING FACTORS FOR MEMBRANE FILTRATION

More stringent disinfection and disinfection by-product (DBP) standards are making membrane filtration a more attractive treatment option. Many state regulatory agencies are giving greater pathogen removal credit for membrane treatment than for conventional treatment, which reduces finished water disinfection requirements. Membrane treatment minimizes the required chlorine dose necessary to meet additional inactivation and residual disinfection requirements.

Lower chlorine doses and contact times as a result of pathogen removal credit by membrane filtration will also reduce the production of DBPs during finished water disinfection. The more stringent Stage 2 DBP Rules, which reduce maximum contaminant level (MCL) of total trihalomethanes from 80 µg/L to 40 µg/L, and total haloacetic acids from 60 µg/L to 30 µg/L, are easier to meet when using membrane filtration.

The finished water quality from a membrane filtration process is relatively constant and independent of feedwater quality, and is not susceptible to upset like conventional treatment. For treating raw water supplies that may experience significant changes in quality, or for applications where a consistent finished quality is required (such as reverse osmosis pretreatment), membrane filtration is the preferred treatment process.

Many utilities require a water treatment plant with a small footprint that can blend in with a surrounding community. Membrane filtration typically requires less chemical addition than conventional processes, which results in smaller bulk chemical storage tanks and feed facilities. Membrane filtration does not require the large structures associated with clarifiers and media filters.

Membrane filtration plants are easier to operate and monitor and require less supervision than conventional plants. Some membrane filtration facilities can be unmanned and monitored from a remote location.

MEMBRANE FILTRATION PROCESS

Membrane filtration processes can be used with minimal pretreatment on good-quality source waters. Figure 1 presents a typical process flow diagram using strainers to protect the membranes.

Membrane filtration is also becoming more popular in applications where high turbidity removal is necessary. An in-line filtration setup can be used where coagulant is added ahead of the membranes to form small-size suspended and colloidal solids that

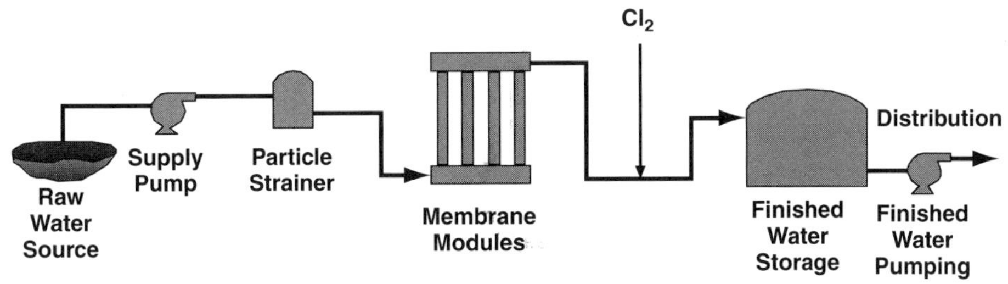

Figure 1 Typical membrane filtration process flow diagram

are then removed by the membrane. These floc particles may also help to reduce fouling and plugging of some types of membranes. Coagulant dosing is optimized during pilot testing to maximize time between membrane backwashing and cleaning.

When organic removal is necessary, pre-treatment by coagulation, flocculation, and sedimentation may be required. In some cases, membranes can be immersed within the filtration tank. Figure 2 shows an immersed membrane configuration.

SURVEYED MEMBRANE FILTRATION PLANTS

Several membrane filtration facilities constructed over the past five years in the United States and Canada that have initial capacities greater than 0.3 mgd are included in this cost study. The included plants treat both raw surface and groundwater supplies with varying water quality characteristics. In all of the plants, membrane filtration is the primary treatment process used for turbidity and pathogen removal. Most of the plants typically use direct filtration and have design flux of approximately 45 to 55 gallons per day per square foot of membrane surface area (gpd/sf). Figure 3 shows the names and locations of the included plants.

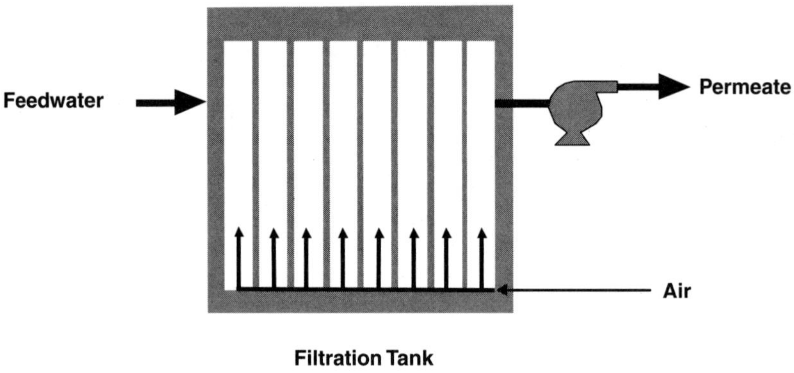

Figure 2 Immersed membrane process configuration

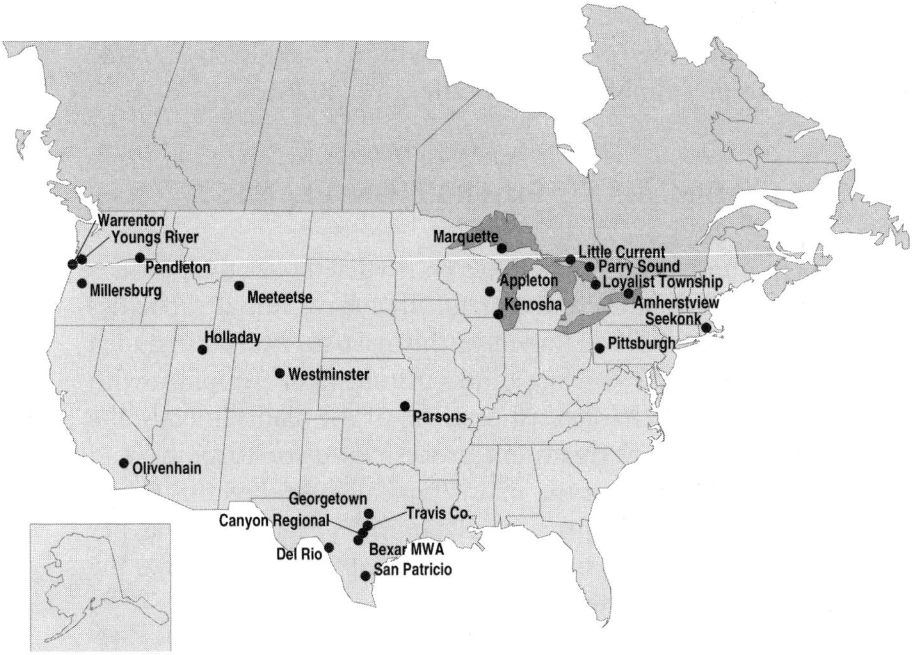

Figure 3 Map of membrane filtration plant locations

INSTALLED MEMBRANE EQUIPMENT COST

Membrane equipment includes some or all of the following components depending on the membrane manufacturer: membrane modules, clean-in-place (CIP) equipment, feed or permeate pumps, backwash pumps, blowers, instrumentation and control equipment, and tanks associated with cleaning and backwashing. While the cost of the membrane modules is proportional to the design capacity and flux of the system, the cost of the other components has a significant economy of scale. One cleaning system and one air system can be used for several process trains. One feed pump can be used for trains of varying size. Therefore, as the train size and system capacity increase, the relative unit cost of these components will decrease. As overall system capacity increases, the actual membrane module cost will be a greater percentage of the overall cost.

The total membrane module cost is primarily affected by flux, or the amount of membrane area needed to produce a given amount of water each day. Individual membrane module cost has little or no economy of scale; therefore additional membrane modules needed to reduce flux will increase membrane cost proportionally.

The length of a membrane replacement warranty will have a significant effect on membrane module cost. Using these warranties, manufacturers guarantee the useful life for the membrane and/or prorate the cost of replacement. Current membrane warranties are typically five years.

Figure 4 shows the unit cost of installed membrane equipment as a function of total water treatment plant capacity. Above 2 mgd, only a small economy of scale is realized as the membrane module cost becomes the most significant portion of the total membrane equipment cost.

The membrane equipment cost curve shown in Figure 4 is based on a typical design flux of 50 gpd/sf. To apply this curve to other plants, the membrane equipment cost must be multiplied by the ratio of the actual flux to the 50 gpd/sf used to develop this curve.

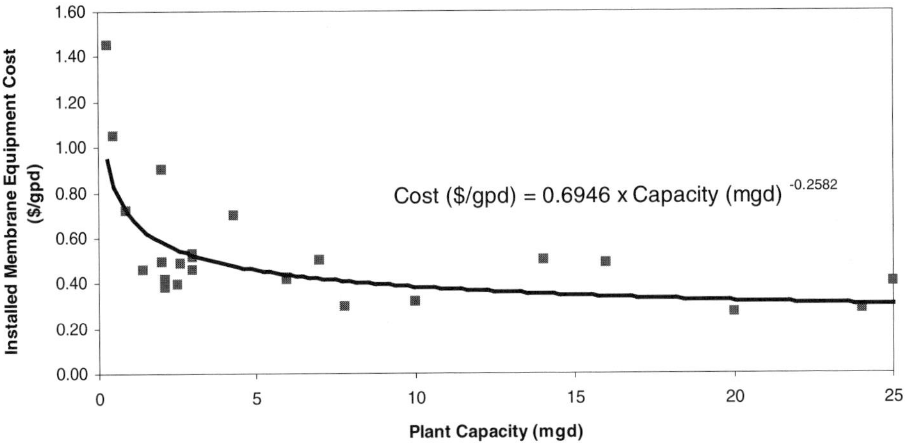

Figure 4 Unit installed equipment cost of membrane filtration plants

The design flux is dependent on feedwater quality and may be determined during pilot testing.

MEMBRANE FILTRATION PLANT CONSTRUCTION COST

Membrane treatment plant construction cost is dependent on the membrane equipment cost, configuration and pretreatment necessary, site constraints, and the building requirements of the facility. The plants surveyed have a range of pretreatment requirements depending on the source water quality. Some of the surveyed membrane plants required more costly site preparation because of existing facilities, while others required special building construction to meet local area needs. These factors have a large impact on the overall unit construction cost of the membrane filtration plants.

Figure 5 shows the total membrane water treatment plant (WTP) construction cost in dollars per gallon of installed membrane capacity ($/gpd). The plant construction data shown in Figure 5 have significantly more scatter than the equipment cost

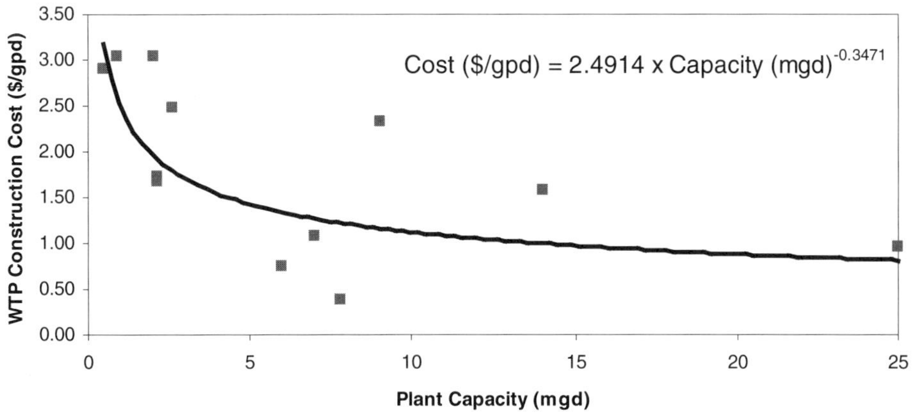

Figure 5 Unit WTP construction cost of membrane filtration plants

data shown in Figure 4. This is due to the wide variety of source water quality and therefore pretreatment processes used in each of the plants. An economy of scale exists for the construction cost of the plants as they increase in capacity to about 5 mgd. Above 5 mgd, the membrane equipment cost becomes greater than 50 percent of the total construction cost of the membrane filtration plant. As plant capacity increases, the economy of scale becomes less significant (see Figure 6).

Figure 6 shows the membrane equipment cost percentage of the total membrane plant construction cost. The membrane equipment cost of small plants below 3 mgd is 20 to 30 percent of the total plant construction cost. As plant size increases, the percentage of the total plant cost from membrane equipment increases to nearly 50 percent of the total construction cost. As mentioned above, as plant size increases, the membrane module cost becomes the most significant portion of the membrane equipment cost. Therefore, the membrane module cost, which has little economy of scale, becomes the driving cost factor for larger plants.

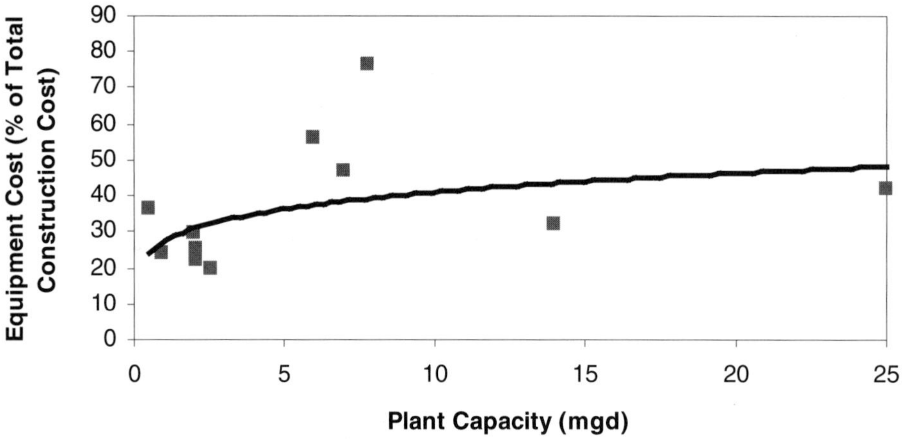

Figure 6 Percent of total WTP construction cost for membrane equipment

Table 1 shows the plant locations and cost data used to create the membrane cost curves shown above.

Table 1 Membrane filtration cost data

Location	Capacity (*mgd*)	WTP Const. Cost (*$*)	MF/UF Equip. Cost (*$*)	Const. Cost (*$/gpd*)	Equip. Cost (*$/gpd*)	Equip. Cost (*%*)
Meeteetse, WY	0.302	–	438,000	–	1.45	–
Youngs River, WA	0.5	1,450,000	525,000	2.90	1.05	36
Little Current, Ont	0.9	2,730,000	650,000	3.03	0.72	24
Gibson Canyon, CA	1.4	–	640,000	–	0.46	–
Travis County, TX	2	–	980,000	–	0.49	–
Millersburg, OR	2	6,063,000	1,800,000	3.03	0.90	30
Amherstview, Ont	2.1	3,500,000	880,000	1.67	0.42	25
Loyalist Township, Ont	2.1	3,640,000	800,000	1.73	0.38	22
Holladay, UT	2.5	–	980,500	–	0.39	–
Parry Sound, Ont	2.6	6,430,000	1,260,000	2.47	0.48	20

Table 1 Membrane filtration cost data *(continued)*

Location	Capacity (*mgd*)	WTP Const. Cost (*$*)	MF/UF Equip. Cost (*$*)	Const. Cost (*$/gpd*)	Equip. Cost (*$/gpd*)	Equip. Cost (*%*)
Canyon Regional, TX	3	–	1,540,000	–	0.51	–
Georgetown, TX	3	–	1,380,000	–	0.46	–
Parsons, KS	3	–	1,590,000	–	0.53	–
Seekonk, MA	4.3	–	3,000,000	–	0.70	–
Pendleton, OR	6	–	2,500,000	–	0.42	–
Warrenton, OR	6	4,520,000	2,525,000	0.75	0.42	56
Marquette, MI	7	7,500,000	3,500,000	1.07	0.50	47
San Patricio, TX	7.8	3,000,000	2,300,000	0.38	0.29	77
Bexar Met. Water Auth., TX	9	21,000,000	–	2.33	–	–
Canyon Regional, TX	10	–	3,170,000	–	0.32	–
Kenosha, WI	14	22,000,000	7,000,000	1.57	0.50	32
Del Rio, TX	16	–	7,777,778	–	0.49	–
Pittsburgh, PA	20	–	5,400,000	–	0.27	–
Appleton, WI	24	–	6,800,000	–	0.28	–
Olivenhain, CA	25	24,000,000	10,000,000	0.96	0.40	42

MEMBRANE FILTRATION O&M COST

Major variable membrane filtration operations and maintenance (O&M) costs are membrane replacement, chemicals, and power. As membrane capacity increases, these costs will increase proportionally. Labor and general maintenance are fixed O&M costs, which will remain relatively constant independent of membrane plant size.

Figure 7 shows estimated O&M cost per 1,000 gallons of finished water as a function of plant capacity. Projected O&M cost is estimated based on a membrane plant with only basket strainers for pretreatment, raw water pumping, membrane feed (or permeate) pumping, finished water pumping, finished water disinfection, general maintenance, and staffing for a continuously operated plant (five full-time operators). As plant size increases, the fixed costs

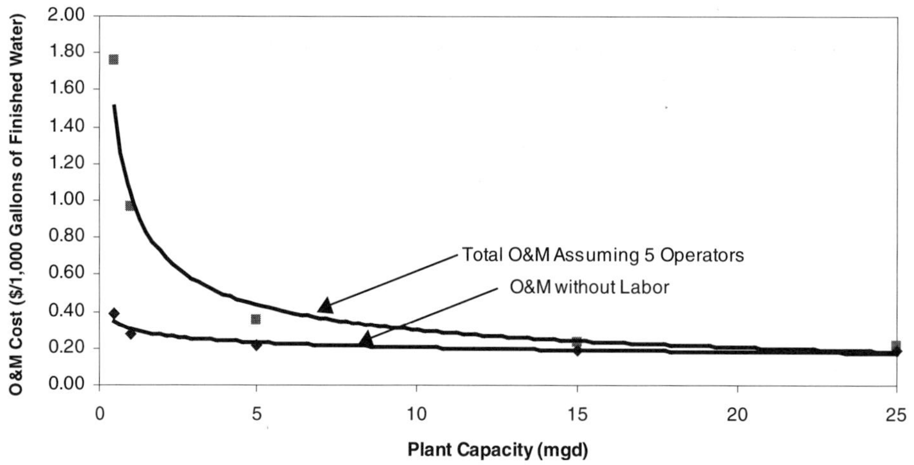

Figure 7 Unit membrane filtration operations and maintenance cost

become less significant and variable costs become a larger percentage of the total O&M cost. The O&M cost for large plants approaches the manufacturer estimates of $0.10 to $0.15 per 1,000 gallons for the membrane filtration variable cost.

TOTAL MEMBRANE TREATMENT UNIT COST

The total WTP construction cost curve estimated was used to develop construction cost for plants of different sizes between 0.5 mgd and 25 mgd. The construction costs were then amortized over 20 years using an interest rate of 8 percent. The total membrane filtration treatment unit cost curve in Figure 8 was then developed by adding the estimated O&M annual cost (without labor cost) to the amortized capital cost to calculate the total treatment unit cost per 1,000 gallons of finished water produced.

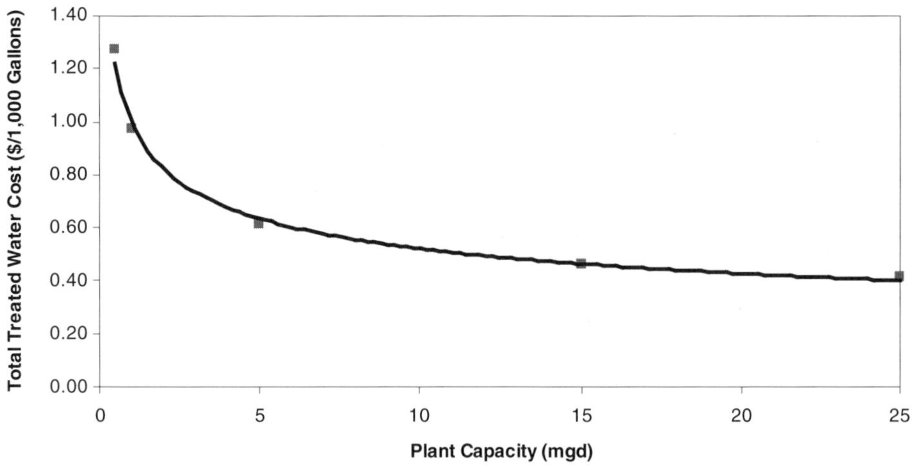

Figure 8 Unit membrane filtration total water treatment cost

CONCLUSIONS

- Site-specific water quality and other construction factors must be considered when estimating plant construction cost.
- Membrane equipment cost can be used to estimate total WTP construction cost for plants above about 5 mgd.
- Membrane filtration cost, especially for larger plants (above 5 mgd), is significantly influenced by total membrane cost. There has been a general decrease in the membrane module cost in the past few years, which has resulted in membrane filtration being more competitive with conventional processes.
- Design flux directly affects membrane equipment cost and, therefore, WTP construction cost. The design flux for a membrane filtration plant is dependent upon the source water quality, finished water quality goals, and how a specific membrane type performs under the design conditions.

ACKNOWLEDGEMENTS

The author thanks the following individuals for their assistance: Martin Gravel, Paul Berg, Tom Engleson, Jamal Awad, James Lozier of CH2M HILL, and Rick Moro from Pall Corp.

REFERENCES

1. Smith, M.C. (1998–2000). *Water Desalination Report.* Vols. 34–36. Tracey's Landing, Md.

CHAPTER · 2

Cost Model for Low-Pressure Membrane Processes

Sandeep Sethi, Project Engineer
Metcalf & Eddy, Inc., Atlanta, Georgia

Mark R. Wiesner, Professor and Associate Dean
Environmental Science and Engineering, Rice University, Houston, Texas

INTRODUCTION

Ultrafiltration (UF) and microfiltration (MF) are finding increased use in water and wastewater treatment. Cost models and analyses are required to assess the expenses incurred in utilizing UF/MF, to compare with alternate treatment technologies, and for process optimization. Cost modeling and analyses can be used to narrow down the scope of pilot studies and orient them to the most promising scenarios.

The primary objective of this work is to present a new model for estimating the capital costs of membrane facilities by establishing separate correlations for major cost components. This approach accounts for the individual economies of scale associated with different equipment and facilities and thereby considers an overall economy of scale for the entire membrane system that is a function of the design mix. Previous approaches for modeling the capital costs of membrane processes typically present a single correlation for total capital costs that does not provide insight into the different economies of scale associated with the different components involved. The complete cost model is obtained by coupling the new capital cost model with equations for operating cost developed in a previous cost model [1]. A second objective is to utilize the newly formulated model to predict and analyze treatment costs for various raw waters and plant capacities.

BACKGROUND

The economics of membrane filtration are determined by the initial investment made towards the membrane modules along with the associated ancillary equipment and facilities, and the various operating costs incurred in running the process. A key parameter describing membrane process performance is the permeate flux, or the clean water permeation rate per unit membrane area. Being a direct measure of productivity, permeate flux can be expected to significantly affect treatment costs. In addition to all the variables that determine permeate flux, treatment costs are affected by module geometry, frequency of the hydrodynamic cleaning, membrane plant characteristics, and other economic factors such as interest rates. Thus, membrane treatment costs are typically a complex function of a large number of variables.

Based on a literature review of published data, capital costs for current membrane systems appear to demonstrate economies of scale characterized by exponents between 0.4 and 0.8, with the lower value corresponding to smaller plants and the upper value corresponding to larger plants.

OPERATING COSTS

In this work, we adopt a previous model [1] for the purposes of estimating the operating costs, which are briefly reviewed here. The operating costs arise from energy requirements, membrane replacement, chemicals, and concentrate disposal. Energy costs are calculated for the feed, recycle, backflush, and fastflush pumping requirements. Membrane replacement costs are amortized over the life of the membrane. Chemical costs were not considered in the simulations performed in this work. Concentrate disposal costs are approximated by accounting for the cost of energy invested in the water.

DEVELOPMENT OF MODEL FOR CAPITAL COSTS

Membrane system costs typically comprise the costs of pumps, pipes and values, instrumentation and controls, tanks and frames, membrane modules, and other miscellaneous equipment and facilities. The capital cost of membranes was calculated as the product of the membrane cost per unit area and the membrane area required to meet the design capacity:

$$C_{mem} = C^{*}_{mem} A_{mem} \tag{1}$$

The membrane area for producing a given design flow of Q_{des} is calculated from the ratio of the required design capacity to the net permeate flux:

$$A_{mem} = \frac{Q_{des} t_{tot}}{\bar{J} t_o - J_{bf} t_{bf}} \tag{2}$$

where $\bar{J}$ = permeate flux, J_{bf} = backflush flux, t_{bf} = backflush duration. The total time for one complete operating and flux enhancement cycle is calculated as $t_{tot} = t_o + t_{bf}$ where t_o is the operating time between two flux enhancement cycles. This expression assumes that fastflushing occurs during backflushing, with a duration less than or equal to the backflush duration.

Power law correlations were established for the major categories of non-membrane equipment and facilities. Membrane area was selected as the design parameter in the correlations for all the non-membrane equipment and facilities except pumps. Pump costs are calculated as a function of flow and pressure. Data from various sources [2–4] were used to establish the factors representing the economies of scale in the power law correlations for the costs of manufactured equipment. The correlations established are:

Pipes and valves

$$C_{PV} = 5926.13\ (A_{mem})^{0.42} \tag{3}$$

Instruments and controls

$$C_{IC} = 1445.50\ (A_{mem})^{0.66} \tag{4}$$

Tanks and frames

$$C_{TF} = 3047.21\ (A_{mem})^{0.53} \tag{5}$$

Miscellaneous

$$C_{MI} = 7865.02\ (A_{mem})^{0.57} \tag{6}$$

Pumps

$$C_{\text{pump}} = I^{*} f_1{}^{*} f_2{}^{*} L^{*} 81.27 (Q^{*} P)^{0.39} \tag{7}$$

The "miscellaneous" category includes process equipment building, electrical supply and distribution, disinfection facilities, treated water storage and pumping, and wash water recovery system. In Equation (7), I = a cost index ratio used to update the cost to the recent year (base year was reported to be 1979, first quarter), f_1 = a factor to adjust for pump construction material, f_2 = a factor to adjust for suction pressure range, and L = a factor used to incorporate labor costs. Based on a January 1979 pumps and compressors index of 269.9 and a March 1996 index of 613.5 obtained from *Chemical Engineering*, a value of I = 2.28 (= 613.5/269.9) is applied to update the pump costs to March 1996, the base year selected for this work. Labor costs of 40% are typically reported for installation of manufactured equipment [5], implying that L =1.4. For simulations performed in this work, type 316 stainless steel surfaces were assumed, implying that f_1 = 1.5 [3]. For suction pressures up to 150 psi (1034.5 kPa), f_2 = 1.0 [3], which was the value assumed in this work.

The exponents in Equations (3) through (7) are seen to vary significantly for different equipment, implying different inherent economies of scale for different cost components, and a changing overall economy of scale for the entire membrane system, as the system design incorporates variable mixes of the components. The leading constants in these correlations, which represent the relative weights of the cost components, were estimated by calibrating [6] the model using recent data on membrane system costs [7] and typical percent contributions of the different equipment to the total cost [8]. The calibrated model was seen to be in good agreement with the correlation reported in [7] over most of the range of the plant design capacity considered [6].

COST SIMULATIONS

Simulations were performed to estimate cost behavior of a typical UF system for water treatment. Based on estimates provided by membrane manufacturers on the costs of UF hollow-fiber membranes, a value of C^*_{mem} = \$100/m^2 (1996 prices) was utilized in this work. Clearly, this value is a function of the type of membrane material, geometry, and process. In addition, this value for any given membrane is likely to be strongly influenced in the near future by the rapid growth of the membrane market.

The membrane systems modeled are assumed to be operated in the mode of constant trans-membrane pressure. A previously developed mathematical model [9] was utilized to estimate the permeate flux. The permeate flux model considers mass transport due to Brownian diffusion, shear-induced diffusion, inertial lift, and concentrated flowing layers. The model is ideally applicable to monodisperse suspensions over a wide range of particle sizes, and accommodates polydisperse suspensions by utilizing an "average" particle size. Details of the permeate flux model are presented elsewhere [9]. A key prediction of the permeate flux model is that particles with diameters on the order of 10^{-1} μm should exhibit unfavorable accumulation on the membrane surface due to a minimum in net back-transport away from the membrane.

The effect on cost of a UF membrane system was investigated by simulating the behavior of the permeate flux and the associated cost parameters for various raw water qualities and plant capacities. Raw water quality was represented by an "average" particle size and concentration in the feedwater. The UF system was assumed to be operated in the feed and bleed mode, in which a waste stream is generated continuously. In this mode the system operates in a pseudo-steady-state condition. Thus, this mode was investigated assuming conditions of steady-state permeate flux. Behavior of treatment costs is presented over the domain of raw water quality represented by particle size and concentration in the feedwater. The baseline design configuration assumed for a 158 m^3/hr (1 MGD) low-pressure membrane plant simulated in this work is given in Table 1.

Table 1 Values of system parameters used in the cost estimates

Parameter	Value
Membrane radius (mm)	0.5
Membrane length (cm)	105
Transmembrane pressure (kPa)	100
Crossflow velocity (cm/s)	60
System recovery (%)	90
Plant design capacity (m^3/hr)	157.73
Plant design life (years)	20
Membrane MWCO (Daltons)	100,000
Membrane cost ($/$m^2$)	100
Membrane resistance (1/cm)	1.46×10^9
Membrane life (years)	5
Backflush duration (s)	45
Backflush flux (L/m^2/hr)	384
Backflush pressure (kPa)	200
Fastflush duration (s)	15
Fastflush velocity (cm/s)	113
Energy cost ($/kwh)	0.07
Cost of capital (%)	10
Earning interest rate (%)	8
Efficiency of pumps (%)	70

Behavior of Treatment Costs with Raw Water Quality

Treatment costs for the baseline design configuration are presented as a function of particle size and concentration in Figure 1 for a 158 m^3/hr (1 MGD) plant. Treatment costs are significantly affected, and are essentially a mirror image of the permeate flux behavior (Figure 2) with the raw water quality. Very small and large particles that are characterized by high net back-transport and/or relatively porous cakes demonstrate high permeate fluxes, which translate into low treatment costs. Particles in the intermediate size range for which the back-transport away from the membrane surface is unfavorable [9] yield low permeate fluxes and are hence characterized by high total treatment costs.

Sensitivity Analyses

Sensitivity analyses were performed to study the effects of plant capacity on treatment costs. All simulations were based on a feed suspension concentration of 200 mg/L, and are presented here for

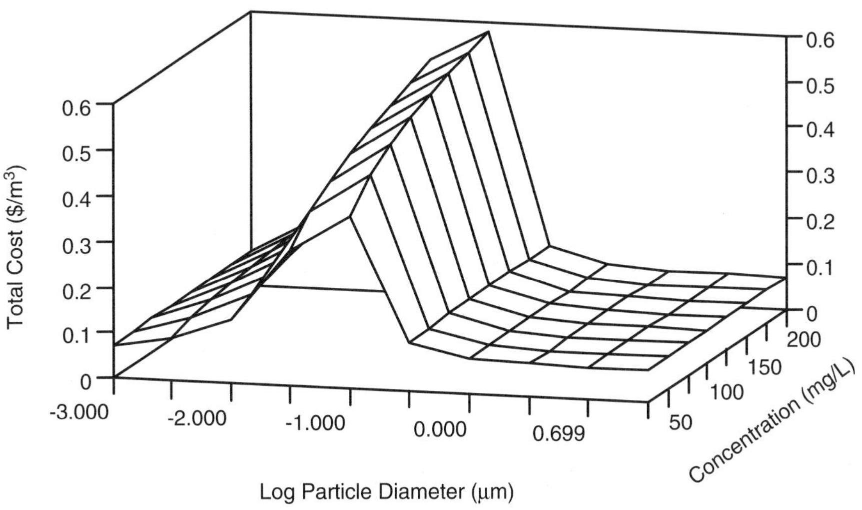

Figure 1 Behavior of treatment costs with particle size and concentration. Plant capacity =157.73 m^3/hr (1-mgd)

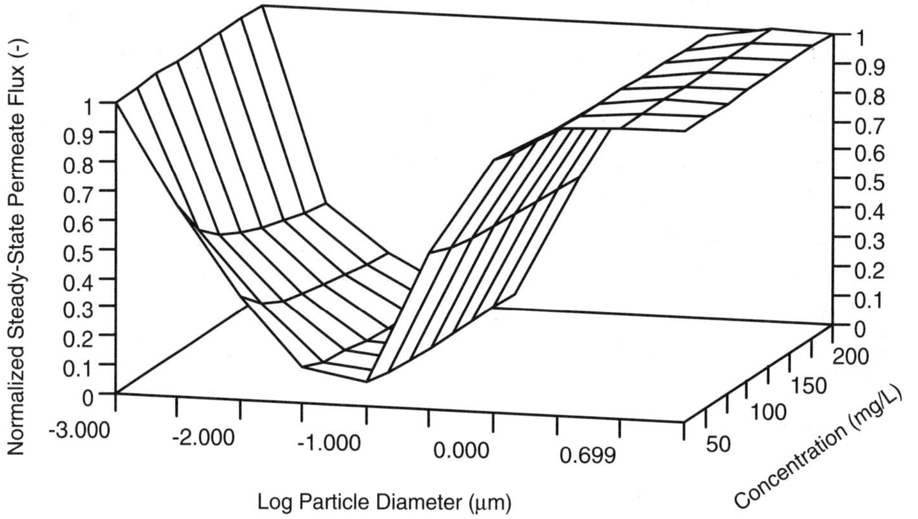

Figure 2 Behavior of steady-state permeate flux with particle size and concentration

three representative particle sizes (small particle: 0.01 μm; intermediate/unfavorable sized particle: 0.1 μm; and large particle: 1.0 μm). Total treatment costs per unit volume of permeate produced are plotted for a range of plant capacities from 1.5773 m^3/hr (0.01 MGD) to 15,773 m^3/hr (100 MGD) in Figure 3. Treatment costs for all the three particle sizes demonstrate a significant decrease with plant capacity due to the economies of scale exhibited by the non-membrane capital costs. At higher capacities the incremental cost of building the plant decreases rapidly. The costs of membranes and energy invested per unit volume of permeate produced remain constant with capacity. Costs are highest at all capacities for the intermediate-sized particles, which are associated with low permeate fluxes.

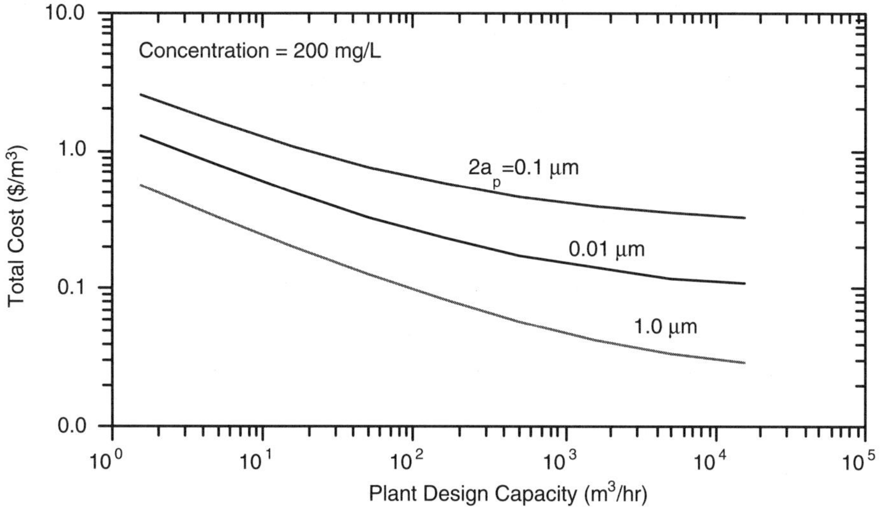

Figure 3 Behavior of treatment costs with plant capacity

CONCLUSIONS

A new model for estimating the capital costs of membrane plants has been developed that incorporates separate cost correlations for several different categories of non-membrane equipment and facilities. This approach thus considers the individual economies of scale associated with different equipment and facilities. It is deemed more suitable for estimating and comparing costs of membrane plants with different design and operating conditions, and for estimating costs of larger facilities for which cost and design data are currently unavailable. The calibrated model compares well with the correlation reported in [7] based on an international survey of UF/MF membrane plants.

Estimates of treatment costs as a function of particle size and concentration, for common ranges of system parameters characterizing typical UF facilities, indicate that costs are significantly affected by the permeate flux. In fact, total costs are essentially a mirror image of the permeate flux behavior with the raw water quality. Thus, membrane fouling is predicted to translate into high capital as well as operating costs by affecting the membrane area and energy requirements. Treatment costs demonstrate a significant decrease with plant capacity. This is attributable to the economies of scale exhibited by the non-membrane capital costs.

ACKNOWLEDGEMENTS

This work was performed at Rice University as part of Sandeep Sethi's doctoral dissertation, and was partially funded with support from the United States Environmental Protection Agency.

REFERENCES

1. Pickering, K.D., M.R. Wiesner (1993). Cost Model for Low-Pressure Membrane Filtration. *J. Envir. Engrg.*, ASCE, 119(5), 772–797.
2. Gumerman, R.C., B.E. Burris, S.P. Hansen (1985). *Estimation of Small System Water Treatment Costs.* US Environmental Protection Agency, Cincinnati, Ohio.
3. Perry, R.H., C.H. Chilton (1991). *Chemical Engineers' Handbook.* McGraw-Hill, New York.
4. Peters, M.S., K.D. Timmerhaus (1991). *Plant Design and Economics for Chemical Engineers.* McGraw-Hill, New York.
5. Holland, F.A., F.A. Watson, J.K. Wilkinson (1984) Process Economics. *Chemical Engineers' Handbook* (ed. R.H. Perry, D. Green) McGraw Hill, New York.
6. Sethi, S. (1997). *Transient Permeate Flux Analysis, Cost Estimation, and Design Optimization in Crossflow Membrane Filtration.* Ph.D. Dissertation, Rice University, Houston.
7. Adham, S.S., J.G. Jacangelo, J-M. Laîné (1996). Characteristics and Costs of MF and UF Plants. *J. AWWA*, 88(5), 22–31.
8. Eykamp, W. (1991). Microfiltration. *Membrane Separation Systems: Recent Developments and Future Directions* (ed. R.W. Baker) Noyes Data Corporation, Park Ridge, N.J.
9. Sethi S., M.R. Wiesner (1997). Modeling of Transient Permeate Flux in Cross-flow Membrane Filtration Incorporating Multiple Particle Transport Mechanisms. *J. Mem. Sci.*, 136, 191–205.

CHAPTER · 3

Comparative Life-Cycle Costs for Operation of Full-Scale Conventional Pretreatment/RO and MF/RO Systems

Wyatt Won
West Basin Municipal Water District, El Segundo, CA

Pat Shields
West Basin Municipal Water District, El Segundo, CA

INTRODUCTION

Microfiltration is a viable, cost-effective alternative to conventional lime clarification/filtration pretreatment for reverse osmosis. This chapter will compare full-scale conventional pretreatment to full-scale microfiltration pretreatment at a water recycling plant in El Segundo, California. The West Basin Water Recycling Plant (WBWRP) affords a unique opportunity for a side-by-side comparison of the two pretreatment processes. The WBWRP includes three 2.5 million gallons per day reverse osmosis trains. Each train uses cellulose acetate membranes. Feedwater to Train Nos. 1 and 2 is pretreated using a conventional process consisting of lime clarification, recarbonation, and filtration. Feedwater to Train No. 3 is pretreated using microfiltration.

The two pretreatment processes will be evaluated with regard to life-cycle costs and pretreated water quality. Run times between reverse osmosis chemical cleanings will also be evaluated as a proxy for the impacts of the two pretreatment processes on the reverse osmosis membranes. Costs that will be considered include:

- Capital costs
- Operation and maintenance labor

- Chemical costs
- Sludge handling and disposal
- Power costs
- Replacement Parts and Supplies

Pretreated water quality parameters that will be considered are:

- Turbidity
- SDI
- Conductivity
- Magnesium

BACKGROUND

The West Basin Municipal Water District (West Basin) is a public agency providing wholesale water to local water utility companies and municipal water departments. West Basin provides imported and recycled water to 17 cities and unincorporated areas of southwest Los Angeles County.

West Basin has implemented a water recycling program that so far includes a 42 million gallons per day (mgd) pump station that pumps secondary effluent through a 3-mile-long, 60-inch-diameter force main to the 37.5 mgd West Basin Water Recycling Plant (WBWRP) in El Segundo, California. The treated water is distributed by a pump station through about 60 miles of pipelines to over 100 recycled water customers and two satellite nitrification facilities.

The WBWRP treats secondary effluent from the City of Los Angeles' Hyperion Wastewater Treatment Plant and produces two types of water for reuse. The first type of water produced meets Title 22 standards. Title 22 refers to the section of the California Code of Regulations that addresses quality standards for water that is recycled. This type of recycled water is typically used for landscape irrigation and industries such as dye houses and oil refineries. The processes used to treat the secondary effluent to Title 22 standards include coagulation, flocculation, filtration, and disinfection. The water is further treated at the satellite nitrification plants to remove ammonia for use in refinery cooling towers.

The second type of water produced at the plant is distributed to the West Coast Barrier, a series of wells that inject a blend of potable and recycled water into the groundwater basin to protect it from seawater intrusion. Barrier water must meet all primary and secondary drinking water standards. Reverse osmosis is used to treat Barrier water. Pretreatment for reverse osmosis includes conventional and microfiltration pretreatment. This chapter focuses on the Barrier portion of the plant, and specifically compares the two types of pretreatment for the reverse osmosis process.

PROCESS DESCRIPTION

A process schematic of the Barrier portion of the plant is shown in Figure 1. Secondary effluent from the City of Los Angeles' Hyperion Wastewater Treatment Plant is fed to both reverse osmosis pretreatment systems.

The Barrier portion of the plant can produce 7.5 mgd of product water. Before reverse osmosis, 6.1 mgd is run through conventional pretreatment. Before reverse osmosis, 3.0 mgd is run through microfiltration. About 85 percent of the reverse osmosis feedwater is recovered as product. The initial construction of the Barrier portion of the WBWRP was completed in June 1995 and consisted of the conventional pretreatment/reverse osmosis process for the Barrier. Microfiltration was pilot tested at the WBWRP for nine months prior to its implementation under a Phase 2 plant expansion. The Phase 2 expansion of the WBWRP using the microfiltration pretreatment/reverse osmosis process was completed in June 1997.

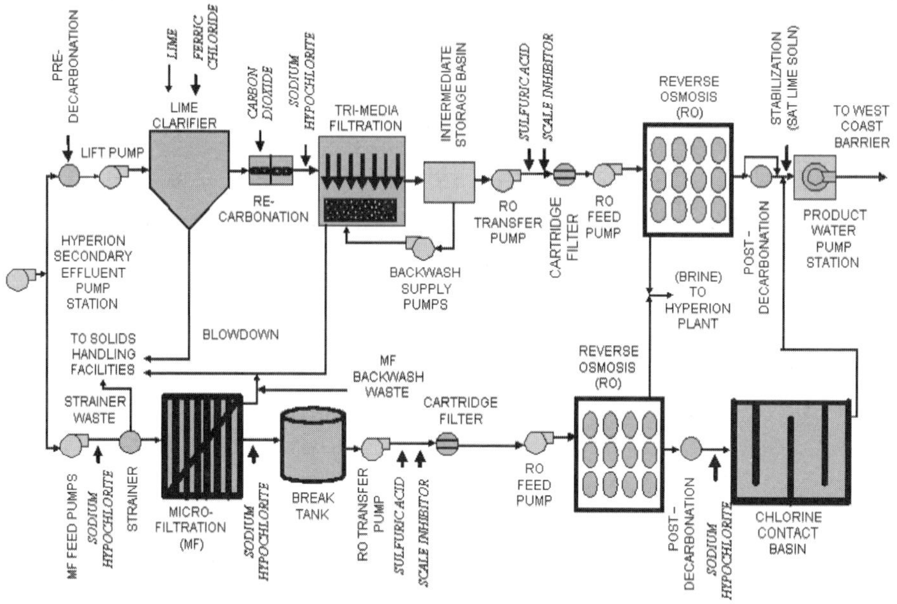

Figure 1 Barrier process flow schematic

CONVENTIONAL PRETREATMENT

As shown in Figure 1, conventional pretreatment for reverse osmosis Train Nos. 1 and 2 consists of pre-decarbonation, lime clarification, recarbonation, and filtration prior to reverse osmosis, post-decarbonation, and pH stabilization. Pre-decarbonation reduces the amount of dissolved carbon dioxide in the secondary effluent and thereby reduces the amount of lime required in the clarifier. This, in turn, reduces the amount of sludge produced in the lime clarifier. In the pre-decarbonation process, dissolved carbon dioxide is stripped from the secondary effluent in a packed media countercurrent force draft stripping tower.

The lime clarifier reduces the amount of suspended solids and hence turbidity in the secondary effluent. It also removes carbonate hardness and non-carbonate magnesium hardness. Pebble lime is slaked and added to the clarifier as a slurry. Sufficient lime is added to raise the pH of the water to about 11.2. Insoluble precipitates are formed and removed by settling within the lime clarifier. There is also the capability to add ferric chloride to the lime clarifier to help settle out the suspended solids during high-turbidity/suspended-solids episodes. The lime sludge is removed from the clarifier at the sludge concentrators and pumped to the sludge thickeners for further processing.

Recarbonation uses gaseous carbon dioxide to lower the pH prior to tri-media filtration to about 8. This is done to protect the tri-media filters from potential long-term calcium carbonate deposition.

The recarbonated water is filtered using tri-media filters to remove floc carried over from the lime clarifier or that developed in the recarbonation process. The filter is made up of 16.5 inches of anthracite coal, over 9 inches of garnet, and over 4.5 inches of fine illumanite. Filtered water is used as the backwash water supply. Waste backwash is discharged to the solids-handling facilities for further processing.

MICROFILTRATION PRETREATMENT

Pretreatment for reverse osmosis Train No. 3 consists of basket strainers and microfilters prior to reverse osmosis, post-decarbonation, stabilization, and disinfection. Microfiltration is a fine filtration process that uses a membrane filter to remove particles greater than approximately 0.2 micron from the secondary effluent feed stream from the Hyperion Wastewater Treatment Plant.

Filtration takes place from the outer surface of the fiber to the hollow inner core. Feedwater passes through the porous wall of the fibers and suspended matter remains on the outside. This filtration process removes solids larger than approximately 0.2 micron. Microfiltration is a barrier filtration system that provides positive mechanical retention of suspended solids and microorganisms. As a guide, bacteria are typically larger than about 1 micron.

As deposits build up on the fibers, filtration flow resistance increases and the membrane has to be backwashed to restore flow performance. During backwash, filtration is stopped and compressed air is used to remove deposits from the outer surface of the fibers. Feedwater is used to transport the deposits to the backwash drain lines. Backwashes are performed on each of the five units every 20 minutes and take approximately 2.5 minutes to complete. Backwash waste is sent to the solids-handling facilities for further processing.

Filtrate quality is consistent despite wide variations in feedstream conditions. Filtrate quality is independent of operator interaction. Operator input is minimal, making remote, unattended operation possible.

The microfiltration system at WBWRP is composed of five units capable of treating 3.4 mgd. Flow to the units is pumped at a constant pressure of approximately 35 psi that is maintained by an adjustable-frequency-drive pump motor. Sodium hypochlorite (NaOCl) is added downstream of the feed pump, prior to the 500 micron basket strainers. The addition of NaOCl is needed to prevent bio-fouling of the microfilters. NaOCl is also added to the microfilter effluent (filtrate) to maintain residual in the transport system to the reverse osmosis elements. The filtrate is collected in a break tank from which it is pumped through 20-micron cartridge filters to the reverse osmosis high-pressure feed pump. Sulfuric acid and scale inhibitor are added to the flow stream immediately upstream of the cartridge filters.

With microfiltration, pretreatment disinfection of the reverse osmosis permeate is required by the regulatory agency and is achieved with the addition of NaOCl to the chlorine contact basin. The basin provides over 90 minutes of contact time.

COMPARISON METHODOLOGY

The two pretreatment processes were compared with regard to cost and water quality. Costs for each of the pretreatment processes were determined using historical cost information specific to each pretreatment process. Operating costs for this direct comparison were gathered for the period July 1997 to October 1998. Costs were gathered for the following categories.

Fixed Costs

Capital Costs

Costs to construct the facilities and purchase the land were gathered from West Basin's financial records. Construction costs were then amortized at 5.5 percent over 20 years to determine debt payments per year. It should be noted that construction costs for conventional pretreatment were based on a capacity about twice that of the microfiltration system. Cost for an equivalent-size microfiltration pretreatment system might be lower per acre-foot than that shown in Table 1 because of economies of scale. Conventional pretreatment capital costs included the pre-decarbonation tower, lift station, lime clarifier, recarbonation basin, tri-media filters, chemical storage and feed facilities, and a major portion of the sludge handling and disposal facilities. Microfiltration pretreatment capital costs included adjustable-frequency-drive microfilter feed pumps, basket strainers, microfilters and associated controls and ancillary equipment, chemical storage and feed, and a chlorine contact basin.

Operation and Maintenance Labor

Labor costs to operate and maintain the pretreatment processes were determined based on labor charges during the fiscal year from July 1997 through June 1998.

Replacement Parts

Because the plant is relatively new, costs for replacement parts were estimated based on life expectancy of individual process components. There are few data on the life of the microfilter elements treating secondary effluent. For this study, microfilter elements were assumed to be replaced every five years. Replacement costs were escalated at 3 percent, brought back to present worth at 5.5 percent and amortized over 20 years.

Variable Costs

Chemical Costs

Chemical usage for each pretreatment process was determined based on plant historical records for the study period. The chemicals and the cost per acre-foot of RO product water are listed in Table 1.

Sludge Handling and Disposal

Sludge handling and disposal costs were determined using plant historical records for the study period.

Power Costs

Power costs were determined using plant power use records for the study period.

Table 1 Chemical usage

Chemical	Cost (*$/AF*)
Conventional Pretreatment	
Sodium Hydroxide	22
Lime	52
Ferric Chloride	10
Carbon Dioxide	25
Microfiltration Pretreatment	
Sodium Hypochlorite	33
Citric Acid	1
Memclean C	1

In order to compare the costs of each pretreatment train, costs are expressed as dollars per acre-foot of reverse osmosis product water. Annual fixed costs were divided by the reverse osmosis system product water capacity in acre-feet per year. Variable costs were divided by the reverse osmosis product water delivered over the study period in acre-feet.

The two pretreatment processes were also compared with regard to quality of the pretreated water and its impacts on the performance of the reverse osmosis units. Performance of the reverse osmosis units was measured in terms of run time between chemical cleanings. The parameters used to compare the two processes include:

- Turbidity
- SDI
- Conductivity
- Magnesium

RESULTS OF THE COMPARISON

The costs are summarized in Table 2. As shown in Table 2, costs for pretreatment of reverse osmosis feedwater using microfiltration are less than for conventional pretreatment for all of the cost categories except replacement parts. Overall, microfiltration costs were about $250 per acre-foot lower than those for conventional pretreatment. Higher microfilter element replacement costs were more than offset by lower capital, operation and maintenance labor, chemical, sludge handling, and power costs.

Monthly average plant influent, conventional pretreatment effluent, and microfiltration effluent turbidities are shown in Figure 2. As shown in Figure 2, monthly average plant influent turbidity varied from about 4.5 ntu to 7.5 ntu from November 1997 to October 1998. Over the same period, conventional pretreatment effluent turbidity varied from about 0.13 ntu to 0.33 ntu and

Table 2 Cost summary

	Conventional Pretreatment			Microfiltration		
Description	(a) (*$*)	(b) (*AF*)	(c) (*$/AF*)	(d) (*$*)	(e) (*AF*)	(f) (*$/AF*)
Fixed Costs						
Capital Costs	1,502,340	5,611	268	442,753	2,806	158
O&M Labor	315,914	5,611	56	78,979	2,806	28
Replacement Parts & Supplies	35,000	5,611	6	78,830	2,806	28
Subtotal Fixed Costs	1,853,255	5,611	330	600,562	2,806	214
Variable Costs						
Chemical Costs	784,930	7,198	109	100,219	2,993	33
Sludge Production & Handling	496,748	7,198	69	12,737	2,993	4
Power	196,666	7,198	27	79,990	2,993	27
Subtotal Variable Costs	1,478,344	7,198	205	192,946	2,993	64
Total Fixed & Variable Costs			536			279

microfiltration effluent turbidity varied from about 0.05 ntu to 0.10 ntu. The turbidity of the conventional pretreatment effluent generally follows the turbidity of the plant influent, rising when plant influent turbidity rises and falling when plant influent turbidity falls. The microfiltration pretreatment effluent was more consistent, generally around 0.05 ntu, with a maximum of about 0.1 when high influent turbidity was sustained over a period of several months.

The silt density indexes (SDI) for conventional pretreatment effluent and microfiltration pretreatment effluent are shown in Figure 3. The SDI is consistently lower for microfiltration pretreatment effluent over the study period.

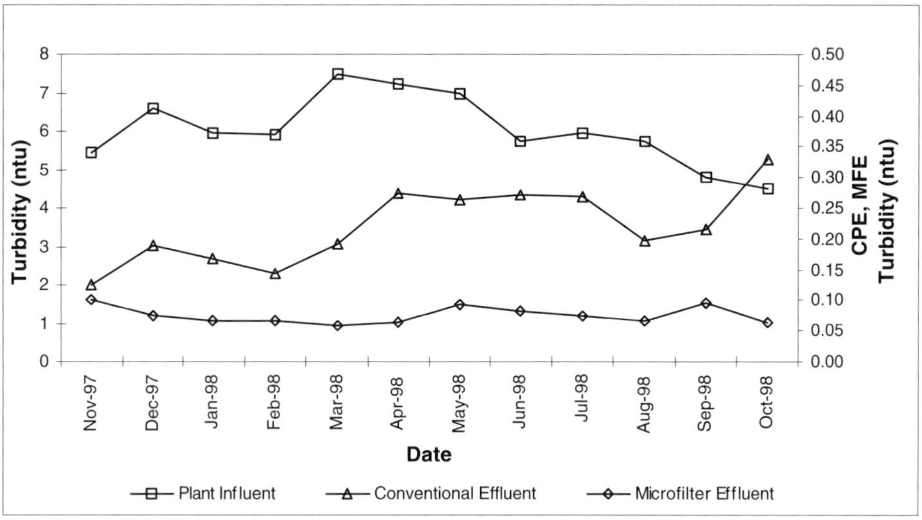

Figure 2 Average monthly turbidity

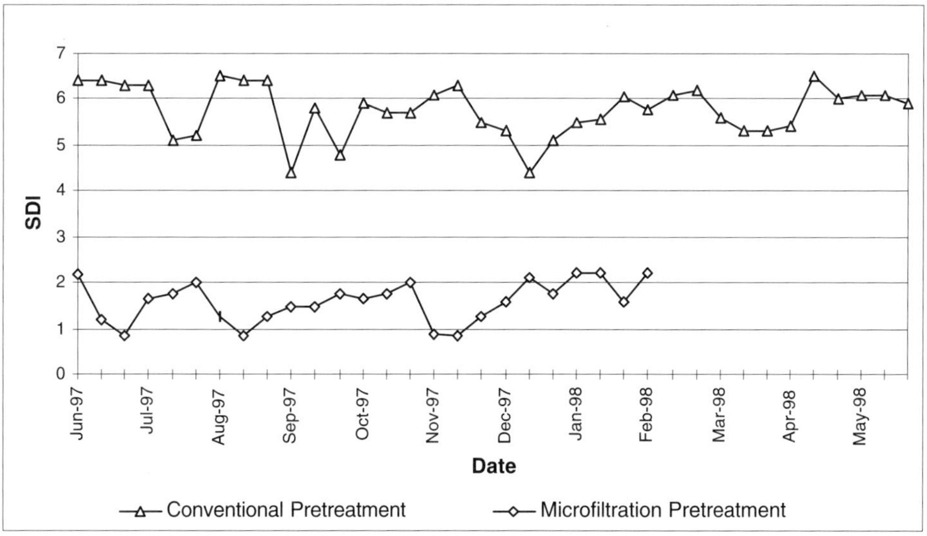

Figure 3 Silt density index (SDI)

The conductivities for conventional pretreatment effluent and microfiltration pretreatment effluent are shown in Figure 4. Conductivity is a measure of dissolved solids in the water. The conductivity for the conventional pretreatment effluent is consistently higher than the microfiltration pretreatment effluent over the study period.

Magnesium levels for conventional pretreatment effluent and microfiltration pretreatment effluent are shown in Figure 5. As shown in Figure 5, magnesium levels from the microfilters were consistently higher than from conventional pretreatment, indicating magnesium is being removed in the conventional pretreatment process.

Run times for reverse osmosis Train Nos. 1, 2, and 3 are shown in Table 3. Train Nos. 1 and 2 use conventional pretreatment prior to the reverse osmosis membranes. Train No. 3 uses microfiltration pretreatment prior to the reverse osmosis membranes. The average times between cleaning for Train Nos. 1, 2, and 3 are 26 days, 27 days, and 31 days respectively. The run times for each train are

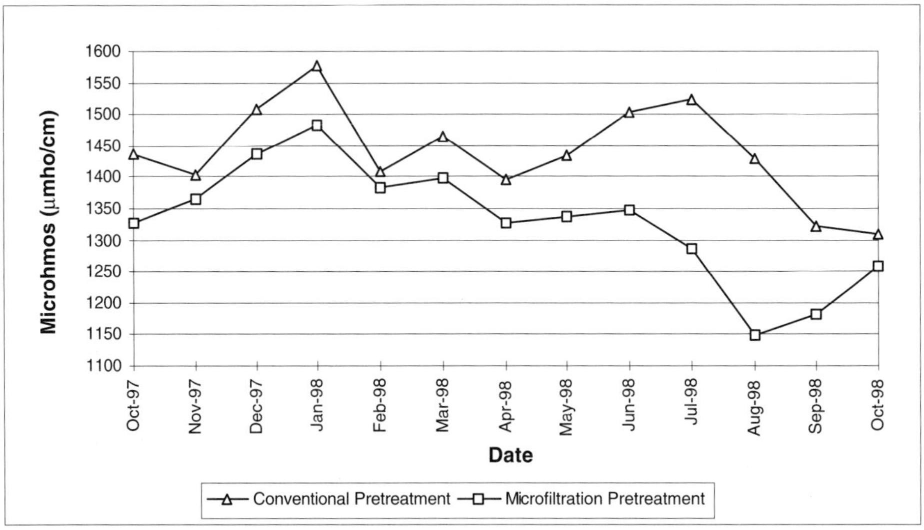

Figure 4 Conductivity

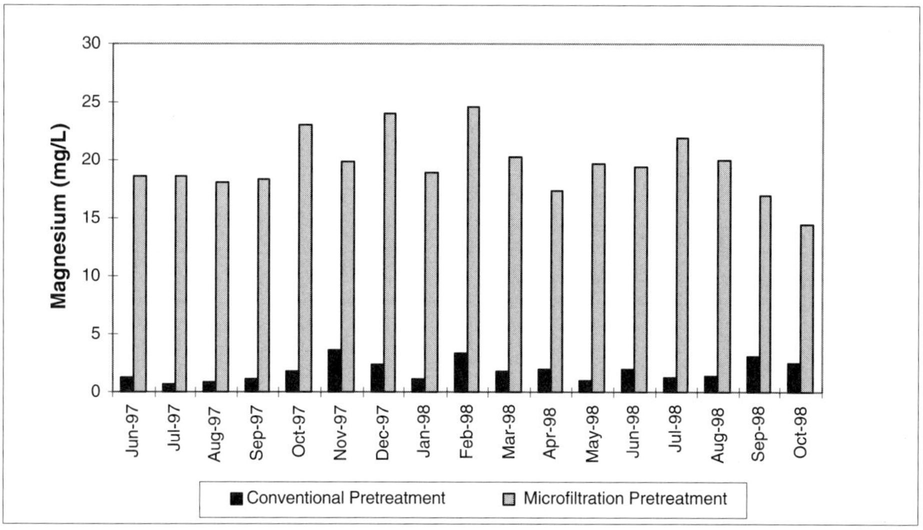

Figure 5 Magnesium levels

generally the same. The slightly longer run time on Train No. 3 could be attributable to the age of the membranes. Train Nos. 1 and 2 are about three years old, whereas Train No. 3 is about one year old.

DISCUSSION

Costs for microfiltration are about 50 percent lower than those for a comparable conventional pretreatment process consisting of pre-decarbonation, lime clarification, recarbonation, and filtration. The higher cost of microfilter elements is more than offset by less capital and other operation and maintenance costs.

The quality of reverse osmosis feedwater from microfilters is better than conventional pretreatment in terms of turbidity, SDI, and conductivity. However, magnesium levels from the microfilters are higher than those from conventional pretreatment. Magnesium is not removed through the microfilters.

Table 3 RO run times

Conventional Pretreatment				Microfiltration Pretreatment	
RO Train No. 1		RO Train No. 2		RO Train No. 3	
Date	Days Since Last Cleaning	Date	Days Since Last Cleaning	Date	Days Since Last Cleaning
7/20/97	42	7/20/97	41	8/17/97	60
10/7/97	79	10/13/97	85	9/22/97	36
11/12/97	36	11/14/97	32	10/21/97	29
12/8/97	26	12/30/97	46	11/9/97	19
12/26/97	18	1/17/98	18	12/15/97	36
1/16/98	21	2/4/98	18	1/12/98	30
2/2/98	17	2/25/98	21	2/19/98	38
2/12/98	10	3/20/98	23	3/18/98	29
3/8/98	24	4/6/98	17	4/16/98	29
3/28/98	20	4/24/98	18	5/3/98	17
4/16/98	19	5/4/98	10	5/28/98	22
5/3/98	17	5/25/98	18	6/15/98	18
5/20/98	17	6/3/98	9	7/20/98	34
6/2/98	13	6/24/98	21	8/31/98	42
6/18/98	16	7/22/98	28		
7/15/98	27	9/8/98	48		
8/27/98	43	10/3/98	24		
9/22/98	26	10/20/98	17		
10/4/98	12	10/27/98	7		
10/26/98	9				
Average Run Time	26		27		31

In spite of the differences in water quality in terms of turbidity, SDI, conductivity, and magnesium for the reverse osmosis feedwater from the two pretreatment processes, there was no measurable difference in the performance of the reverse osmosis membranes.

Run times between membrane cleanings were approximately the same for the two pretreatment systems. However, the long-term impacts of the different reverse osmosis feedwaters are not certain at this time.

CONCLUSIONS

Microfiltration is a cost-effective alternative to conventional pretreatment for reverse osmosis membranes. A side-by-side cost comparison shows that microfiltration costs are about 50 percent lower than conventional pretreatment costs. Capital, chemical, and sludge handling and disposal costs are significantly lower for microfiltration than for conventional pretreatment.

The water qualities of the two pretreatment processes were different in terms of turbidity, SDI, conductivity, and magnesium. However, there was no measurable difference in the performance of the reverse osmosis membranes from the two different pretreatment processes as measured by time between chemical cleanings. Impacts of the two pretreatment processes on the long-term performance of the RO membranes are uncertain at this time.

ACKNOWLEDGEMENTS

The authors would like to acknowledge and thank Jude Perera of United Water Service for his assistance and support in the preparation of this chapter.

Part 2
Fouling

Critical Recovery Concept in Constant Flux Microfiltration and Effects of Operational Mode on Microfilter Performance

Aerobic Versus Anaerobic Nanofiltration: Fouling of Membranes

Solving Membrane Fouling for the Boca Raton 40-mgd Membrane Water Treatment Plant: The Interaction of Humic Acids, pH and Antiscalants with Membrane Surfaces

Morphological Properties of Fouling Cakes Formed from Heterogeneous Suspensions

Characterization of UF and NF Membrane Fouling by Different Biopolymer Fractions Using Dead-End Filtration and Atomic Force Microscopy

CHAPTER · 4

*Critical Recovery Concept in Constant Flux Microfiltration and Effects of Operational Mode on Microfilter Performance**

Shankararaman Chellam, Senior Engineer
Montgomery Watson, Herndon, VA

Joseph G. Jacangelo
Montgomery Watson, Herndon, VA

BACKGROUND

Microfiltration (MF) is an ultra-low-pressure membrane process that has been shown to be effective for the removal of particles, turbidity, *Giardia,* and *Cryptosporidium* [1]. MF has also been demonstrated to be an effective pretreatment technique for higher-pressure membrane processes such as nanofiltration [2, 3, 4].

Pressure-driven membrane filters can be operated in either constant pressure or constant flux, as depicted in Figure 1. Drinking water MF units are frequently operated in constant flux mode with periodic backwashing. Thus, the filtration pressure is increased to maintain a predetermined flow rate across the membrane as it gets fouled. Upon reaching a specified terminal pressure (usually based on manufacturers' recommendations), the unit is shut down and cleaned. With increasing pressure operation, system productivity

*Presented at the 1999 AWWA Membrane Technology Conference, Long Beach, Calif.

could be compromised because of membrane compaction, cake compression, and/or decreasing backwash effectiveness.

Alternatively, membranes can also be operated in a constant pressure, declining flux mode. This mode of operation may result in a larger membrane area necessary for a given plant capacity, but is expected to decrease chemical and labor costs (because of less frequent cleanings). This may be particularly important for smaller installations where skilled labor may not be easily available to perform frequent membrane cleanings. Therefore, differences in fouling patterns and operating philosophies for MF units will have important implications for plant design and cost. To the authors' knowledge, there are no data in the drinking water literature that have carefully compared these two modes of operation.

This paper reports on our continuing work in optimizing MF for water treatment and builds on previous publications [e.g., 2, 6]. A more complete description of the findings and conclusions have been subject to peer review and published elsewhere [7]. The principal objective of this presentation is to compare MF fouling rates and filtrate water quality under conditions of constant flux and constant pressure operation. The implications for MF plant design and operation will also be analyzed.

FOULING ANALYSIS

MF specific flux (J_S) profiles normalized to the initial specific flux (J_{SO}) are commonly analyzed using time (t) as the independent variable (Equation 1). Subsequently, statistical fits to such temporal flux profiles are obtained using linear regression and chemical cleaning intervals calculated by extrapolation. Even though such an approach yields valuable information on anticipated run times for various experimental conditions, it does not normalize for the differences in volume filtered when membranes are operated at different flux and backwashing intervals. Hence, to better compare MF fouling rates when operating conditions are changed, it is proposed that the cumulative volume filtered (V) per unit membrane area (A_m) be employed as the independent variable (Equation 2).

$$\frac{J_s}{J_{so}} = Ae^{-bt} \tag{1}$$

$$\frac{J_s}{J_{so}} = Ae^{-bV/A_m} \tag{2}$$

where A is a pre-exponential factor, and b is similar to a first-order decay term. The parameter b can be interpreted as a fouling rate.

A recently published mechanistic model [5] based on compressible cake filtration was also used to derive backwash effectiveness parameters under various fluxes and recoveries investigated. This approach was used to verify the empirical analysis of MF fouling rates as outlined in the previous paragraph in addition to deriving site-specific values of the backwash effectiveness parameter.

EXPERIMENTAL METHODS

A detailed description of the experiments, protocols employed, source water quality, membranes, etc. has recently been published elsewhere [7]. However, for the sake of completeness, a brief summary of the experimental methods are given next.

Membranes and Membrane Cleaning

Two hollow fiber membrane units were operated in parallel at a water filtration plant. The effective filtration area for each unit was 8.8 m^2. These units were operated in direct flow mode and backwashed periodically using compressed air at a pressure of 621 kPa. Following this step, a fast flush using raw water was employed to sweep away the materials dislodged by compressed air. At the end of each experiment, membranes were cleaned using a two-step procedure. First, a 0.5% (v/v) phosphoric acid solution (pH = 2) was used in an attempt to remove metallic and silica-based foulants. Following the acid cleaning procedure, a 3% (v/v) solution of sodium hydroxide mixed with surfactants (pH = 12) was used to removed adsorbed organic materials. Typically, the membranes were cleaned when the transmembrane pressure reached approximately 138 kPa.

Experimental Protocols

Three constant flux experiments were conducted by maintaining the flux and backwash interval at predetermined values to assess differences in the performance of the two membrane skids. The backwashing interval was maintained constant in constant flux experiments so as to maintain a constant recovery for each experiment. However, in constant pressure experiments, the backwash interval was increased manually as the flux dropped so that the same volume of water was filtered between two backwash events. Experiments were conducted in a wide range of fluxes ($49 \leq J \leq L/m^2/h$) and recoveries ($75 \leq R \leq 100\%$).

Source Water

Untreated water from a reservoir having a relatively low turbidity and suspended solids content was used. This was a slightly alkaline water with total organic carbon (TOC) concentration near 2 mg/L. Manganese concentrations were less than 0.05 mg/L except during turnover, when they increased to 1.2 mg/L. The MF feedwater also contained HPC and coliform bacteria. The median silica concentration was measured at 6 mg/L. Table 1 provides a summary of the important raw water quality parameters during the pilot testing period.

Table 1 Summary of important raw water quality parameters

Parameter	Units	Median	Range	# of Observations
Turbidity	NTU	1.47	0.68–7.37	189
TOC	mg/L	2.2	1.8–2.8	68
Temperature	°C	11.4	3.8–22	188
Alkalinity	mg/L as $CaCO_3$	46.5	26–59	46
pH	–	7.49	6.68–8.28	188
TSS	mg/L	2	1–6	33
Silica	mg/L	6	3–8	37
Manganese	mg/L	0.03	0.01–1.2	206
Iron	mg/L	0.25	0.1–0.8	6
Total coliforms	#/100 mL	260	30–1,600	24
HPC bacteria	CFU/mL	220	60–1,300	35

RESULTS AND DISCUSSION

Reproducibility Between MF Units

Constant flux experiments were conducted by operating the two MF units at the same flux and backwash interval to investigate differences in specific flux profiles. Several such experiments were conducted during the course of the study. Results from these experiments are shown in Figures 2a and 2b by linearizing Equation 1:

$$\mathrm{Ln}\left[\frac{J_s}{J_{so}}\right] = \mathrm{Ln}(A) - b\frac{V}{A_m} \tag{3}$$

The error bars in Figure 2 denote 95% confidence intervals calculated using linear regression. Figure 2a denotes the non-dimensional pre-exponential factor A, and Figure 2b depicts the fouling rate parameter b expressed in m^2/L. These data suggest no differences in operation at 95% confidence level and that specific flux profiles obtained from these two MF units could be compared.

Membrane Compaction

The clean membrane resistance (R_m) was measured at various pressures using Darcy's law (Equation 4) following each chemical cleaning event.

$$J = \frac{\Delta P}{\mu R_m} \qquad (4)$$

where μ is the absolute viscosity of water at ambient temperature and ΔP is the instantaneous transmembrane pressure. The membrane resistance was empirically observed to follow a straight-line increase with the applied pressure as given in Equation 5 [7, 8].

$$R_m = 74.75 \times 10^{10} + 4.39 \times 10^{6} \Delta P \qquad (5)$$

where R_m is expressed in m^{-1}, and ΔP is in Pa.

Cake Compressibility

The specific cake resistance (α) was also measured at various pressures. These experiments were performed using flat disc membranes rated at 0.1 μm. The specific resistance of cakes formed by colloidal materials in the source water was found to obey a power law relationship as given in Equation 6 [9].

$$\alpha = \alpha_o (\Delta P)^n \qquad (6)$$

where α_o is the specific resistance at unit pressure and n is the compressibility index. These data showed that cakes formed by natural colloidal materials were highly compressible ($n = 0.93$).

Effect of Backwashing Interval and Recovery on Fouling

Constant flux experiments were conducted to investigate the effect of recovery (backwash interval) on specific flux profiles as shown in Figure 3. These data obtained at a flux of 99 L/m^2/h and three different backwash intervals show that decreasing the recovery retarded fouling. Without backwashing ($R = 100\%$), membranes had to be cleaned after filtering only 500 L/m^2. However, even at a recovery of 97.4%, approximately 6,600 L/m^2 were filtered before necessitating chemical cleaning. As the recovery was further decreased to 76%, more than 14,000 L/m^2 were filtered before cleaning was warranted.

Changes in MF Fouling with Operational Mode

MF specific flux profiles were generated under constant flux and constant pressure operation. These operational modes were compared under conditions of constant initial flux and recovery. In all cases, a constant pressure operation resulted in decreased fouling. For example, Figure 4 depicts specific flux profiles at 117.5 L/m^2/h flux and 97.8% recovery. It is observed that constant flux operation resulted in accelerated fouling compared to constant pressure operation. This can be explained in part due to membrane compaction and cake compression. It is also observed that the initial portion of the specific flux profiles is similar to each other. Hence, fouling mechanisms may also be similar at the start of these runs [10]. However, backwashing appears to be more effective toward the later part of the constant pressure run. This may be in part due to changes in the mechanism of fouling. It is speculated that a constant pressure operational mode may result in decreased pore penetration (because of decreased flux) compared to constant flux operation. This decrease in pore penetration may be expected to increase the effectiveness of backwashes. Similar observations were made at recoveries in the range $75 < R < 100\%$, and at flux of 99 L/m^2/h.

Water Quality

Alternating from constant flux to constant pressure modes of operation is warranted only if there are no changes in filtrate water quality. Therefore, turbidity, true color, UV^{254}, and concentrations of particles, TOC, alkalinity, total hardness, SDSTHM, and SDHAA precursors were monitored in the raw water, and MF filtrate during constant pressure and constant flux experiments. Non-parametric statistical analysis similar to that employed in some previous research on water quality comparisons [11] revealed that there were no differences in filtrate water quality with operational mode [7].

SUMMARY AND CONCLUSIONS

Fouling in direct flow MF systems was shown to be reduced by periodic backwashing. However, a critical recovery was identified above which fouling rates were more sensitive to backwashing frequency. Below this critical recovery point, fouling rates and backwash effectiveness were relatively insensitive to the frequency of backwashing. Longer chemical cleaning intervals and lower fouling rates were shown to be attained by operating potable water MF systems at constant pressure rather than constant flux. Increasing chemical cleaning intervals by constant pressure operation may be particularly valuable to small systems where skilled labor for cleaning may not be readily available. However, a complete cost analysis needs to be conducted before further recommendations can be made.

REFERENCES

1. Jacangelo, J.G. et al. Mechanism of *Cryptosporidium*, *Giardia* and MS2 Virus Removal by MF and UF. *Journal AWWA*, 89 (9), 107.
2. Chellam, S. et al. (1997). Effect of Pretreatment on Surface Water Nanofiltration. *Journal AWWA*, 89 (10), 77.

3. Lozier, J.C. et al. (1997). Integrated Membrane Treatment in Alaska. *Journal AWWA*, 89 (10), 50.
4. Kruithof, J.C. et al. (1996). Integrated Multi-Objective Membrane Systems for Control of Microbials and DBP Precursors. *AWWA Membrane Technology Conference Proceedings*, February 23–26, 1997, New Orleans, La., p. 307.
5. Chellam, S., et al. (1998). Modeling and Experimental Verification of Pilot-Scale, Direct Flow, Hollow Fiber Microfiltration with Periodic Backwashing. *Environ. Sci. Technol.*, 32 (1), 75.
6. Chellam, S., J.G. Jacangelo (1998). Fouling of Microfilters in Direct Flow: Comparison of Constant Flux and Constant Pressure Operation. Presented at the AWWA Annual Conference and Exposition, Dallas, Tex.
7. Chellam, S., J.G. Jacangelo (1998). Existence of Critical Recovery and Impacts of Operational Mode on Potable Water Microfiltration. *Jour. Envi. Engrg.*, 124 (12), 1211.
8. Persson, K.M., V. Gekas, G. Tragardh (1995). Study of Membrane Compaction and Its Influence on Ultrafiltration Water Permeability. *Jour. Memb. Sci.*, 100, 155–162.
9. Nakanishi, K., T. Tadokoro, R. Matsuno (1987). On the Specific Resistance of Cakes of Microorganisms. *Chem. Eng. Commun.*, 62, 187–201.
10. Nagata, N., K.J. Herouvis, D.M. Dziewulski, G. Belfort (1989). Cross-Flow Membrane Microfiltration of a Bacterial Fermentation Broth. *Biotech. Bioeng.*, 34, 447–466.
11. Krasner, S.W., M.J. McGuire, J.G. Jacangelo, N.L. Patania, K.M. Reagan, E.M. Aieta (1989). The Occurrence of Disinfection By-products in US Drinking Water. *Journal AWWA*, 81 (8), 41.

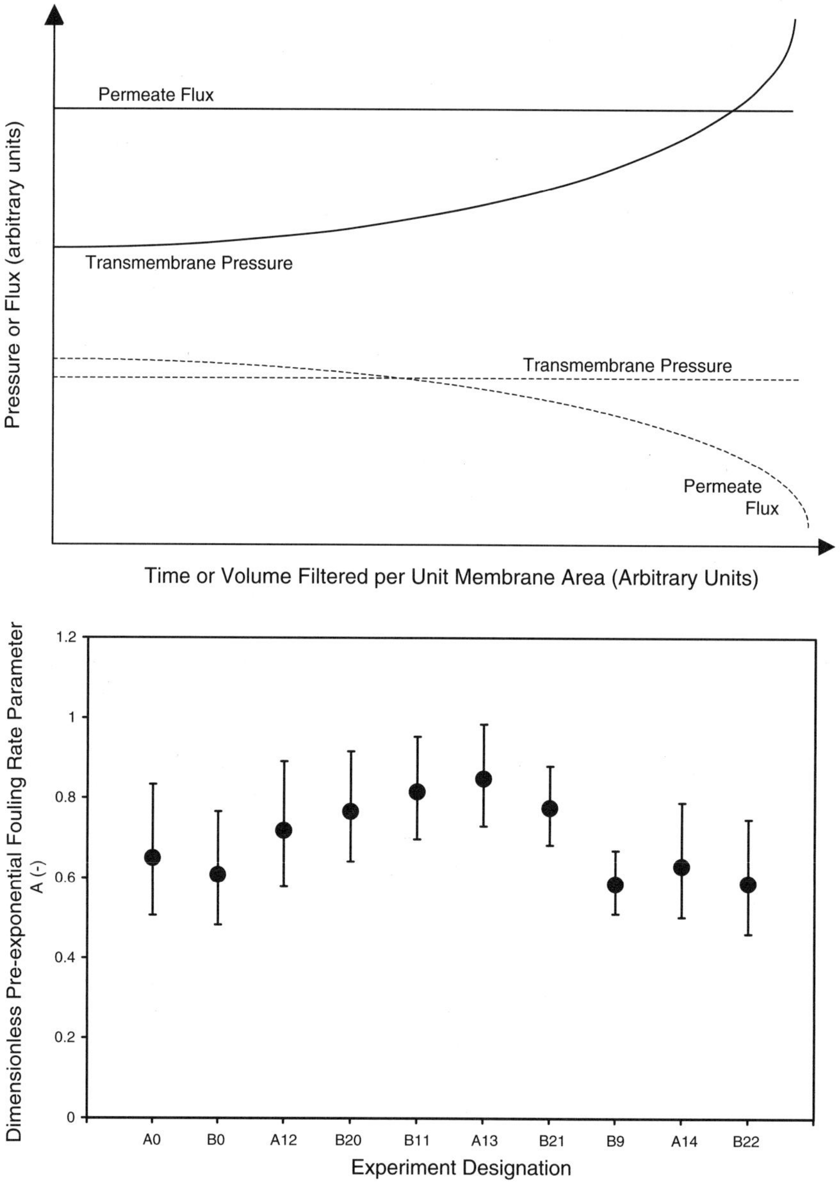

Figure 1 Different operational modes for pressure-driven membrane processes. The solid lines depict constant flux (increasing pressure) operation, whereas the dotted lines depict constant pressure (declining flux) operation.

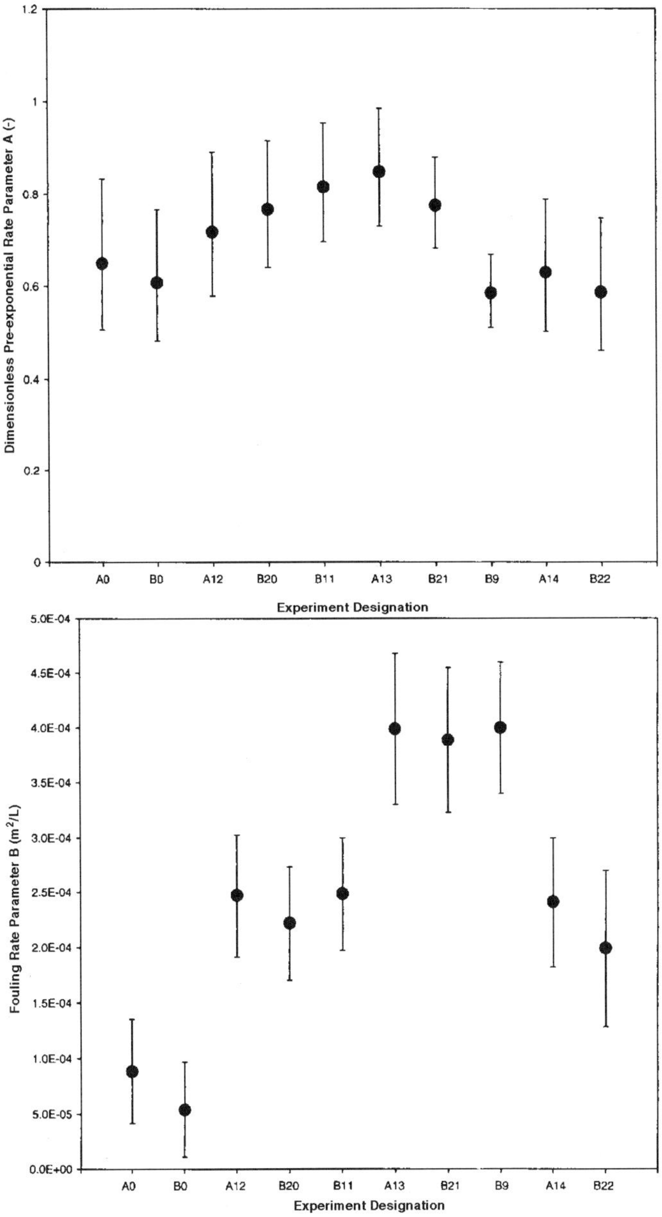

Figure 2 Comparison of experimental reproducibility during the study. Figure 2a depicts the pre-exponential parameter A, whereas Figure 2b depicts the fouling rate parameter b.

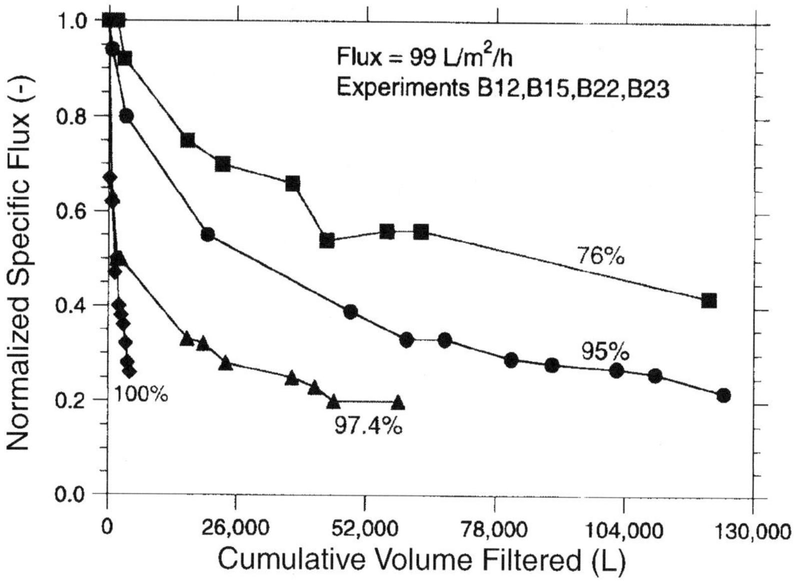

Figure 3 Effect of recovery on MF specific profiles

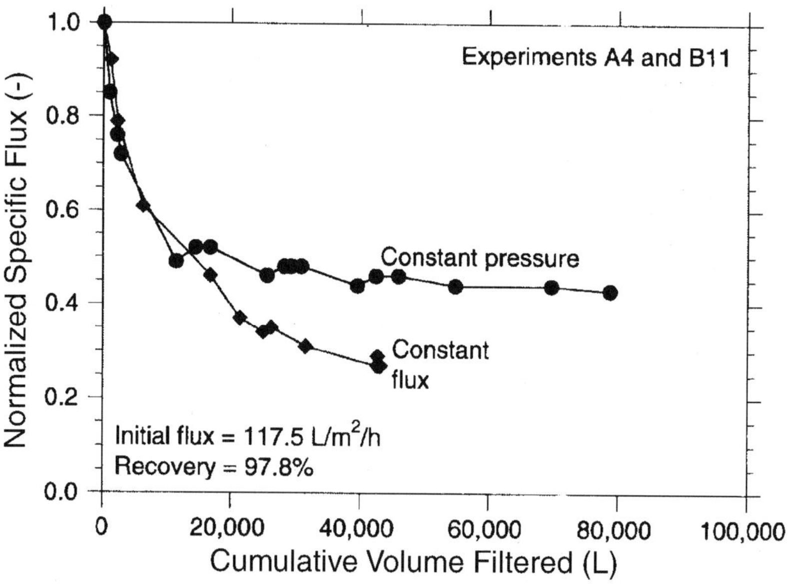

Figure 4 Effect of operational mode on MF specific flux profiles

CHAPTER · 5

Aerobic Versus Anaerobic Nanofiltration: Fouling of Membranes

Peter Hiemstra, M.Sc.
Overijssel Water Supply, Zwolle, The Netherlands
Witteveen + Bos Consulting Engineers, Deventer, The Netherlands

Jacques Van Paassen, B.Sc.
Overijssel Water Supply, Zwolle, The Netherlands

Bas Rietman
Overijssel Water Supply, Zwolle, The Netherlands

Jil Verdouw, B.Sc.
Kiwa NV, Research and Consultancy, Nieuwegein, The Netherlands

BACKGROUND

The consumers of tap water in the Netherlands desire a hardness between 1 and 1.5 mmol/l and a color of less than 10 mg Pt/l. For a number of treatment plants, these objectives cannot be reached with standard groundwater treatment, like aeration and filtration.

For this reason the Water Supply Company of Overijssel (WMO) started research with nanofiltration in 1994 for the future application at 11 of the 27 treatment plants. Extended pilot plant investigation with nanofiltration has been carried out on four groundwater treatment plants:

1. Hammerflier
2. Vechterweerd
3. Boerhaar
4. Engelse Werk

In the United States nanofiltration generally is the first step in the treatment. The difference with the situation in the Netherlands is the iron content of the groundwater. Anaerobic conditions and a high iron content (respectively 25 and 11 mg/l) can characterize the groundwater of Hammerflier and Vechterweerd.

To avoid problems with iron fouling due to the possible introduction of small amounts of oxygen WMO started research with nanofiltration at the end of the conventional groundwater treatment (aeration, 1^e RSF, aeration, 2^e RSF). The consequence of this choice is that the nanofiltration-installation is fed with aerobic water. The MFI-value of the feedwater is 3–4 s/l^2 (in both cases).

The research program in Hammerflier showed that the main causes of membrane fouling were deposition of iron and microbiological regrowth. Microbiological fouling was hard to control. One of the tested membrane types turned out to be sensitive for adsorption of humic acids (NOM) resulting in irreversible fouling.

The research program in Vechterweerd showed comparable results as in Hammerflier. Better cleaning results were obtained with anaerobic treatment of the membranes. The control of microbiological fouling improved by a strict monitoring program for every new batch of chemicals. Nevertheless the cleaning frequency of the membranes is still high.

Evaluating the results of Hammerflier and Vechterweerd WMO made the choice for Boerhaar to place the nanofiltration as the first step in the treatment. The groundwater of Boerhaar is anaerobic and contains only 4 mg/l iron. The MFI of the anaerobic feedwater is < 0.5 s/l^2. The results of the pilot plant are very good. During a period of 9 months no decline of the MTC-value occurred. Also the NPD showed no difference.

To confirm the better performance of the anaerobic operation of nanofiltration both systems (aerobic and anaerobic) are tested at the treatment plant Engelse Werk. The membrane with the anaerobic feedwater showed much better results than the aerobic-operated membrane (in terms of NPD).

Due to the change from aerobic to anaerobic feedwater WMO made a big step in the control of membrane fouling. This is caused by two phenomena:

1. Microbiological fouling can be neglected under anaerobic conditions. Keeping the stable anaerobic groundwater under anaerobic conditions has a very good influence on the prevention of membrane fouling.

2. Deposition of iron can be avoided having anaerobic conditions. The aerobic pretreatment changes the oxidation state of the iron (iron-II to iron-III). Although most of the iron is removed during the aerobic pretreatment, very small amounts of iron-III cause more problems in aerobic nanofiltration than big amounts of iron-II in anaerobic nanofiltration.

INTRODUCTION

In 1998 the Overijssel Water Supply (WMO) delivers drinking water to about 1.2 million consumers (461,000 connections). The annual consumption is 83 million m^3/year (94 MGD). This water is produced by twenty-seven treatment plants (sources: twenty-five groundwater treatment plants, one surface water treatment plant and one treatment plant with bank infiltration).

A growing number (11) of treatment plants of WMO have two or more of the following quality problems (Table 1):

1. Color, caused by humic acids, above 15 mg Pt/l

2. Hardness above 2.0 mmol/l

3. Chloride above 100 mg/l, which can cause corrosion of cast iron distribution pipes

4. Pesticides in the raw water above 0.1 μg/l, if the source is surface water or bank-infiltrated surface water

The parameters in Table 1 are not removed with conventional groundwater treatment, like aeration and filtration. Until now, pesticides were removed with the use of activated carbon filtration. In the Netherlands insight is growing that, if a treatment plant has

Table 1 Overview of raw water quality of 11 treatment plants of WMO

Treatment Plant WMO (1997)	Color >15 mg Pt/l	Hardness >2 mmol/l	Chloride >100 mg/l	Pesticides >0,1 μg/l
Diepenveen	22	1.6	120	not present
Ceintuurbaan	20	1.7	130	not present
Zutphenseweg	18	1.4	105	not present
Sint Jansklooster	21	2.4	37	not present
Witharen	21	2.4	31	not present
Staphorst	21	2.3	39	not present
Boerhaar	20	3.5	32	not present
Rodenmors	20	2.1	36	not present
Hammerflier	18	2.2	20	not present
*Vechterweerd *)*	17	2.3	20	present
*Engelse Werk *)*	7	2.2	100	present

SOURCE: Groundwater (9 plants), bank infiltrated surface water *) (2 plants)

two or more of the mentioned water quality problems, nanofiltration is a very promising technology. However, until 1999 nanofiltration was not in use in full-scale installations for the production of drinking water in the Netherlands. This is caused by three main factors:

1. The costs of operation of nanofiltration
2. The challenge to get permission for concentrate disposal to surface water
3. Membrane fouling

During the last ten years, the operation costs of nanofiltration have been declining due to lower operation pressure and lower membrane investments. The challenge to get permission for concentrate disposal to surface water can be solved in cooperation with the local water authorities. The remaining problem is the membrane fouling. In this chapter an overview is presented to control the membrane fouling at four locations.

Objectives

The outlines of the WMO water treatment philosophy are:

- Physical processes are favorable above chemical processes.
- To produce biologically stable drinking water without disinfectant residual [1]
- To produce chemical stable drinking water (prevention of corrosion in the distribution system)

WMO research for the application of nanofiltration distinguishes several objectives:

1. To minimize membrane fouling and scaling. The maximum cleaning frequency of nanofiltration membranes should be twice a year, preferably less.
2. To maximize the recovery of the nanofiltration in combination with a composition of the concentrate that is acceptable for disposal to local surface water.
3. To maximize the rejection of hardness, color and other parameters that need improvement.
4. To minimize the investments and the costs of operation and maintenance of nanofiltration.

In this chapter the results of the research for the first objective will be presented.

Productivity Loss of Nanofiltration

Productivity loss of (nanofiltration) membranes can occur due to the following mechanisms:

1. *Biological fouling:* Biofouling is the growth of biological species on the membrane surface, or in the case of spiral-wound membranes on the feed spacer as well as on the membrane surface. Biofouling reduces productivity and increases the feed concentrate pressure drop. High tem-

peratures, the presence of easily assimilable organic carbon (AOC) and the presence of an antiscalant may promote biofouling.

2. *Colloidal fouling:* Deposition of suspended or colloidal particles on the membrane surface will result in a loss of flow through the membrane. In anaerobic groundwater, iron is in solution (Fe-II). In conventional groundwater, treatment water is aerated to oxidize iron (Fe-II to Fe-III). The oxidized iron is insoluble and many colloids are formed. The rapid sand filters (RSF) remove most of the iron, but after RSF in the filtrate a few thousand particles $> 1\ \mu m$ per ml may still be present. Installing cartridge filters in the feedwater may reduce iron deposition onto the membranes. The cartridge filters of 10 μm must be replaced frequently, but will not remove all the iron.
3. *Organic fouling:* Organic fouling (adsorption to the membrane surface) may occur and can hardly be predicted. It depends on the nature of the organics in the raw water and on the type and nature (charge) of the membrane surface. Organic fouling results in a constant flux decrease and an increase of the feed concentrate pressure drop of the membrane process [2].
4. *Scaling:* Generally scaling is defined as the formation of mineral deposits precipitating from the feed stream to the membrane surface. The most important deposits are calcium carbonate, calcium sulphate and barium sulphate. The scale formation is caused by supersaturation of one or more of the poor soluble salts. The precipitation will take place in the concentrate phase, especially in the final phase of the installation. Scaling is controlled by the addition of an acid to remove the carbonate anion or an antiscalant to complex the metal involved in the formation of the salt or to inhibit the crystal growth.

METHODS

To determine the membrane fouling and scaling a number of specialized tools have been used.

Biological Fouling

The AOC content gives an indication about the level of (easily) assimilable organic nutrients in the water. Aerobic water with an AOC content below 10 μg Ac-C/l is regarded as biologically stable for water in the distribution system [3].

The biofilm formation rate (BFR) gives an indication of the biofilm growth on rings of glass in a biofilm monitor. BFR values below 10 pg ATP/cm^2.day are considered to be safe for regrowth of aeromonads in the distribution system [4]. The biofilm monitor can also provide information about the iron accumulation rate (IAR).

Colloidal Fouling

With the membrane filtration index (MFI) an impression about the colloidal fouling can be obtained. Membrane feedwater with a MFI 0.45 μm less than 3 s/l^2 is regarded as low fouling [5]. WMO did adopt an Fe(III) content of < 0.03 mg/l as a tentative standard for nanofiltration with aerobic feedwater.

Organic Fouling

Experiments with the water to be treated can give information whether adsorption of organics on the membrane surface will occur or not. Testing several types of membrane elements in a test rig before pilot plant experiments are started can do this.

Scaling

Through projection calculations of the membrane suppliers, scaling may be predicted. Scaling can be prevented by dosage of acids and/or antiscalants.

Specific Parameters for Membrane Filtration

The calculation of the mass transport coefficient (MTC) and the normalized pressure drop (NPD) is presented in Appendix 1. The criteria for membrane cleaning are:

1. A normalized flux (MTC) decline of 15%, compared to the initial MTC value.
2. An increase of the normalized feed concentrate pressure drop (NPD) over a pressure vessel (with 4 elements 4 × 40") of 40 kPa (applied from 1997).

Autopsy of Membrane Elements

In case of operational problems or at the end of a test period elements were taken from the membrane filtration to determine a number of parameters:

- Visual inspection.
- XPS (X-ray Photoelectron Spectroscopy) to determine the adsorption of organic substances on the membrane surface (organic fouling).
- Samples to determine the total amount of biomass and iron present on the membrane, the feed spacer and the product spacer. These results give information about the extent of biofouling and colloidal fouling.

PILOT PLANTS

In Table 2 an overview is presented of the several research periods, the way of operation of the membrane filtration (aerobic and/or anaerobic), the number of stages of the pilot plant and the recovery of the installation. In all the pilot plants spiral-wound nanofiltration elements (thin film composites) are used.

Table 2 Overview of treatment plants and research periods

Treatment Plant	Research Period	Operation	Stages	Recovery
Hammerflier	Nov 94 / Apr 97	aerobic	5/3	90/80%
Vechterweerd	Aug 96 / Dec 98	aerobic	3	80%
Boerhaar	Sep 97 / May 98	anaerobic	1	18%
Engelse Werk	Jun 98 / Oct 98	aerobic/anaerobic	1	10%

Hammerflier

The research program started at the treatment plant Hammerflier in November 1994. The aim was to reduce the color and the hardness of the groundwater. In the case of Hammerflier great fouling problems were expected by the application of nanofiltration on the anaerobic groundwater due to the high iron content (25 mg/l). Therefore it was decided to apply nanofiltration after extended pretreatment (plate aeration, 1^e RSF, aeration tower, 2^e RSF) [2]. As a consequence of this decision the feedwater of the pilot plant is aerobic.

First Research Period Hammerflier

Data pilot plant (Table 3): The iron content in the feedwater was < 0.03 mg/l and the $MFI_{0.45\ \mu m}$ 3.5 s/l^2. From Figure 1 it can be seen that the initial flux is very high (2.7×10^{-8} m/s.kPa). Within 10 days of operation there had been a loss in flux of more than 20%. The first cleaning took place after one month of operation, when the flux decline was > 30%. The cleaning at pH 2 had hardly any effect on MTC. The cleaning at pH 10.5 after 50 days of operation was more effective. The MTC increased with 15%; still a loss in MTC of 25% remained. There was no evidence that scaling took place.

The membrane fouling was caused by:

1. Adsorption of NOM on the membrane surface (organic fouling). It was demonstrated by XPS analysis of the membrane surface that adsorption of humic acids took

Table 3 Data pilot plant Hammerflier (first period)

Stages	five-stage array, configuration: 5, 3, 2, 1, 1 pressure vessels per stage
Pressure vessel	length 160", each vessel contains 4 membrane elements, 4" diameter, 40" length
Feed capacity	8 m3/hr (0.05 MGD)
Product flux	20 $l/m^2.h$ (at 10°C)
Recovery	90%
Dosage	sulphuric acid (to pH 6.6) and 2 mg/l antiscalant (Flocon 260, polyacrilic acid)
Membrane type	Dow Filmtec NF 70 (second batch)

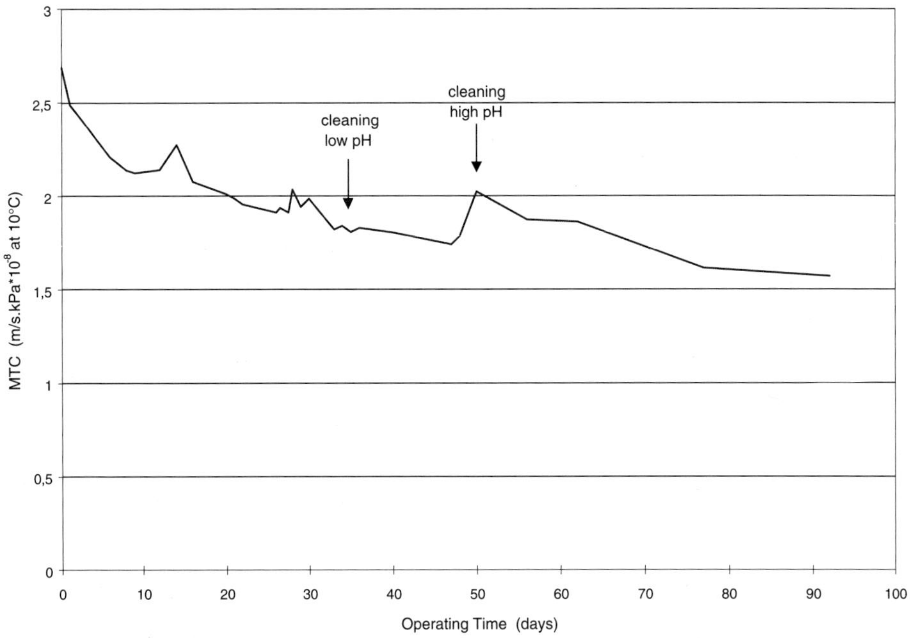

Figure 1 MTC of the five-stage nanofiltration plant in Hammerflier

place. The adsorbed humic acids could not be removed by cleaning. The adsorption of humic acids was irreversible.

2. Biological fouling, mainly caused by the antiscalant. The antiscalant (see Table 3) increased the AOC content of the feedwater from 10 to 55 μg Ac-C/l [5].
3. The deposition of iron (colloidal fouling). Though the level of iron in the feedwater is very low, less than 0.05 mg/l, deposition of iron took place and was responsible for part of the increase of pressure drop and MTC decline.

Second Research Period Hammerflier

In the second research period the configuration, the recovery and the dosage to the feedwater were changed to reach a better composition of the concentrate. With the improvement of the quality of the concentrate WMO got permission for disposal of the concentrate to local surface water.

Even though permeate quality was satisfying [2], the process operation and maintenance were not. Figure 2 shows the MTC as a function of time. The first cleaning had to be carried out after 55 days of operation. The cleaning at pH of about 11 and with alkaline detergents turned out to be very insufficient and the running time was shortened after each cleaning session.

Next to the organic fouling (see the first research period Hammerflier) biological fouling was an important cause for membrane fouling. Measurements pointed out that the AOC content of the feedwater was increased from 10 to 150 μg Ac-C/l by the technical-grade hydrochloric acid that was used.

Third Research Period Hammerflier

The third experiment was carried out with two membrane elements, type Hydranautics PVD 1, in a small test unit (1 stage, 20% recovery). The pH of the feedwater was lowered to 6.3 by dosing chemical-pure-grade (instead of technical-grade) hydrochloric acid.

Table 4 Data pilot plant Hammerflier (second period)

Stages	three-stage array, configuration: 5, 3, 2 pressure vessels per stage
Pressure vessel	length 160", each vessel contains 4 membrane elements, 4" diameter, 40" length
Feed capacity	8 m^3/hr (0.05 MGD)
Product flux	20 l/m^2.h
Recovery	80%
Dosage	hydrochloric acid (to pH 6.3)
Membrane type	Dow Filmtec NF 70 (second batch)

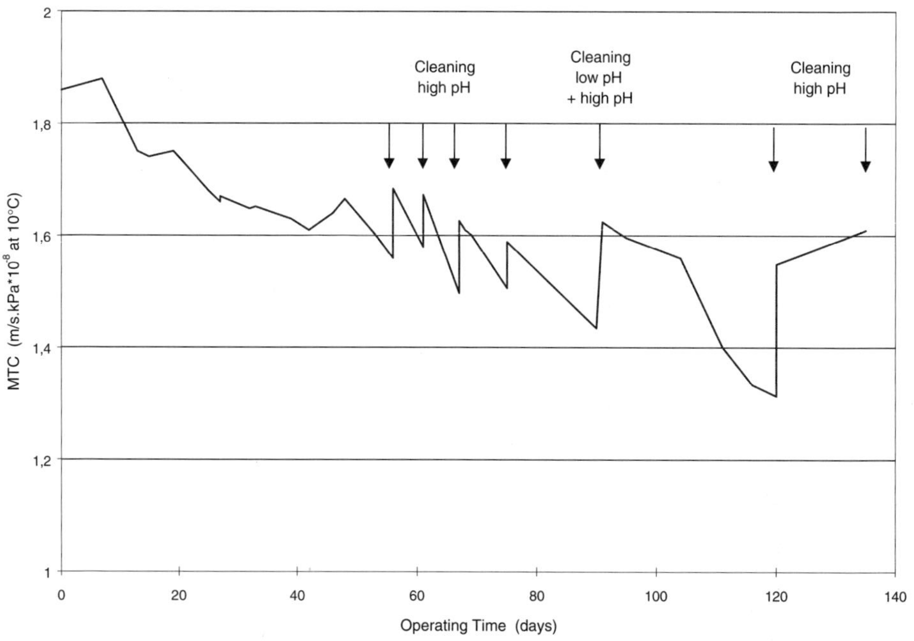

Figure 2 MTC of the three-stage nanofiltration plant in Hammerflier

These elements showed no organic fouling, due to adsorption of NOM. A cleaning frequency of once every three months was sufficient. This result was the best obtained at Hammerflier. The results of this period are shown in Table 9 (Hf = Hammerflier/period 3).

Vechterweerd

At the site of the future production plant Vechterweerd surface water is abstracted via bank filtration. The raw water has a high color and hardness. Moreover the water contains a number of synthetic organic chemicals originating from the river. The iron content of the raw water is 11 mg/l. In place of the nanofiltration in the treatment scheme the same approach as in Hammerflier is chosen. The river bank filtrate is pretreated by a double aeration and two rapid sand filtration steps. Like Hammerflier the feedwater of the nanofiltration is aerobic [6]. The pH of the feedwater is lowered from 7.5 to 6.1 by dosing chemical-pure-grade hydrochloric acid.

Table 5 Data pilot plant Vechterweerd

Stages	three-stage array, configuration: 5, 3, 2 pressure vessels per stage
Pressure vessel	length 160", each vessel contains 4 membrane elements, 4" diameter, 40" length
Feed capacity	8 m^3/hr (0.05 MGD)
Product flux	23 l/m^2.h (stage 1: 23.5/stage 2: 21.1/stage 3: 19.0 l/m^2.h)
Recovery	80%
Dosage	hydrochloric acid (to pH 6.1)
Membrane type	Hydranautics PVD 1 (16 months)/TriSep TS 80 (10 months)

Aug 96/Nov 97: Hydranautics PVD 1 Membranes

The pilot plant was operated 480 days with the Hydranautics PVD 1 membranes. Hydrochloric acid was applied to avoid scaling. Use of antiscalants was avoided to restrict biofouling. The initial MTC-value was about $0.86*10^{-8}$ m/s.kPa [6, 7].

For the first 160 days MTC values showed a gradual decrease in combination with a gradual increase of the normalized feed concentrate pressure drop. The AOC of the feedwater was stable at 8 μg Ac-C/l. The results of this period are shown in Table 9 (Vw1 = Vechterweerd/period 1).

From day 180 to day 260 an exponential increase in the feed concentrate pressure drop of the first stage was observed. The BFR of the feedwater, after dosage of hydrochloric acid, showed a sharp increase to 310 pg ATP/cm^2.d in the same period. Autopsy proved that a strong biofouling buildup onto the feed spacer of the membrane modules caused the exponential increase in NPD. The increase in NPD of the first elements was three times higher than of the other elements of stage 1. It is observed that strong biofilm formation leads to enhanced iron accumulation [8]. The results of this period are shown in Table 9 (Vw2 = Vechterweerd/period 2).

The biofouling proved to be related to the quality of the chemical-pure hydrochloric acid (36%). The AOC of the feedwater rose in this period to 27 μg Ac-C/l. It was evident that the hydrochloric acid was polluted with biodegradable compounds. The AOC measurement takes too much time for the quality control of new batches of strong acids. For the quality control a new protocol was developed that the acid was acceptable with a DOC content < 5 mg/l.

Thirteen cleanings have been carried out during the 480 days. Cleaning of the membranes did not restore the original levels of MTC and pressure drop. One cleaning was not a standard procedure (high pH, low pH) but an anaerobic treatment. The temporary anaerobic conditions were meant to inactivate the biomass. This approach was more successful in terms of MTC and NPD; however, the biomass was not removed from the membrane surface.

Jan 98/Dec 98: TriSep TS 80 Membranes

After a number of measures to minimize fouling, especially biofouling, the research continued in January 1998 with new membrane elements (TriSep TS 80). The measures that were taken are:

1. To prevent biofouling: quality control of new batches of chemical-pure acid (DOC < 5 mg/l) is added to AOC measurements of feedwater.
2. The flux of the first and second stages is controlled to about 20 $l/m^2.h$.
3. Optimization of the pretreatment to minimize the deposition of iron (colloidal fouling).
4. Adjusted cleaning criteria: cleaning at a pressure drop increase of only 5% of the initial pressure drop.

In Figure 3 the results are shown. At day 62 a cleaning was carried out.

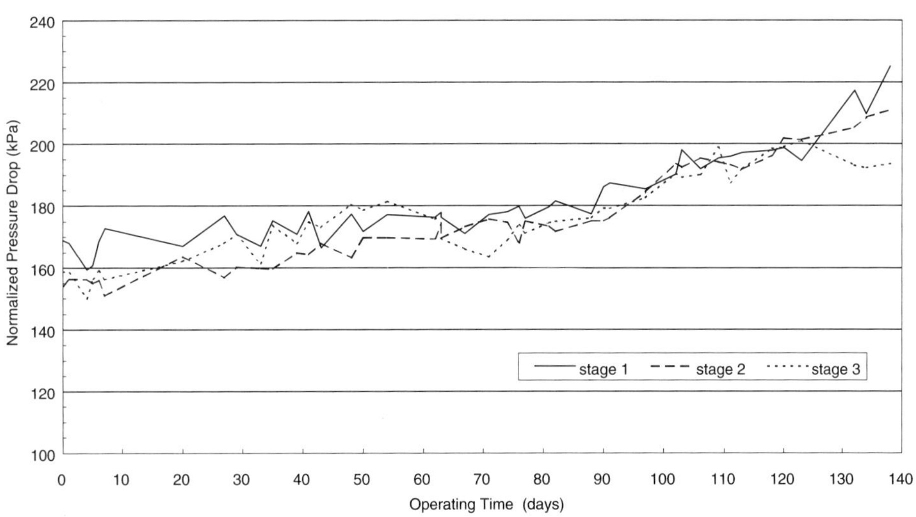

Figure 3 Normalized pressure drop of the three stages of the nanofiltration plant in Vechterweerd (after modification)

The measures taken to minimize membrane fouling lead to a cleaning frequency of once per 3 months. The prevention of membrane fouling remains difficult: in Vechterweerd membrane fouling is mainly caused by biogrowth.

Boerhaar

Within two years the color and hardness of the raw water of the treatment plant Boerhaar will rise to values of 20 mg Pt/l and 3.5 mmol/l. This is due to a change in the groundwater catchment. The aim of the research is to investigate the addition of nanofiltration in order to reduce the hardness and the color of the water.

Evaluating the pilot plant investigations at Hammerflier and Vechterweerd WMO was not satisfied with the results of the prevention of membrane fouling. Biogrowth in aerobic feedwater leads to cleaning frequencies of up to four times a year.

In the USA nanofiltration is often applied to the untreated groundwater [9, 10]. Typical cleaning frequencies for full-scale nanofiltration systems treating groundwater in the USA are 3 months to 2 years, with an average of 6 months [11].

Until Boerhaar direct application of nanofiltration to the groundwater was not considered because of the high iron content of the groundwater. The groundwater of Boerhaar is anaerobic and contains only 4 mg/l iron. The $MFI_{0.45\ \mu m}$ of the groundwater is $< 0.5\ s/l^2$. Due to membrane fouling with the aerobic operation at Hammerflier and Vechterweerd, WMO decided to investigate anaerobic operation of the nanofiltration installation at Boerhaar.

To compare the performance of two membrane types under anaerobic conditions a bench-scale experiment was carried out. The bench scale unit was equipped with two pressure vessels. See Table 6.

The results (Figure 4) are very good. The flux over the first 45 days was 16.9 $l/m^2.h$. From day 45 the flux was increased to a level of 20.7 $l/m^2.h$. From day 172 an antiscalant (3 mg/l) was dosed to the feedwater, to examine possible effects on membrane fouling. No problems with maintaining the anaerobic conditions of the membranes are observed [12].

Table 6 Data pilot plant Boerhaar

Stages	one stage (2 parallel pressure vessels)
Pressure vessel	length 80", each vessel contains 2 membrane elements, 4" diameter, 40" length
Feed capacity	2 × 1.6 m^3/hr
Product flux	vessel 1: 20 l/m^2.h/vessel 2: 22 l/m^2.h
Recovery	17.9%
Dosage	3 mg/l antiscalant (Permatreat 191) from day 172
Membrane type	pressure vessel 1: TriSep TS 80/pressure vessel 2: Hydranautics ESNA

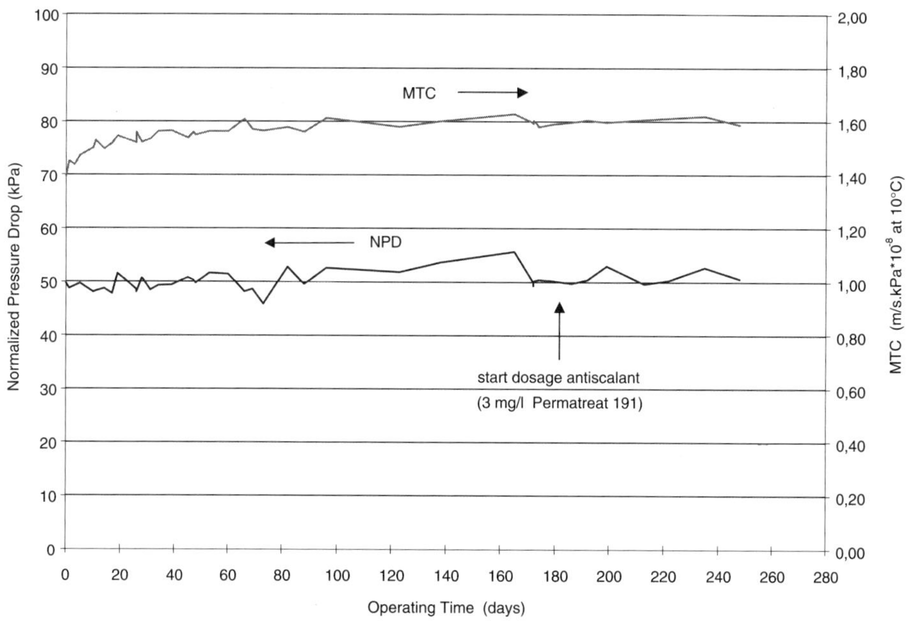

Figure 4 NPD and MTC of anaerobic nanofiltration in Boerhaar (pressure vessel 1)

During a period of 9 months the MTC-value slightly rose from 1.39 to 1.6*10-8 m/s.kPa. A change in the NPD is not observed, thus a cleaning procedure is not necessary during this period. In this case membrane fouling can be neglected. It can be expected that the cleaning frequency of the full-scale installation will be less than 2 times a year.

Some data (pressure vessel 1 = pv1) are presented in Table 9. Measurements of the BFR are not carried out due to the anaerobic conditions of the water. AOC measurement of the anaerobic feedwater didn't show reliable results.

Engelse Werk

At the treatment plant Engelse Werk, surface water is abstracted via bank filtration. After extended pretreatment (aeration, 1^e RSF, aeration and 2^e RSF), activated carbon filtration is applied. Due to consumers' complaints about hardness and a slight taste, WMO decided to investigate the improvement of water quality with an additional step, nanofiltration.

The research objective was: what is the best place for nanofiltration in the treatment scheme to prevent membrane fouling?

For the position of the nanofiltration in the treatment scheme two possibilities were considered:

1. First step in treatment. Consequence: anaerobic feedwater (iron 7 mg/l).
2. After extended pretreatment. Consequence: aerobic feedwater (iron 0.02 mg/l).

A bench-scale unit is used to compare both possibilities. The bench-scale unit is equipped with two pressure vessels, so both experiments are carried out on the same time. Also, the same membrane type is used (Table 7). There is no dosage of antiscalant to the feedwater, because Boerhaar showed that the antiscalant (Permatreat 191) didn't contribute to membrane fouling. No cleanings are carried out during the 120 days of the experiments.

Table 7 Data pilot plant Engelse Werk

Stages	one stage (2 parallel pressure vessels: 1 anaerobic, 1 aerobic operation)
Pressure vessel	length 40", each vessel contains 1 membrane element, 4" diameter, 40" length
Feed capacity	1.5 m^3/hr (each vessel)
Product flux	25 l/m^2.h
Recovery	10%
Dosage	none
Membrane type	TriSep TS 80

The changes in MTC and NPD for the two pressure vessels are summarized in Table 8:

Table 8 Changes in MTC and NPD for the two pressure vessels

Engelse Werk operation // day	MTC (10^{-8} *m/s.kPa*)		NPD (*kPa*)	
	0	120	0	120
anaerobic	1.24	1.39	35	35
aerobic	1.43	1.57	30	57

The MTC of both membranes increased slightly (same as Boerhaar). The NPD of the membrane with the aerobic feedwater increased 90%. Autopsy showed that biofouling (slimy layer) and deposition of iron (brown) occurred on the surface of the aerobic membrane. The results of the change in NPD of both pressure vessels are shown in Figure 5.

The membrane with the anaerobic feedwater showed no change in NPD. Autopsy showed that the membrane surface is partly black-colored, but no biofouling or deposition of iron is observed.

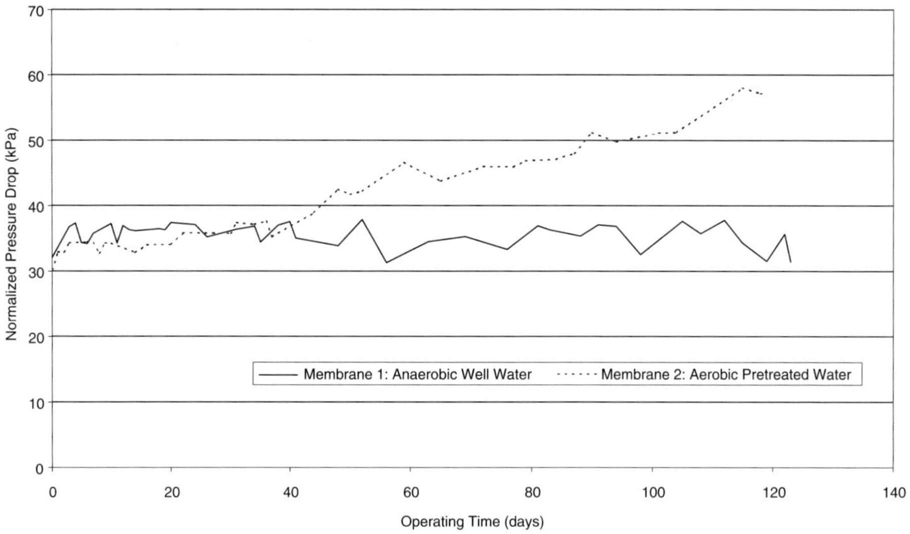

Figure 5 Normalized pressure drop of anaerobic and aerobic nanofiltration in Engelse Werk

The results of this bench-scale study are shown in Table 9 (EW = Engelse Werk).

DISCUSSION AND CONCLUSIONS

Prediction of Membrane Fouling

In the presented experiments biofouling and colloidal fouling are the main mechanisms for membrane fouling. There is no evidence for scaling (Hammerflier, Vechterweerd). Organic fouling only occurred at Hammerflier with a membrane type that is sensible for adsorption of NOM (humic acids).

Parameters that may predict biofouling (AOC and BFR) and colloidal fouling (MFI and Fe) are summarized in Table 9. Data about membrane fouling of two Dutch surface water plants, where

Table 9 Parameters that indicate membrane fouling for the different studies

Treatment Plant	Feedwater	AOC (*µg Ac-C/l*)	BFR (*pg ATP/cm²*)	MFI (*s/l2*)	Fe (*mg/l*)	NPD-change (*kPa/day*)	Temp (°C)
Hammerflier	aerobic/period 3	10	20	3.5	<0.03	0.22	10.5
Vechterweerd	aerobic/period 1	8.5	2.7	3	0.025	0.33	11.3
Vechterweerd	aerobic/period 2	27	310	4	0.025	7.8	11.3
Boerhaar	anaerobic pv 1	–	–	<0.5	3.9	~0.01	10.6
Engelse Werk	anaerobic	8	–	<0.5	7	~0.01	13
Engelse Werk	aerobic	7	~10	0.75	0.02	0.23	13
Andijk/RO	aerobic/after UF	8	max. 10	0.2	<0.04	max. 0.8	2–22
Leiduin/RO	aerobic/ref. scheme	3.8	0.92	1.5	<0.01	~0.01	12–25

research with *reverse osmosis* membranes is performed, are added to Table 9 [7, 13, 14, 15].

From the results it can be concluded:

- In the case of *anaerobic* feedwater the low MFI-value and the anaerobic conditions give an indication that membrane fouling is not expected. In the pilot plant it was no problem to maintain the anaerobic conditions of the membranes.
- Values for *aerobic* feedwater with a $MFI_{0.45\ \mu m} < 3$ s/l^2, an Fe(III)-content < 0.03 mg/l and an AOC < 10 μg Ac-C/l don't give a guarantee that membrane fouling of nanofiltration will not occur. The meaning of the MFI seems to be limited [16]. In this research the contribution of biofouling to membrane fouling is a dominant factor. To analyze the contribution of the MFI to membrane fouling it is necessary to exclude biofouling.
- Measurement of the biofilm formation rate (BFR) for *aerobic* feedwater gives a better indication for the prediction of membrane fouling caused by biogrowth (measured as NPD). See Figure 6.

In Figure 6 the assumption is made that membrane cleaning is necessary if the increase of the NPD is 40 kPa.

Prevention of Membrane Fouling

Prevention of membrane fouling should be focussed on two issues: the quality of the feedwater and the quality of the dosages.

Aerobic Feedwater

- Membrane fouling due to biogrowth can be prevented if the aerobic feedwater has a low BFR. An indication from Figure 6 can be obtained that a BFR < 1.7 pg ATP/cm^2 seems to be safe for a cleaning frequency lower than 2 times a year. A BFR < 5 pg ATP/cm^2 seems to be safe for a

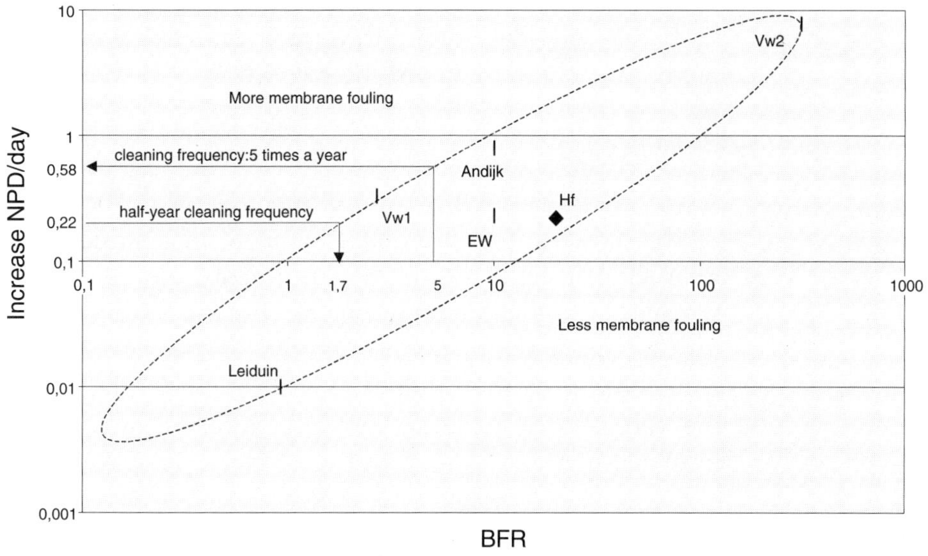

Figure 6 Relation between BFR and increase NPD per day for nanofiltration and RO systems with aerobic feedwater

cleaning frequency of less than 5 times a year. These tentative guidelines are very strict and need further foundation. Higher product fluxes of the first stage (> 23 $l/m^2.h$) will lead to higher cleaning frequencies.

- At Vechterweerd the biofilm monitor shows that biofouling promotes iron deposition. Values for $MFI_{0.45\ \mu m} \approx 3\ s/l^2$ and an Fe(III) content < 0.03 mg/l, which seem to be more or less safe at low BFR-values (< 1.7 pg ATP/cm^2), can be unsafe at higher BFR-values (> 10 pg ATP/cm^2).

Anaerobic Feedwater

- Keeping the stable anaerobic groundwater under anaerobic conditions has a very good influence on the prevention of membrane fouling. Microbiological fouling can be neglected under anaerobic conditions. Anaerobic groundwater has a low MFI-value (< 0.5 s/l^2) due to the good solubility of Fe(II), so colloidal fouling through iron deposition on the membrane surface can also be neglected.

Quality of the Dosages

- A part of the biofouling proved to be related to the quality of the dosages. For the quality control of chemical-pure hydrochloric acid (36%) a new protocol is developed that the acid is acceptable with a DOC content < 5 mg/l. Also, for the antiscalants quality control is needed. For the contribution of the dosages to membrane fouling a BFR measurement of the feedwater (after addition) will provide information.

Anaerobic or Aerobic Feedwater?

Due to the change from aerobic to anaerobic feedwater, WMO made a big step in the control of membrane fouling. This is caused by two phenomena:

1. Microbiological fouling can be neglected under anaerobic conditions.
2. Deposition of iron can be avoided by having anaerobic conditions. The aerobic pretreatment changes the oxidation state of the iron (iron-II to iron-III). Although most of the iron is removed during the aerobic pretreatment, very small amounts of iron-III cause more problems in aerobic nanofiltration than big amounts of iron-II in anaerobic nanofiltration.

Advantages of anaerobic operation of nanofiltration membranes are:

1. The increase of the NPD is lower and better to control.
2. The cleaning frequency is lower (biofouling can be neglected), less than 2 times/year.

3. Due to the application of nanofiltration as the first treatment step in groundwater treatment, less conventional treatment is needed (Fe and Mn are removed with nanofiltration, post-treatment is only needed to add O_2 and to remove CH_4, CO_2 and NH_4^+).

Disadvantages of anaerobic operation of nanofiltration membranes are:

1. Special precautions must be taken to keep the installation anaerobic during production stops.
2. If the iron content of the concentrate is too high (the Netherlands: Fe > 5 mg/l), additional treatment before discharge may be needed.

Application of Results in Design Treatment Plant Boerhaar

WMO applied the results of the research program in the design of the treatment plant Boerhaar. The capacity of the treatment plant is 2 million m^3/year drinking water (2.3 MGD). The choice is made that:

- Nanofiltration is the first step in water treatment (after cartridge filtration). The feedwater is anaerobic. The recovery is restricted to 80% to prevent $BaSO_4$-scaling. An antiscalant dosage of about 2 mg/l is needed (Permatreat 191).
- Split treatment is applied to maintain a minimum hardness. It results in a hardness of 1.5 mmol/l and a color of 8.5 mg Pt/l.

Use of results in design treatment plant Boerhaar

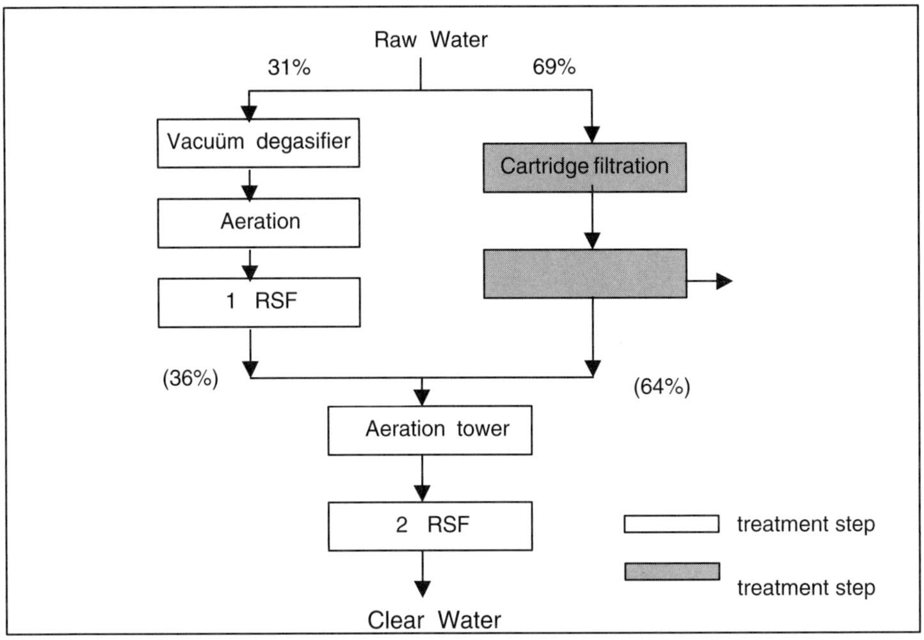

ACKNOWLEDGEMENTS

The investigations at the location Vechterweerd have been performed as part of the Integrated Membrane Systems project (IMS) conducted by three water supply companies in the Netherlands and three in the USA, Kiwa and the University of Central Florida (UCF). The American Water Works Association Research Foundation (AWWA-RF) and the US Environmental Protection Agency, Drinking Water Research Division, cofunded the project. The IMS project is part of the Joint Research Program of the Netherlands Water Works Association (VEWIN) and the Research Program of the American Water Works Association Research Foundation (AWWA-RF).

The authors thank Frank Schoonenberg Kegel of Kiwa and Simon Bakker and Simon in 't Veld of WMO for their contribution to the study.

REFERENCES

1. Van der Kooij, D., Van Lieverloo, H., Schellart, J. and Hiemstra, P. (1998). Distributing drinking water without disinfectant: highest achievement or height of folly? *Proceedings specialized conference on drinking water distribution*, Mülheim an der Ruhr, Germany, September 28–30, 1998, pp. VII 1–13.
2. Hiemstra, P., Van Paassen, J.A.M., Rietman, B.M., Nederlof, M.M. and Verdouw, J. (1997). Reduction of color and hardness of groundwater with nanofiltration — Is pilot plant investigation really important?, *Proceedings AWWA Membrane Technology Conference*, New Orleans, February 23–26, 1997, pp. 857–881.
3. Van der Kooij, D. (1992). Assimilable organic carbon as an indicator of bacterial regrowth. *Journal American Water Works Association* (84): February 1992, pp. 57–65.
4. Van der Kooij, D., Vrouwenvelder, J.S., Veenendaal, H.R. and Van Raalte-Drewes, M.J.C. (1994). Multiplication of aeromonads in groundwater supplies in relation with the biofilm formation characteristics of drinking water. *Proceedings AWWA WQTC*, San Francisco California, November 6–10, 1994, pp. 1349–1363.
5. In 't Veld, S. (1998). Voorspellen van biologische vervuiling van membranen met AOC-metingen en biofilmmonitoring (in Dutch). *H_2O* (31) 1998, nr. 21, pp. 22–25.
6. Van Paassen, J.A.M., Kruithof, J.C., Bakker, S.M. and Schoonenberg Kegel, F. (1998). Integrated multi-objective membrane systems for surface water treatment: pre-treatment of nanofiltration by riverbank filtration and conventional ground water treatment. *Desalination* 118 (1998), pp. 239–248.
7. Vrouwenvelder, J.S., Van Paassen, J.A.M., Folmer, H.C., Hofman, J.A.M.H., Nederlof, M.M. and Van der Kooij, D. (1998). Biofouling of membranes for drinking water production. *Desalination* 118 (1998), pp. 157–166.
8. Vrouwenvelder, J.S., In 't Veld, S. and Van Paassen, J.A.M. (1998). Biofilmvorming in nanofiltratiemembranen van de proefinstallatie te Vechterweerd (in Dutch). Nieuwegein, Zwolle, The Netherlands, *KOA* 98.017, 1998, pp. 1–51.
9. Morin, O.J. (1994). Membrane plants in North America. *Journal American Water Works Association* (86), December 1994, pp. 42–54.

10. Bergman, R.A. (1996). Cost of membrane softening in Florida. *Journal American Water Works Association* (88), May 1996, pp. 32–43.
11. Taylor, J.S., Mulford, L.A., Durenceau, S.J., Barrett, W.M. (1989). Cost and performance of a membrane pilot plant. *Journal American Water Works Association* (81), November 1989, pp. 52–60.
12. Rietman, B.M. and Verdouw, J. (1998). Nanofiltratie op anaëroob grondwater te Boerhaar (in Dutch). *WMO report*, Zwolle, The Netherlands, 1998, pp. 1–26.
13. Kamp, P.C. (1995). Integral approach of surface water treatment using ultrafiltration and reverse osmosis. *Proceedings AWWA Membrane Technology Conference*, Reno, Nevada, August 13–16, 1995, pp. 31–38.
14. Kruithof, J.C., Schippers, J.C., Kamp, P.C., Folmer, H.C. and Hofman, J.A.M.H. (1998). Integrated multi-objective membrane systems for surface water treatment: pretreatment of reverse osmosis by conventional treatment and ultrafiltration. *Desalination* 117 (1998), pp. 37–48.
15. Van der Hoek, J.P., Bonné, P.A.C., Van Soest, E.A.M. and Graveland, A. (1997). Fouling of reverse osmosis membranes : effect of pretreatment and operating conditions. *Proceedings AWWA Membrane Technology Conference*, New Orleans, February 23–26, 1997, pp. 1029–1041.
16. AWWA Membrane Technology Research Committee (1998). Membrane processes. *Journal of American Water Works Association* (90), June 1998, pp. 91–105.

APPENDIX 1. SPECIFIC PARAMETERS FOR MEMBRANE FILTRATION

MTC = Mass Transport Coefficient

Normalized flux [m/s.kPa*10¯8 at 10°C]

$$MTC = \frac{Q_p * TCF}{(((P_v + P_c)/2 - (\pi_v + \pi_c)/2) - P_p) * A * 3600}$$

[m/s.kPa*10^{-8} at 10°C]

MTC	=	mass transport coefficient [m/s.kPa*10^{-8} at 10°C]
Q_p	=	permeate flow [m^3/h]
P_v	=	feed pressure [kPa]
P_c	=	concentrate pressure [kPa]
P_p	=	permeate pressure [kPa]
π_v	=	osmotic feed pressure [kPa]
π_c	=	osmotic concentrate pressure [kPa]
A	=	membrane area [m^2]
TCF	=	e^x = temperature correction factor

where $x = U (1 / (t_v + 273) - 1 / (t_r + 273))$

U = membrane constant (depend on membrane type)

t_v = actual temperature feedwater [°C]

t_r = reference temperature feedwater [=10°C]

Normalized Pressure Drop (NPD)

$$\Delta P_n = \Delta P_a * \frac{(Q_r)^{1.6}}{\{(Q_v + Q_c)/2\}^{1.6}} * (1.03(t_v - t_r))^{0.4}$$

ΔP_n	=	normalized pressure drop [kPa]
ΔP_a	=	actual pressure drop (= feed pressure – concentrate pressure) [kPa]
Q_v	=	feed flow [m^3/h]
Q_c	=	concentrate flow [m^3/h]
Q_r	=	reference flow [= 1.5 m^3/h]
t_v	=	temperature feedwater [°C]
t_r	=	reference temperature feedwater [=10°C]

CHAPTER · 6

Solving Membrane Fouling for the Boca Raton 40-mgd Membrane Water Treatment Plant: The Interaction of Humic Acids, pH and Antiscalants with Membrane Surfaces

April Richards, E.I., Environmental Engineer
Camp Dresser & McKee, Miami, Florida

William Suratt, P.E., Vice President
Camp Dresser & McKee, Ft. Lauderdale, Florida

Harvey Winters, Ph.D.
Fairleigh Dickinson University, Teaneck, New Jersey

Don Kree, Chemist
City of Boca Raton, Florida

INTRODUCTION

A 40-mgd membrane softening process is currently in design for the City of Boca Raton, Florida, and will be the largest nanofiltration plant in the world.

In recent years, the City of Boca Raton, Florida, has experienced a gradual but steady increase in the levels of color in the raw and finished water at its Glades Road Water Treatment Plant. In addition, increasing levels of dissolved organics in the raw water have made compliance with more stringent regulations for disinfection by-products (DBPs) more difficult with the existing conventional lime-softening process. While the lime process does remove some color, DBP precursors, and associated constituents, it is not capable of meeting both DBP and color standards simultaneously. In addition, the utility is receiving increasing customer complaints about high color levels in the finished water. With the passage of more stringent regulations for DBPs under the United States Environmental Protection Agency (USEPA) and State of Florida Safe Drinking Water Acts (SDWAs), the City has recognized the need to pursue other treatment process alternatives to provide continued compliance with drinking water quality regulations.

DESIGN

Once membrane technology was chosen to meet the City's multiple water quality objectives, other decisions followed, including the treatment capacity of the membrane process, the design recovery rate, the number and design of the membrane skids, membrane selection, and pumping configurations. In addition to design alternatives directly associated with the membrane process, pre- and post-treatment alternatives were evaluated, including multimedia pressure filters, strainers, cartridge filtration, chemical feed, degasification and odor control, and concentrate disposal. Continued testing of design alternatives and operating parameters is being performed using the City's membrane pilot plant. The basic components of the full-scale membrane plant design are summarized in Table 1.

Raw Water Quality

The City's existing raw water facilities consist of six wellfields that draw from the Biscayne Aquifer. There are a total of 56 wells with a total design pumping capacity of approximately 94 mgd. The wells are constructed to depths ranging from approximately 105 to 220 feet below land surface, with casing diameters ranging from 14 to 24 inches. This wellfield has historically supplied raw water to the existing lime-softening process, and constitutes the proposed source for the membrane-softening process. Table 2 presents the average raw water quality for the membrane process.

Table 1 Membrane plant design parameters

Parameter	Stage 1 & 2 Units	Stage 3 Units (Concentrators)
Capacity	36.76 mgd	3.24 mgd
Recovery Rate	85%	50%
Raw Water Capacity	43.2 mgd	6.48 mgd
No. Skids	10 × 3.68 mgd	2 × 1.62 mgd
Array	72:36	54 single stage
No. Elements/Vessel	7	7
Flux Rate	12.2 gfd	10.7 gfd
Transmembrane Pressure	85 psi	70 psi

Recovery Rate

The majority of membrane softening plants in Florida are two-stage plants with a recovery rate of 85 percent. This value is a practical upper limit for a two-stage process because of the requirement to maintain a minimum crossflow velocity in the last membrane element in each stage to prevent fouling. At this recovery rate, 47 mgd of well water would be required to produce 40 mgd of permeate, and 7 mgd of concentrate would be generated requiring disposal.

The City can dispose of the concentrate through their existing ocean outfall if the volume of concentrate can be reduced to approximately 3.5 mgd. To accomplish this, the design is based on a two-stage system with a recovery rate of 85 percent, followed by an independent third stage with a recovery rate of 50 percent, for an overall recovery rate greater than 90 percent. This combination would convert the 7 mgd of concentrate described above to 3.24 mgd of concentrate. Instead of 15 percent of the feedwater remaining as concentrate, now only 7.5 percent would remain, allowing the City to utilize the existing ocean outfall.

Table 2 Average raw water quality

Parameter	Units	Average
Aluminum	mg/L	< 0.025
Barium	mg/L	0.021
Calcium	mg/L	93
Copper	mg/L	< 0.026
Iron	mg/L	0.15
Magnesium	mg/L	3.81
Manganese	mg/L	< 0.007
Potassium	mg/L	3.11
Sodium	mg/L	25.1
Strontium	mg/L	0.57
Zinc	mg/L	< 0.028
Alkalinity	mg/L as HCO_3	265
Bromide	mg/L	< 4.8
Chloride	mg/L	38.8
Fluoride	mg/L	< 8
Nitrate	mg/L	< 5
Sulfate	mg/L	17
pH	units	7.2
Conductivity	mhos	562
TOC	mg/L	12.0
TDS (by evaporation)	mg/L	314
Silicon	mg/L as Si	5.0
Phosphorus	mg/L as P	0.77

PILOT PLANT TESTING

The City has owned and operated a membrane pilot plant for over five years. During this time the pilot plant was used by the City to comply with the USEPA's Information Collection Rule (ICR), which was promulgated to help generate a critical mass of data for use in the development of upcoming regulations. From the start of the ICR, the pilot plant experienced rapid fouling of its 5-micron cartridge filters, which required replacement approximately every 7 days. In order to alleviate this problem, a multimedia filter, was installed upstream of the cartridge filters to lengthen their run time. After the installation of the multimedia filter, a significant improvement was seen in the operating life of the cartridge filters, although the pilot plant operating data still showed a continuous decline in membrane specific flux (mass transfer coefficient) due to organic foulants that were not being removed by the multimedia filter. These specific flux decline trends in all three stages of the membrane pilot plant are shown in Figure 1. Fouling at this rate would require cleaning approximately every three months.

The most significant problem encountered during pilot testing has been the rapid membrane fouling by the high humic acid content of the feedwater. The fouling manifests itself in continuously falling permeability and increased feed pressures. Initially, testing with conventional acid and antiscalant addition resulted in fouling that would require frequent membrane cleaning. The pilot was then operated at an 85 percent recovery without acid using only an antiscalant, and no carbonate scaling was observed. There was some improvement in performance, but organic fouling persisted.

The pilot plant was reconfigured to allow "side-by-side" comparative tests of membranes, antiscalants and dispersants. The following comparative tests were conducted:

1. Antifoulant (dispersant) addition versus no antifoulant addition
2. One brand of antiscalant versus another brand
3. One type of membrane versus another type
4. Antiscalant addition versus no antiscalant addition

The final pilot test eliminated the use of all chemicals including antiscalants. The pilot plant performance improved dramatically and it is believed that the naturally occurring humic acids in the raw water are acting as natural antiscalants. To date, stable operation from 85 to over 90 percent recovery has been observed with no apparent fouling.

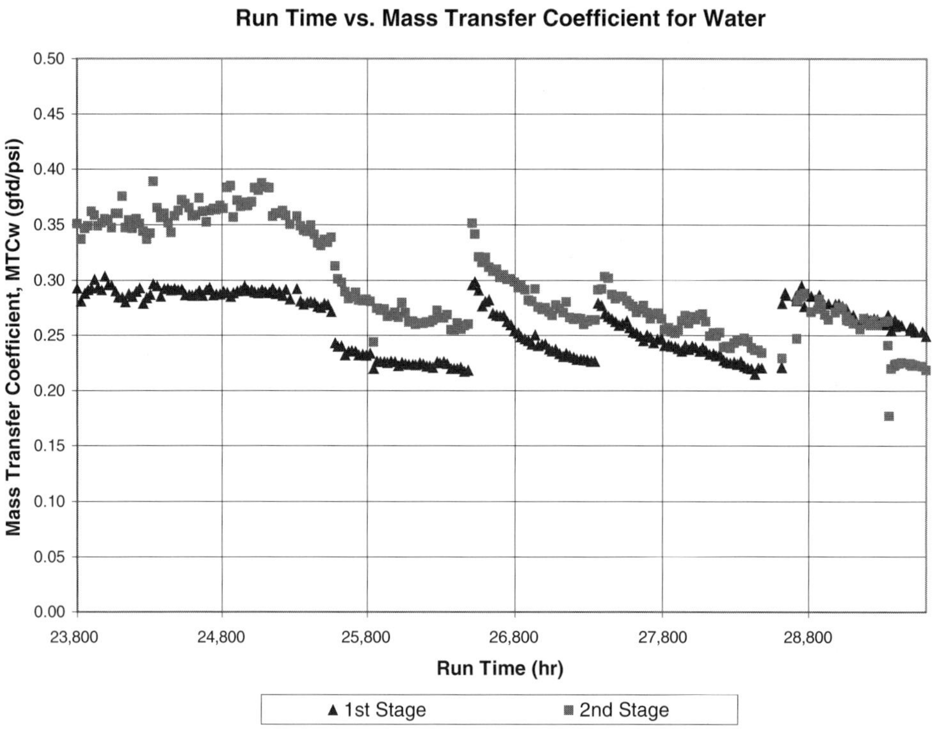

Figure 1 Initial fouling trends of membrane pilot plant

NATURE OF FOULANT

The quality of the raw water from the Boca Raton wellfields has been known to fluctuate over time and there are several areas of concern. The problematic characteristics of this raw water with respect to the membrane process include:

- Natural Organic Matter. Very high dissolved, natural organic matter, mostly humic acids, that impart high color to the water. These substances are adsorbed onto the membrane surface and are a foulant that affects membrane performance significantly. They generally can be cleaned off the membranes with a high-pH cleaner. A small amount of chlorine in the cleaning solution is very effective and does not appear to have any significant short-term effect on membrane performance.

- Clay and silts. Significant amounts of clayey silts and sand with particle sizes less than 5 microns. Some of the silt is colloidal (less than 1 micron) and will pass through media filters and cartridge filters, and deposit on the membranes. The silts typically can be removed with a high-pH cleaner.
- Entrained air. Well inspections have revealed that there are holes in the column pipes of several of the wells and raw water transmission piping, which may be allowing air to enter the raw water system. In addition, there are indications that several of the well pumps may be periodically "breaking suction" and pumping air into the raw water transmission system. This presents the potential for oxidation of dissolved iron and hydrogen sulfide that will foul the membranes.
- Corrosion by-products. These include rust particles and iron sulfide. Most of the wells are equipped with carbon steel casings and column pipes, and the raw water transmission system and wellheads are constructed of ductile iron. Corrosion by-product particles should be filtered out by the cartridge filters, but large quantities may reduce the life of the cartridge filters. Colloidal-sized particles will foul the membranes.
- Hydrogen sulfide. The average concentration in the raw water is approximately 0.25 mg/L, with a maximum of approximately 0.5 mg/L. While dissolved hydrogen sulfide should not cause problems with the membranes, any aeration of the raw water upstream of the membrane process could result in oxidation of the sulfide to colloidal elemental sulfur, which will foul the membranes.

- Dissolved Iron. The average concentration in the raw water is about 0.15 mg/L. Dissolved (ferrous) iron will be rejected by the membranes and not cause a fouling problem. However, if the well water is aerated, the ferrous iron will be oxidized to the ferric form, which exists as ferric hydroxide or ferric carbonate, both of which are precipitates that foul membranes very quickly. Iron hydroxide is a colloidal material that can pass through cartridge filters and plug the membranes. Iron hydroxide is easily removed from the membranes with a low-pH cleaning.
- Bacteria. These include iron bacteria, sulfate-reducing bacteria, and slime-forming bacteria, including Pseudomonas A. These bacteria, if allowed to colonize in the piping and membrane units, could seriously reduce the performance of the membranes and require frequent cleaning and membrane replacement.

Effect of pH

Most of the membrane-softening plants operating in South Florida lower the feedwater pH to 6.0 or lower. This is much lower than necessary for just carbonate scale control if an antiscalant is used with the acid. In fact, carbonate scale can be controlled by antiscalants alone with no acid addition. This has been demonstrated at the City's membrane pilot unit. However, the operators of most plants treating surficial aquifer waters have found that membrane fouling is lower when they operate with a pH of 6.0 or less. Some plants are operating at a pH of 5.8, while others have stopped using antiscalants completely and are controlling carbonate scale with acid addition only. They have experienced much less fouling than the other plants that continue to mix acid with antiscalants. If the pH of the feedwater for the new 40-mgd facility is lowered to 6.0, it will require one 4,000-gallon truckload of sulfuric acid per day.

Operators of plants in South Florida have found that mixing conventional polymer antiscalants with sulfuric acid forms a precipitate that fouls the membranes. The very high amount of humic acid in the feedwater at Boca Raton, and other plants in South Florida, acts as a natural antiscalant. Research has also found that lowering the pH destabilizes the natural organic matter in the feedwater, causing it to coagulate more easily in the presence of polymer antiscalants and contribute to fouling. Furthermore, destabilization by acid may make the organics more assimilable by bacteria, thus encouraging bacterial growth. Considering the safety issues associated with the daily delivery of truckloads of acid, pilot testing without acid has successfully demonstrated that fouling can be controlled with antiscalants alone and will allow the City to avoid handling large volumes of acid. An acid system, however, will be designed and installed because periodic operation at low pH for a few hours per week will probably be necessary as a maintenance routine. Additionally, acid addition may be required for operation of the third-stage units.

Effect of Antiscalants and Dispersants

Side-by-side testing of two membrane pressure vessels with one receiving antiscalant and one receiving both antiscalant and dispersant showed that the vessel receiving both pretreatment chemicals fouled more rapidly than the one receiving just antiscalant. These trends can be seen in Figure 2. During the first 22 days of testing both antifoulant and antiscalant were added to the feedwater. For the following 20 days (until approximately day 40) only antiscalant was used. During the final 60 days, neither antiscalants nor dispersants were added. Discontinuation of antiscalant addition had no apparent effect on performance.

In order to corroborate results of pilot-scale tests, Professor Harvey Winters of Fairleigh Dickinson University conducted several bench-scale tests to measure the fouling potential of various membranes in conjunction with antiscalant and antifoulant addition using the City's feedwater. Zeta potential measurements at

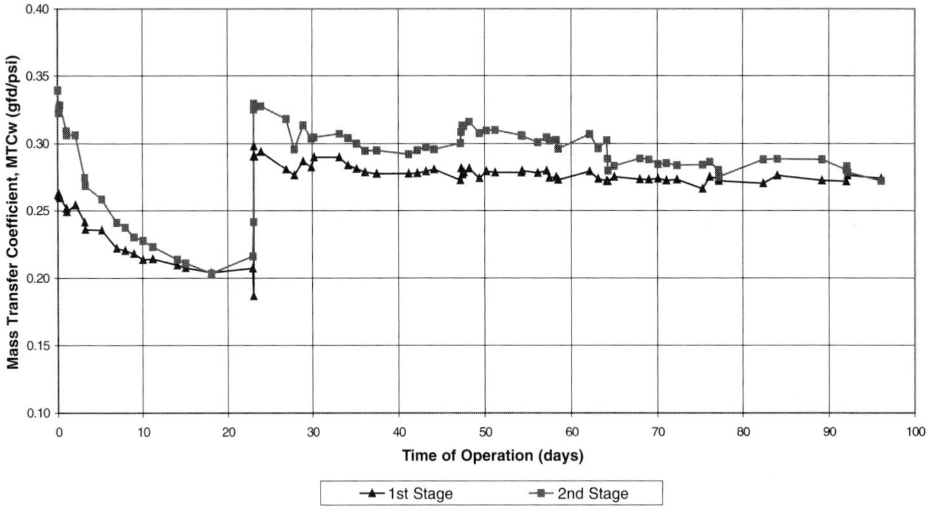

Figure 2 Fouling trends with antiscalants and dispersants

different pH indicated significant differences in the attractive forces of the humic acids in the presence of various antiscalants. Other studies involved electrophoretic measurements, which showed that some antiscalants and dispersants actually enhanced the adsorption of humic acids onto the membrane surfaces. Differences in a membrane surface's ability to attract the organic foulants were also detected.

Zeta Potential and Electrophoresis Tests

The design team engaged Dr. Harvey Winters to conduct laboratory-scale tests aimed at measuring the fouling potential of different membranes and antiscalants and dispersants in combination with Boca Raton's feedwater. Zeta potential measurements of the repulsion between negatively charged humic acids and the membrane were made and revealed significant differences between membranes. Most membranes showed a positive zeta potential at pH levels below 7, which would decrease the repulsion between humic acids and membranes and promote humic acid fouling.

Electrophoretic measurements with polyacrylamide gels were also made by Dr. Winters that showed that some antiscalants and dispersants enhanced the adsorption of humic acids onto the membrane surfaces. The pilot results shown in Figure 2 confirm the bench-scale electrophoretic tests.

Pilot testing has shown that, with the high amount of humic acids in the raw water acting as natural antiscalants, it is possible to operate successfully without acid or antiscalants.

STRATEGIES TO REDUCE FOULING

Wellfield Remediation

Based on the fouling observed during the membrane pilot testing, implementation of a comprehensive, aggressive rehabilitation program for the wellfield and raw water transmission mains, along with an ongoing well and transmission system maintenance program, is considered critical to the success of the project. This program will include rehabilitation of the hydrogeological production characteristics of the well itself, including well disinfection, chemical treatment, and redevelopment. In addition, mechanical improvements to the well casing, pump, column, and wellhead are included, such as lining of the casing with a nonferrous material, replacement of the column pipe with nonferrous pipe, installation of foot valves on any wells that are not so equipped, and rehabilitation of the wellhead piping. In addition, all the raw water transmission piping will be pigged, and provisions will be made to allow regular pigging and other line maintenance in the future. These improvements are aimed at improving the capacity of the wellfield and raw water quality. It is expected that the sand and silt problems can be substantially reduced, and that problems associated with corrosion by-products, aeration of the raw water, and oxidation of the above-discussed constituents can be minimized.

Acid Addition

The pilot plant has been run for several extended periods without acid addition, with no more fouling experienced than with acid addition. This can be accomplished because of the level of humic acids in the raw water, which act as natural antiscalants. It is possible that acid addition will be required in the third stage of membranes due to the concentrated nature of the feedwater entering the third stage. The acid would control carbonate scaling and help to keep iron in solution.

Multimedia Filtration

It has been proven through pilot testing that multimedia filtration is successful at increasing the life of the cartridge filters. The current course of action includes rehabilitation of the existing wellfield to improve the raw water quality entering the membrane plant and pretreatment with multimedia pressure filters to extend the life of the cartridge filters and to protect the membranes.

CONCLUSIONS

The results of pilot-scale and bench-scale testing reveal the following information for design of the new 40-mgd nanofiltration plant.

1. The membrane system at over 90 percent recovery can operate without using acid to control carbonate scale, and operation at a natural pH reduces membrane fouling caused by the dissolved humic acids.
2. Certain commercially available antiscalants and dispersants increase the rate of membrane fouling by humic acids.
3. It may be possible to operate without antiscalants (or dispersants) at over 90 percent recovery because of the natural antiscalant properties of the dissolved organics in the feedwater.

4. Nanofilters have different zeta potential in the presence of the feedwater and also display differing tendencies to be fouled by the dissolved organics. Furthermore, the fouling tendencies increase at lower pH.

5. Zeta potential and electrophoretic measurements are a useful tool for screening membranes and antiscalants.

This project is being conducted by Camp Dresser & McKee in collaboration with CH2M Hill.

CHAPTER · 7

Morphological Properties of Fouling Cakes Formed from Heterogeneous Suspensions

Volodymyr V. Tarabara and Mark R. Wiesner
Department of Environmental Science and Engineering
Rice University, Houston, TX

INTRODUCTION

Formation of fouling layers on membrane surfaces remains a major obstacle in application of membrane filtration as a water treatment unit process [1–3]. The hydrostatic resistance of the fouling layer is often the factor that controls the permeation rate and, yet, is hard to predict quantitatively. The detailed understanding of the link between mechanisms of formation and resulting morphologies of the fouling layers is of critical importance for efforts to minimize fouling by controlling operational parameters and to optimize the performance of a membrane filtration unit.

There is growing evidence that the morphology of colloidal deposits is determined by the mode of long-range transport of particles to the vicinity of a deposit, short-range chemical interactions between the particles and the deposit, and the morphology of the underlying substrate [2, 4–7]. The focus of the research reported on in this chapter is the effect of surface chemistry of particles on the morphology of fouling layers. Different surface chemistries were simulated using a "lump" parameter—collision efficiency α. An on-lattice Monte Carlo model was employed to study the effect of α on the morphology of fouling layers formed from a monodisperse suspension in the course of dead-end filtration. Here, the authors report on results of these simulations and discuss implications of suspension heterogeneity for membrane filtration.

DEPOSITION ALGORITHM

The Monte Carlo simulation of deposition from a monodisperse suspension was performed using 2D on-lattice algorithm described elsewhere [5]. Briefly, particles were introduced into the simulation domain one by one at a release line above the deposit at a random horizontal position along the substrate. A pseudo-Péclét number was used to estimate the relative importance of diffusive and deterministic forces acting on a particle and to determine the distribution of probability of movement in one of 8 directions on the 2D grid for the particle: $N_{Pe} = \frac{V_g + V_{pf}}{V_d}$, where V_g, V_{pf}, and V_d are particle velocities due to gravity, permeate flow and diffusion, correspondingly. The simulation workspace was a square 1,000 × 1,000 lattice with periodical boundary conditions imposed in the horizontal direction to minimize the influence of boundaries.

The collision efficiency α was used to model the propensity of particles in suspension to attach to the deposit upon a contact. Collision efficiency is defined as the probability that an approaching particle is not immediately attached to another particle or cluster but is reflected or detached. For example, the collision efficiency of 0.1 implies that it would take, on average, 10 collisions for the particle to be attached. In this work a collision that did not result in attachment was followed by further migration of the particle in a direction determined by a local structure of the deposit and a force balance on the particle. Nearest neighbors of occupied sites were considered to be active sites, i.e., the sites where attachment may occur. The following assumptions were made in the model: a) suspension was monodisperse, b) particles were spherical, c) no post-attachment was allowed and e) detachment to an active site was forbidden.

To model simultaneous deposition of two sorts of particles, each sort having its own surface chemistry, the particles were assigned a "color" and, corresponding to this, a "color" collision efficiency α. The fraction of each sort of particles was considered to be the same. This was done to factor out effects of surface chemistries of particles.

RESULTS AND DISCUSSION

Effect of Collision Efficiency Versus Effect of Long-Range Transport

The deposition model was run for cases of different N_{Pe} with a fixed value of α and for cases of different α with a fixed value of N_{Pe}. Changes in short-range structure induced by changes in α were found to differ from those induced by changes in N_{Pe} (Figure 1). In the former case, a decrease in α results in a thickening of the deposit branches, which is in sharp contrast to the latter case, where an increasingly ballistic character of deposition leads to formation of an increasingly dense, web-like structure, more homogeneously distributed in space.

To characterize the modeled deposits quantitatively, the fractal dimension of deposits was calculated from the upper surface properties by plotting average height of the deposit $\overline{h}$ versus the number of deposited particles for the growing deposits. For the dependence $\overline{h} \propto N^{\phi}$, the power ϕ is related to fractal dimension D_f by $D_f = d - d_s + \frac{1}{\phi}$. Here, $d = 2$ and $d_s = 1$. As expected, an increase in N_{Pe} and a decrease in α lead to formation of deposits with higher fractal dimensions. It was observed that at large values of N_{Pe}, compact deposits were formed and $D_f \rightarrow 2$ for all values of α. In the case of small N_{Pe} values, though, substantial differences in the morphology were observed. For $\alpha = 1.0$, for example, D_f converged to ~1.77 (1.7 being the theoretical value for diffusion-limited

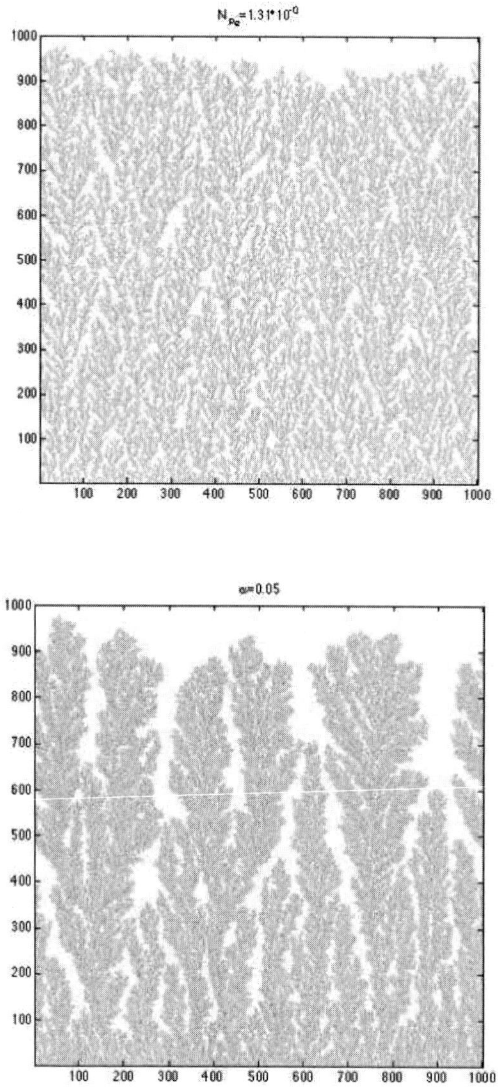

Figure 1 Simulated deposits formed from particles with a) $\alpha^{(1)} = 1$, ${N_{Pe}}^{(1)}$ and b) $\alpha^{(2)} = 0.03$, ${N_{Pe}}^{(2)} \cdot {N_{Pe}}^{(1)} > {N_{Pe}}^{(2)}$

growth), while for $\alpha = 0.1$, D_f was considerably higher and converged to ~1.84 (Figure 2).

It was also found that deposits formed from particles with different α could have the same fractal dimension as corresponding deposits formed from particles with different N_{Pe} while having obviously different structure, as described above. This observation implies that a single fractal dimension is not enough to characterize the deposit fully and uniquely, and that multifractal analysis is necessary [8].

Deposition from Heterogeneous Suspensions

In approaching the problem of modeling the common for many natural and industrial systems situations, when more then one sort of chemically different particles is deposited, the deposition algorithm

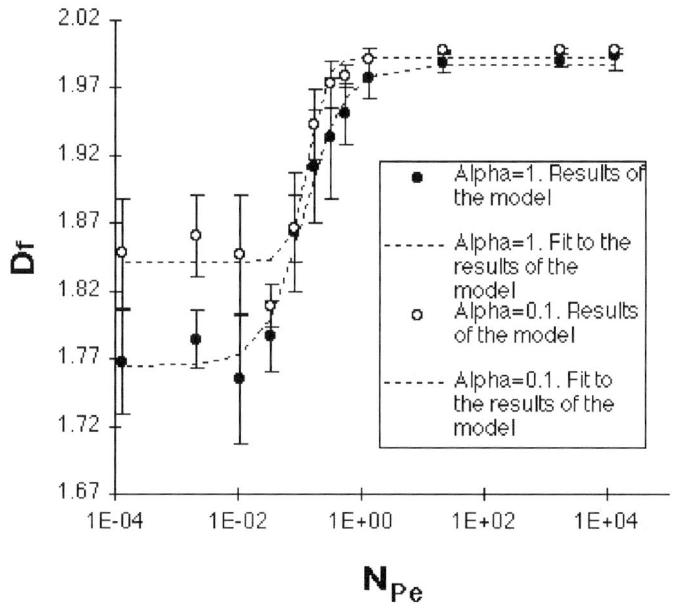

Figure 2 Fractal dimension of deposit as a function of Péclét number for two different collision efficiencies: $\alpha = 0.1$ and $\alpha = 1$

was modified to account for heterogeneity of suspension [9]. The first obvious step in this direction was to have two sorts of particles and, hence, 3 values of collision efficiency— α_{11}, α_{22}, and α_{12}, characterizing collisions between two particles of type 1, between two particles of type 2, and between two particles of types 1 and 2, correspondingly. The case $\alpha_{12} > \alpha_{11}, \alpha_{22}$ is trivial—particles of type 1 attach to particles of type 2 and vice versa and the chessboard-like pattern results. Thus, the non-trivial case $\alpha_{12} < \alpha_{11}$, α_{22} is of most interest. This combination of collision efficiencies corresponds to the suspension consisting of two sets of particles carrying a charge of the same sign and/or having different hydrophobicities.

The model predicts that the structure of the deposit is mostly determined by particles with larger values of α (cohesive fraction), while particles with smaller values of α (non-cohesive fraction) dominate the surface chemistry of deposits. Depositing cohesive particles form networklike structure that is filled by the non-cohesive fraction. The more strict the $\alpha_{12} < \alpha_{11}, \alpha_{22}$ condition, the more pronounced the segregation of particles from different sets into substructures elongated in the direction normal to the membrane surface. This deposit morphology has been observed previously in experiments modeling sand/mud mixture sedimentation in estuaries [10] and when directly studying the structure of membrane fouling cakes with transmission electron microscopy [11].

An important implication of this result for membrane filtration is that the usually employed model of resistances in series for a fouling layer's resistance to the permeate flow should be reconsidered. In fact, the slight deviation from chemical homogeneity in the suspension to be filtered may result in the formation of fouling cakes that have the morphology described above. Clearly, substructures formed from particles of different collision efficiencies have different specific resistances to permeate flow. In this situation, some pathways for the permeate flow through the fouling layer appear to be more favorable than others. In terms of the equivalent scheme of resistances, it means that now a more complex circuit of resistances connected in series and in parallel is formed, a total resistance being less than predicted if suspension is considered to be chemically homogeneous.

CONCLUSIONS

Morphology of fouling layers is a key unknown in modeling permeate flux in membrane filtration. The morphology determines the resistance of the fouling layer to the permeate flux, and is a function of particle transport and particle surface chemistry. It may be predicted based on the knowledge of the properties of particles that form the cake.

The structure of fouling layers formed from chemically heterogeneous suspensions is characterized by a varying degree of segregation of particles from different depositing fractions into substructures elongated in the direction normal to the membrane surface. Generally, cohesive particles determine the underlying structure of the fouling layer while particles from noncohesive fraction fill in the gaps in and, thereby, support otherwise potentially fragile, networklike structure formed from cohesive particles. Substructures formed from particles from different sets have different specific resistances, making some pathways for the permeate flow more favorable than others. As a result, the total resistance of the fouling layer is less than predicted when suspension is considered to be chemically homogeneous.

ACKNOWLEDGEMENTS

This work was supported in part by funding through the Gulf Coast Hazardous Substances Research Center.

REFERENCES

1. Belfort, G. Davis, R. H., Zydney, A. L. The behavior of suspensions and macromolecular solutions in crossflow microfiltration. *J. Membrane Sci.* 96 (1994), 1–58.
2. Wiesner, M. R. Morphology of particle deposits. *J. Environ. Eng.* 125, 12 (1999), 1124.
3. Bowen W. R., Jenner F. Theoretical descriptions of membrane filtration of colloids and fine particles: assessment and review. *Adv. Colloid and Interface Sci.* 56 (1995), 141–200.
4. Veerapaneni, S., Wiesner, M. R. Particle deposition on an infinitely permeable surface, *J. Colloid Interface Sci.* 162 (1994), 110–122.
5. Tarabara V. V., Wiesner, M. R. Effect of Collision Efficiency on the Evolution of the Surface of Diffusion-limited Deposits. Submitted to *J. Colloid and Interface Sci.*
6. Huisman, I. H., Elzo, D., Middelink, E., Trägårdh, C. Properties of the cake layer formed during crossflow microfiltration. *Colloids Surfaces A: Physicochem. Engng Aspects* 138 (1998), 265–281.
7. Riedl, K., Girard, B., Lencki, R. W. Influence of membrane structure on fouling layer morphology during apple juice clarification. *J. Membrane Sci.* 139 (1998), 155–166.
8. Lee, S.-L., Lee, C.-K. Heterogeneous reactions over fractal surfaces: a multifractal scaling analysis, *Int. J. Quantum Chem.* 64 (1997), 337–350.
9. Tarabara, V. V. Collision Efficiency of Colloidal Particles and Morphology of Deposits: Implications for Membrane Filtration. MS Thesis, Dept. Environmental Science and Engineering, Rice University (2000).
10. Toorman, E. A., Berlamont, J. E. Settling and consolidation of mixtures of cohesive and non-cohesive sediments. In *Advances in Hydro-Science and Engineering*, ed. S. S. Y. Wang, v. 1 (1993), 606–613.
11. Parron, C., Pierrisnard, F. Autopsie des Membranes Colmatées de la Première Tranche d'Essais. Rapport. Centre Européen de Recherche et d'Enseignement de Géosciences de l'Environnement, April 2000.

CHAPTER · 8

Characterization of UF and NF Membrane Fouling by Different Biopolymer Fractions Using Dead-End Filtration and Atomic Force Microscopy

Erin Mackey
Carollo Engineers, Boise, ID

Mark R. Wiesner
Dept. of Environmental Science & Engineering, Rice University, Houston, TX

Jean-Yves Bottero
Laboratoire des Géosciences et l'Environnement, Aix-en-Provence cedex, France

INTRODUCTION

Although ultrafiltration (UF) and nanofiltration (NF) are attractive technologies, a major impediment to their optimal performance is fouling potential, which is strongly correlated to raw water quality. Fouling is the reduction of membrane permeability by phenomena such as the accumulation of solutes in the pores and on the surface of the membrane. Fouling is operationally manifested as a drop in permeate flux per unit of applied pressure. It can also result in a change in the rejection characteristics of the membrane (resulting from a change in pore size distribution due to blockage of pores or changes in the charge characteristics of the surface). There are two types of membrane fouling: reversible (foulants that can be removed by hydrodynamic or chemical cleaning) and irreversible (those that cannot). Cleaning the membrane will recover flux lost to reversible fouling, but not irreversible fouling. Therefore, once a membrane is irreversibly fouled, there is a permanent, incremental increase in operating cost for the lifetime of that membrane.

One of the chief foulants encountered in membrane filtration of surface waters is dissolved organic matter (DOM) (1–7). Lainé et al. found the most important parameter in UF membrane fouling by a lake water was the degree of DOM adsorption (2). This was true for both hydrophilic and hydrophobic membranes. DOM interacts with the membrane surface through hydrophobic interactions, hydrogen bonding and other chemical interactions, reversibly and irreversibly fouling the pores. This fouling tendency is strongly influenced by the transmembrane pressure, solution chemistry and membrane character.

Surface waters contain a highly diverse spectrum of DOM molecules. As such, quantifying and qualifying the fouling effect of different types of dissolved organic matter are difficult. One approach is to break DOM into biopolymer groups: proteins, polysaccharides (PSs) and polysaccharides (PHAs), and examine their fouling potentials separately. Previous work using pyrolysis-GC/MS has suggested that organic matter rich in PSs and PHAs might be the dominant source of membrane fouling in surface water treatment (1, 8). But pyrolysis-GC/MS is a rough technique and can only give an approximate value of the DOM distribution. Better characterization of the way different classes of organic molecules foul UF and NF membranes is the ultimate goal of this work.

MATERIALS AND METHODS

This work sought to supplement standard dead-end filtration (DEF) fouling studies with surface imaging using atomic force microscopy (AFM) on freshly fouled samples to better characterize the interaction between proteins, PSs and PHAs and filtration membranes in the NF and the UF regimes. The model biopolymers selected were bovine serum albumin (BSA), polygalacturonic acid and rosolic acid, a protein, a PS, and a PHA respectively. The membranes used were two Spectrum (Laguna Hills, CA) cellulose ester membranes with MWCOs of 100 Da (NF) and 100 kDa (UF). In some cases, the ionic strength was elevated to 0.01 M by addition of NaCl.

Dead-End Filtration

DEF experiments using DOM solutions of varying compositions were prepared. Also, the experiment used various DOM surrogates and the ionic strength solutions. The pH was maintained at 6 for all experiments. Filtration rate and TOC concentrations were monitored over time to determine how the solution composition affected the rate of flux. DEF also provided the fouled membranes for AFM analysis.

DEF is a simple bench-scale method for simulating filtration in a dead-end mode (the bulk flow is perpendicular to the membrane surface). The DEF setup selected was a 200 mL Sartorius Ultrafiltration System. The mechanism is very simple—high-pressure air provides the driving force to push the water through the membrane.

Flat membranes were cut into discs to fit the test cells (48 mm in diameter). The prepared vessels were attached to a zero air cylinder and sealed. A constant pressure of 40 psi (276 kPa) was applied. To create conditions conducive to fouling, the solution was not stirred. This allowed maximum accumulation of solutes near the surface (i.e., maximum concentration polarization). This permeate volume was measured over time and analyzed for total organic carbon (TOC) content using a Shimadzu 5050A TOC analyzer.

In these experiments, membrane fouling followed a three-step process:

1. Clean water flux. An initial clean water flux was run to determine the maximum water flux of the membrane.

2. Filtration. The foulant solution was added and a second filtration run was made. Recovery was 50–75%.

3. Rinse. After the filtration step the membrane was cleaned hydraulically and another clean water flux test was run. Any reduction in clean water flux from step 1 was attributed to irreversible fouling of the membrane matrix.

In this work, the percent rejection (%R) is defined as the percent fraction of the initial mass of material retained by the membrane as concentrate in the cell. Flux reduction is the difference between the first and the second clean water flux rates (%-ΔJ). Fouling effect is the reduction in flux per unit TOC adsorbed ($\%\text{-}\Delta J/M_{ads}$).

Atomic Force Microscopy

AFM is a technique in which a fine tip (made of silicon or silicon nitride) is brought into contact with a surface (contact mode) or atomically close contact with a sample surface without actually touching it (tapping mode).

The AFM apparatus used in these experiments was a Digital Instruments MultiMode™ Scanning Probe Microscope with NanoScope® IIIa V4.23r3 software. The cantilever probes used were Digital Instruments Model NP-S silicon-nitride Nanoprobes™. The membrane samples were imaged wet using a fluid cell.

In AFM, the sample is placed under the mounted cantilever and brought into close proximity (less than 300 μm) to the cantilever probe tip. In these cases, the AFM was operated in "tapping" mode. The tip is lowered atomically close to the surface and lightly taps it. Tapping mode is best for imaging accumulated materials deposited on surfaces (9, 10). It contacts the surface with much less force than contact mode, reducing shear (lateral) forces that can compact the sample and raise the limit of resolution.

The cantilever is set to oscillate at a resonance frequency and is rastered over the surface. As it moves over the surface, the tip is deflected due to changes in the surface contour. Tip deflection is measured by a laser diode, which bounces a laser beam off the oscillating tip. The distance between the tip and the laser can be determined from the simple equation (11):

$$z = z_o + A \sin(\Omega t)$$

where z_o is the initial distance, A is the amplitude of the wave, Ω is the frequency and t is the time. The AFM operational schematic is illustrated in Figure 1.

In past work, AFM images of membranes have been taken in both the wet and the dry mode. Ceramic membranes do not swell when they are wetted and so may be more easily imaged dry (e.g., 12, 13). On the other hand, organic polymer membranes and fouled membranes in general do not image as accurately under dry conditions (14). Organic polymer membranes swell when they are hydrated. These types of membranes are best examined wet—simulating in situ conditions as much as possible. Previous researchers, such as Centoni et al., have observed that samples imaged in water result in pore size images more closely related to their true in situ values, and the amount of bound water is analogous to that which would be found in a typical filtration process (14). Drying the samples can also significantly alter the shape of the pores and the foulant cakes. With these facts in mind, liquid cell AFM was chosen to examine the clean and fouled membranes.

RESULTS AND DISCUSSION

The three different biopolymers formed very different foulant layers and had very different fouling effects on both the UF and the NF membranes, even when their characteristic rejections were similar. The largest fouling effect was caused by the largest molecules (BSA and long-chain PgAs) in both the UF and the NF regimes. In most cases, fouling effect was greater in the UF range than in the NF range. For example, fouling with 1,000 mg/L BSA had very different effects on the NF and the UF membrane performances (Table 1), though it did not have an appreciably different effect on their surface topologies (Figures 2 and 3). This

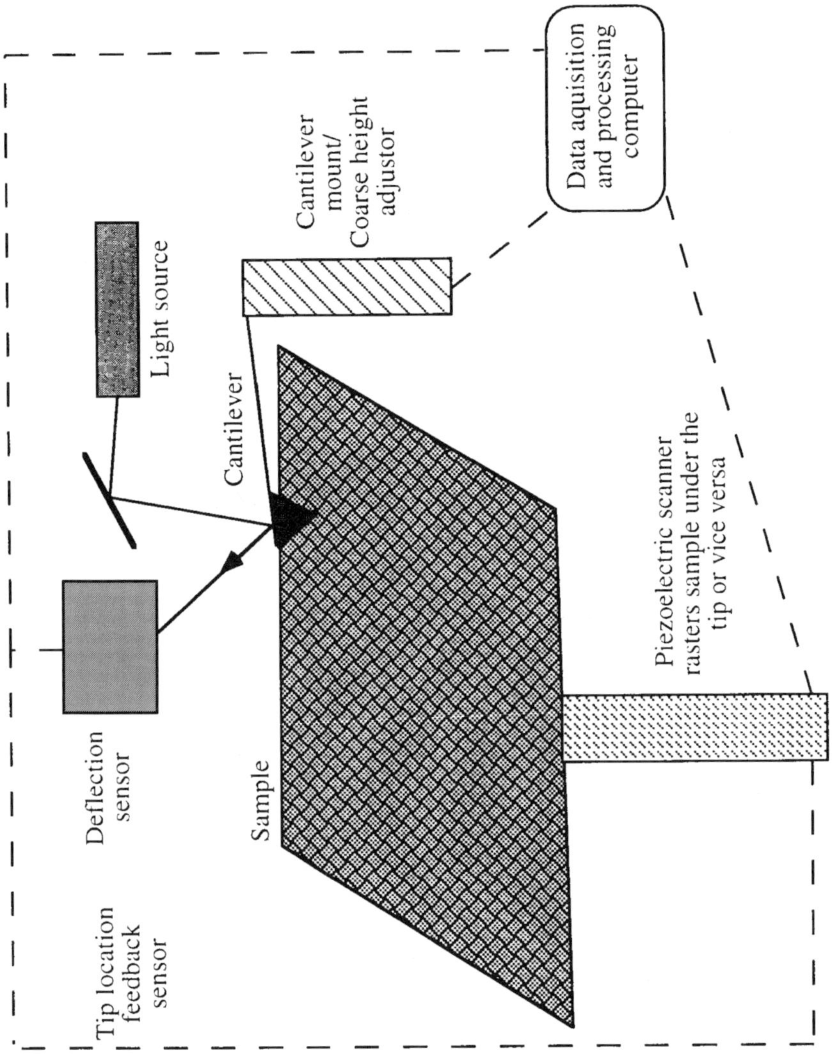

Figure 1 AFM schematic

Table 1 TOC rejection and permeate flux reduction and fouling per mass of BSA at low IS and a feed concentration of 1,000 mg/L BSA

Membrane Type	% Rejection	%-ΔJ	%-ΔJ/$M_{adsorbed}$ (%/mg TOC)
CE100	100.0	11.2	0.5
CE100k	98.0	54.5	2.6

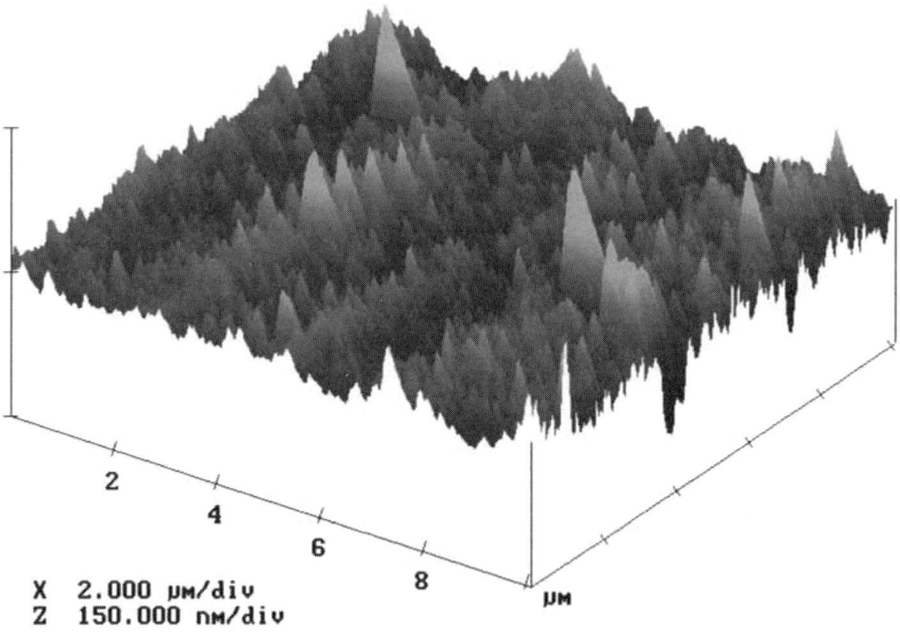

Figure 2 A 10 μm × 10 μm AFM image of a BSA-fouled CE100 membrane at low IS

suggests much more significant irreversible pore fouling in the UF regime. Evidence of surface deposition did not necessarily correspond to a decrease in permeate flux and vice versa. The contributions of the various fouling mechanisms are likely quite different depending upon MWCO.

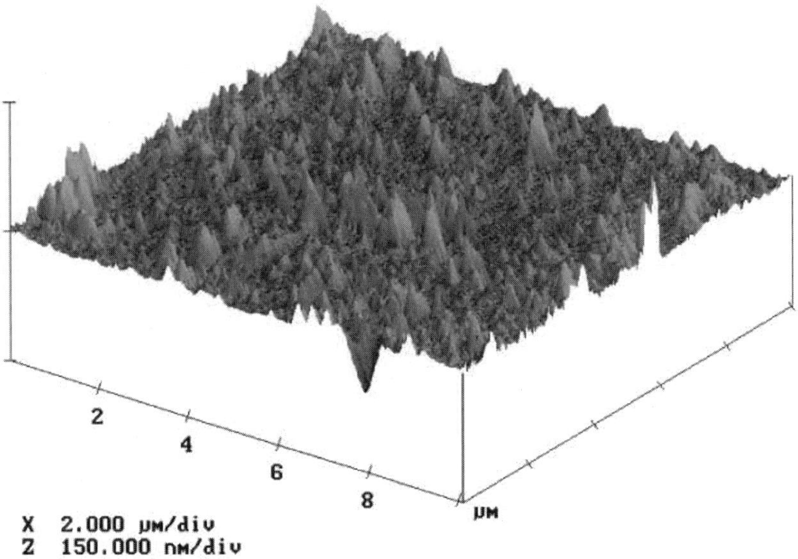

Figure 3 A 10 μm × 10 μm AFM image of a BSA-fouled CE100k membrane at low IS

Ionic strength (IS) effects were a function of the biopolymer type and the MWCO. Elevation of IS enhanced aggregation of each of the foulants into large, globular deposits. In most cases, the effect of IS on the NF and the UF membranes was markedly different. For example, fouling with 1,000 mg/L BSA with 0.01M IS had very different effects on the NF and the UF membrane performances (Table 2). Elevating IS increased flux recovery after fouling, but it decreased the "fouling effect" (reduction in flux per milligram TOC adsorbed) in the NF range and increased it in the UF range. Both surfaces showed a marked increase in deposit size (Figures 3 and 4). This further supports more significant pore fouling in the UF regime. The larger, aggregated proteins could not pack as densely on the membrane surface, or be held as tightly (less NF surface fouling), but would block even the large pores in a looser membrane, into which large molecules might partially or fully penetrate (more UF pore fouling).

Table 2 TOC rejection and permeate flux reduction and fouling per mass of BSA at low and high IS and a feed concentration of 1,000 mg/L BSA

Membrane Type	% Rejection	%-ΔJ	%-$\Delta J/M_{adsorbed}$ (%/mg TOC)
CE100, low IS	100.0	11.2	0.5
CE100, high IS	100.0	8.2	0.3
CE100k, low IS	98.0	54.5	2.6
CE100k, high IS	99.3	12.3	6.5

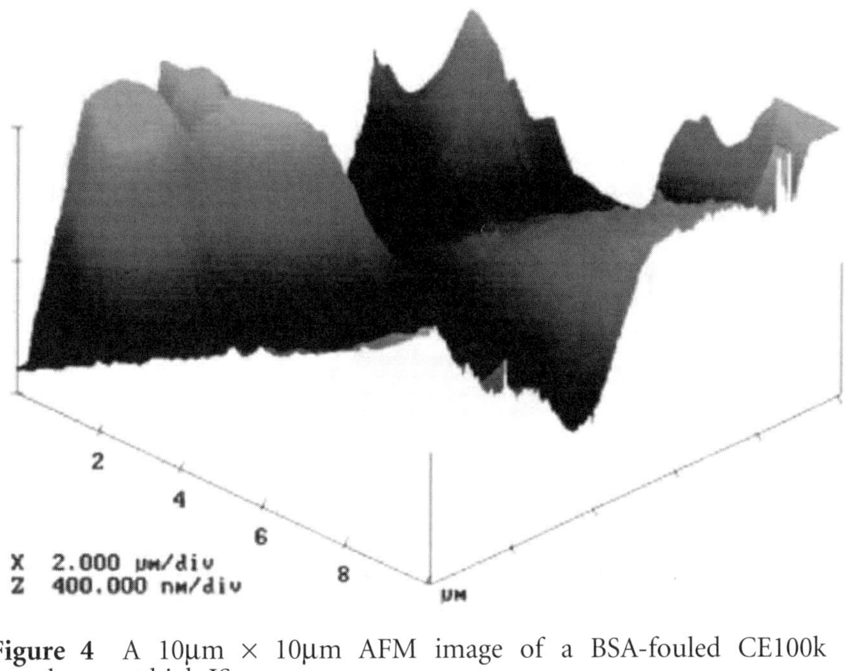

Figure 4 A 10µm × 10µm AFM image of a BSA-fouled CE100k membrane at high IS

The effect of this increased aggregation on fouling followed no consistent trend. Its influence on fouling effect (irreversible reduction of flux per unit mass adsorbed) varied depending upon MWCO and foulant type. This suggests that the effect of ionic strength is a case-dependent phenomenon that is a function of both the nature of the foulants and the nature of the membrane material.

AFM is a powerful technique for the investigation of membrane fouling, but cannot be used alone. Rather, AFM can supplement more standard investigative techniques like dead-end or cross-flow filtration of organic molecules to support or repudiate various fouling scenarios suggested by filtration results.

BIBLIOGRAPHY

1. Mallevialle, J., Anselme, C., Marsigny, O. (1989). Effects of Humic Substances on Membrane Processes. Aquatic Humic Substances. *Proceedings of the 193rd meeting of the American Chemical Society*, Denver, CO, 749–767.
2. Lainé, J.-M., Hagstrom, J.P., Clark, M.M., Mallevialle, J. (1989). Effects of Ultrafiltration Membrane Composition. *Journal of the American Water Works Association*, 81(11), 61–67.
3. Fu, P., Ruiz, H., Thompson, K., Spangenberg, C. (1994). Selecting membranes for removing NOM and DBP precursors. *Journal of the American Water Works Association*, 86(12), 55–72.
4. Allgeier, S. C., Summers, R.S. (1995). Evaluating NF for DBP control with the RBSMT. *Journal of the American Water Works Association*, 87(3), 87–99.
5. Nilson, J.A., DiGiano, F.A. (1996). Influence of NOM composition on nanofiltration. *Journal of the American Water Works Association*, 88(5), 53–66.
6. Speth, T.F., Gusses, A.M., Summers, R.S. Evaluation of Nanofiltration Pretreatments for Flux Loss Control. *American Water Works Association Membrane Technology Conference*, Long Beach, CA, 7(2).
7. Lahoussine-Turcaud, V., Wiesner, M.R., Bottero, J.-Y. (1990). Fouling in Tangential-Flow Ultrafiltration: The Effect of Colloid Size and Coagulation Pretreatment. *Journal of Membrane Science*, 52(1), 173–190.
8. Bersillon, J.L. (1989). *Fouling Analysis and Control, in Future Industrial Prospects of Membrane Processes*. L. Cecille, Toussaint, J.C., ed., Elsevier Applied Science, New York.

9. Maurice, P.A. (1996). Applications of atomic-force microscopy in environmental colloid and surface chemistry. *Colloids and Surfaces A: Physicochemical and Engineering Aspects*, 107, 57–75.
10. Wong, T.M.H., Descouts, P. (1995). Atomic force microscopy under liquid: a comparative study of three different AC mode operations. *Journal of Microscopy*, 178(1), 7–13.
11. Sarid, D. (1994). *Scanning Force Microscopy with Applications to Electric, Magnetic and Atomic Forces*. Oxford University Press, New York.
12. Bottino, A., Capannelli, G., Grosso, A., Monticelli, O., Cavalleri, O., Rolandi, R., Soria, R. (1994). Surface characterization of ceramic membranes by atomic force microscopy. *Journal of Membrane Science*, 95, 289–296.
13. Bowen, W. R., Hilal, N., Lovitt, R.W., Williams, P.M. (1996). Atomic force microscope studies of membranes: Surface pore structures of Cyclopore and Anopore membranes. *Journal of Membrane Science*, 110, 233–238.
14. Centoni, S.A., Vanasupa, L.S., Tong, P.S. (1997). Atomic Force Microscopy for Ultrafiltration Membrane Imaging. *Scanning*, 19, 281–285.

Part 3 Membrane Monitoring

Monitoring Membrane Integrity Using Ultra-High-Sensitivity Laser Light Scattering

Assessing the Integrity of Reverse Osmosis Spiral-Wound Membrane Elements with Biological and Non-Biological Surrogate Indicators

Development of an Innovative Method to Monitor the Integrity of a Membrane Water Repurification System

Membrane Integrity Monitoring at the UF/RO Heemskerk Plant

CHAPTER · 9

Monitoring Membrane Integrity Using Ultra-High-Sensitivity Laser Light Scattering

Ashim Banerjee
Hach Company, Loveland, Colorado

Kenneth Carlson
Colorado State University, Fort Collins, Colorado

James Lozier
CH2M Hill, Tempe, Arizona

BACKGROUND

Membrane filtration plants for drinking water typically use pressure decay testing in conjunction with particle counting and turbidity to monitor membrane integrity. Pilot plants offer the capability of monitoring permeate quality with both intact and intentionally compromised membranes. We compare data from a particle counter, a pressure decay test and a laser turbidimeter on pilot plants from two different manufacturers of microfiltration membranes.

INTRODUCTION

Very significant capital cost reduction has, in recent years, placed membrane systems in a position to significantly impact the drinking water industry. Since these systems offer an absolute barrier (pore sizes between 0.01 and 0.2 micron) that is much smaller than the pathogens of interest in drinking water, they can legitimately claim disinfection credit for cryptosporidium and giardia. In the case of ultrafiltration membranes with pore sizes down to 0.01 micron, that same argument can be extended to viruses. Lower disinfection requirements can significantly impact the cost of chemicals, and cost of storage and handling, not to mention the advantages of reduced disinfection by-products (DBPs). The contrary argument is that membranes offer a single barrier, and in the case of a compromised membrane, no barrier at all.

Membrane manufacturers have incorporated a pressure decay test to verify membrane integrity.[1] However, this test is not online and requires shutdown of part or all of the membrane plant. In practice, pressure decay tests cannot be performed more frequently than once every four to six hours, resulting in concerns regarding water quality in the intervening period. Particle counters and turbidimeters are online instruments that have been employed to measure water quality in multimedia filter-based drinking water plants for many years. Membrane pilot plants typically have these instruments measuring both the feed and permeate water. We added a laser turbidimeter[2] recently introduced by Hach Company, the FilterTrak 660 (Figure 1), to this instrumentation. In this chapter we will compare data from all these different techniques.

EXPERIMENTAL SETUP

A particle counter and a laser turbidimeter were installed on the permeate water from membrane pilot plants (Mem1 and Mem2) provided by two different manufacturers. Both membranes were in the microfiltration class with pore sizes around 0.1 micron. Both membranes were equipped with pressure decay tests. Baseline data from particle counts, turbidity and pressure decay were recorded under conditions of intact membranes, before both membranes were intentionally compromised (by severing one out of the ~5,000 fibers in the membrane cartridge). Data were then collected again.

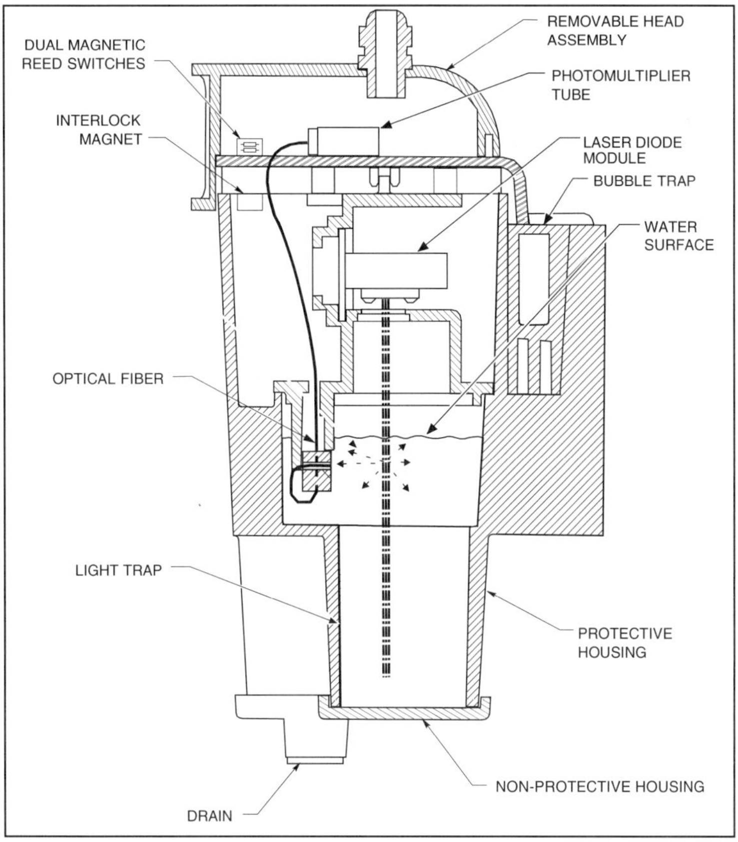

Figure 1 Schematic drawing of the Hach FilterTrak (FT660) laser turbidimeter

RESULTS AND DISCUSSION

Figure 2 shows the pressure decay tests for Mem1 before and after the integrity problem was introduced. Figure 3 shows the same data for Mem2. Mem1 shows a decay of 23 to 18 PSI for a breached membrane (a fall of about 22%) while Mem2 decays from 4 to 2.3 PSI, a drop of 42.4%. It in interesting to note, however, that Mem2 exhibits a slight decay (4 to 3.8 PSI) even under conditions of an

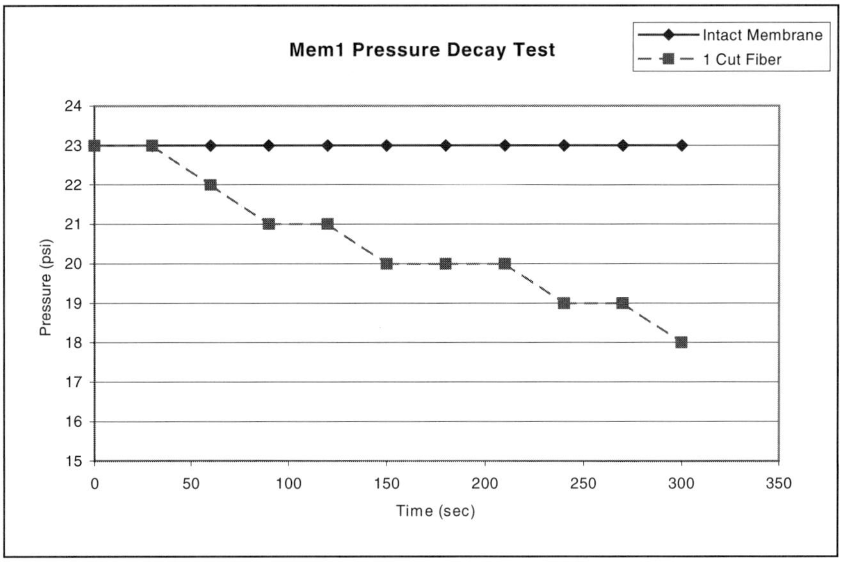

Figure 2 Pressure decay tests for Mem1

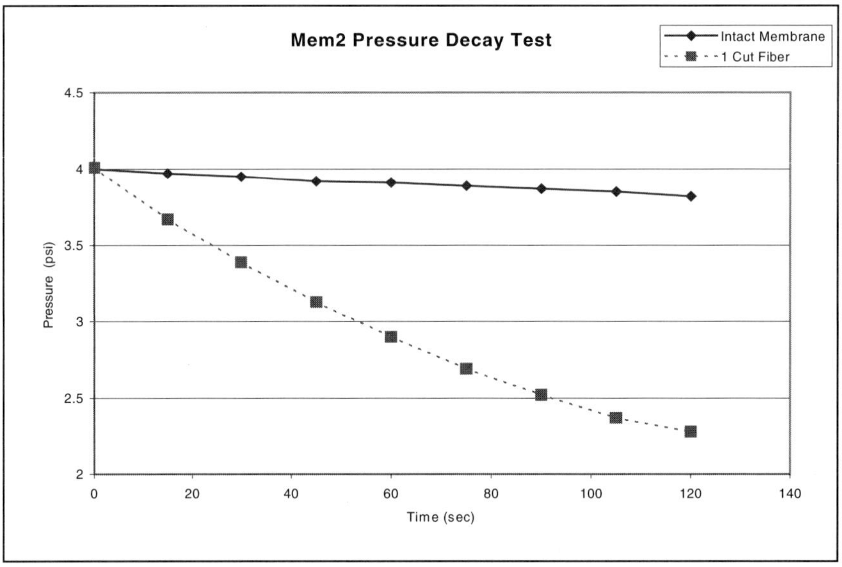

Figure 3 Pressure decay test for Mem2

intact membrane. Thus, the real pressure decay is from 3.8 to 2.3 PSI, or 39%. In either case, the pressure decay tests clearly demonstrate the ability to detect the break in a single fiber in pilot plant testing.

Figures 4 and 5 show the particle count data from Mem1 and Mem2 for both intact and compromised membranes. Both membranes show a very significant degradation in particle counts under conditions of a compromised membrane. While Mem1 shows a rise in counts from 0 to about 30 (increasing to about 100 during a backwash spike), Mem2 jumps up dramatically from less than 1 to over 1,000.

Finally, Figures 6 and 7 show the laser turbidity data for Mem1 and Mem2 for intact and compromised membranes.

For Mem1 the laser turbidity increases from about 14.5 mNTU (intact membrane) to about 18.5 mNTU when the integrity problem is introduced. For Mem2, the increase is from about 14 mNTU to over 250 mNTU. The laser turbidimeter is, therefore, capable of detecting a break in a single fiber in both membranes used in this study. A few interesting points remain that need additional study. These data were collected under conditions of raw water turbidity of around 12 NTU. Clearly, if this influent turbidity were to decrease, so would the change measured by the instruments in the event of an integrity problem (the pressure decay test would remain unaffected). Furthermore, all the data we have seen so far have been collected on pilot-scale studies. In full-scale models, instruments monitoring the permeate from a rack of about 50 membrane cartridges would suffer from very significant dilution. It is not clear if either the particle counter or the laser turbidimeter would in fact be able to detect a break in one fiber in a rack of 50 membrane cartridges.

Another interesting point brought to light by these data is that the laser turbidimeter and the particle counter both sense and recognize a backwash event. It is also clear that, in the event of an integrity problem, the backwash events are amplified far in excess of the increase in the baseline operation value for both laser

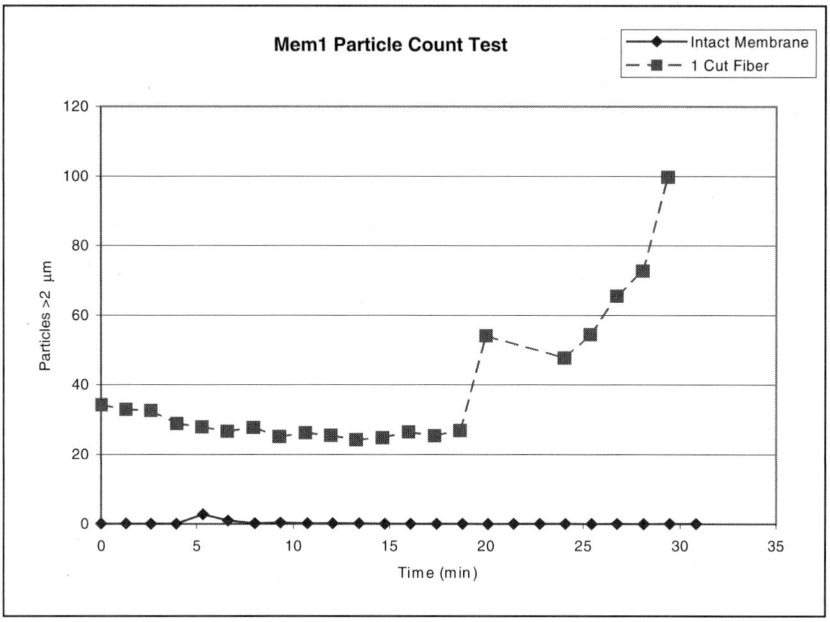

Figure 4 Particle counts for Mem1

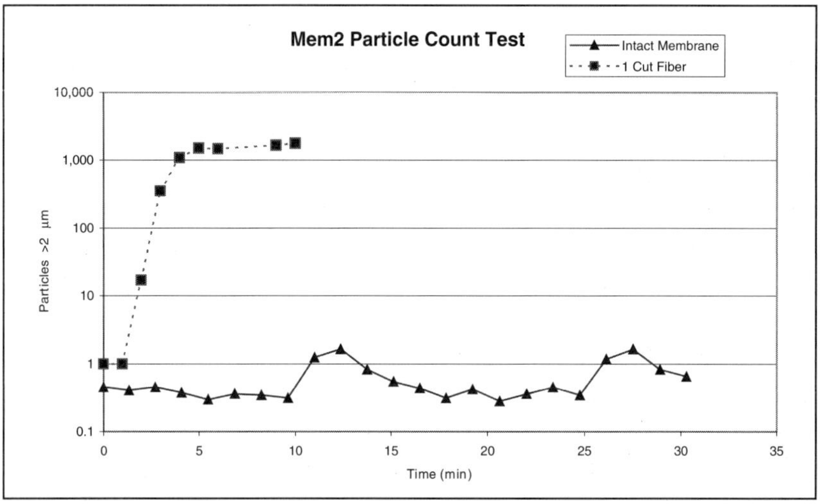

Figure 5 Particle counts for Mem2

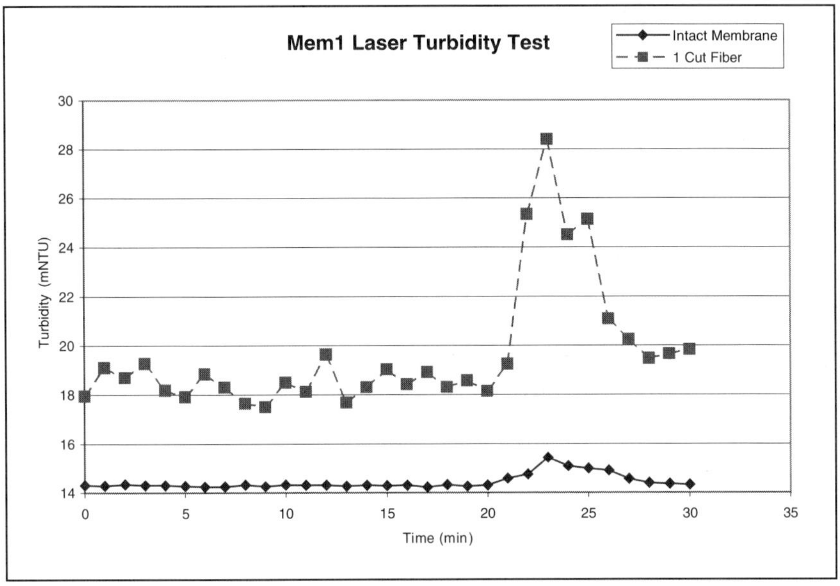

Figure 6 Laser turbidity data for Mem1

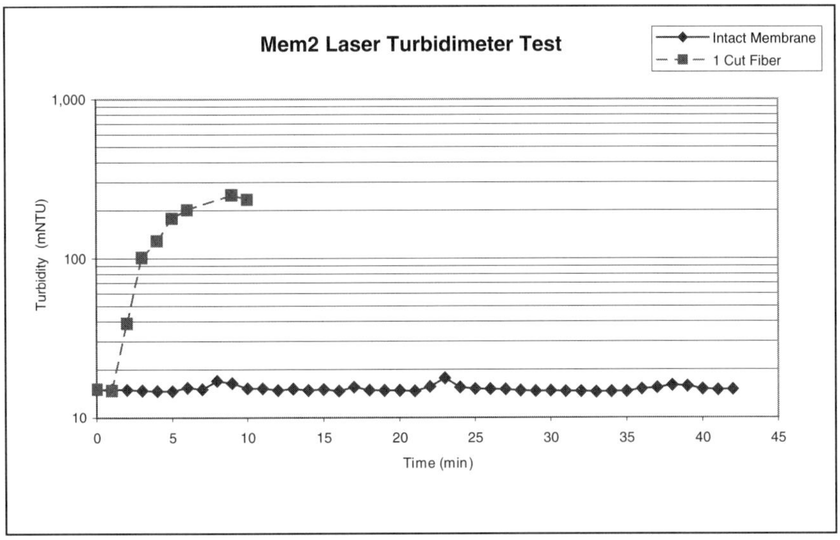

Figure 7 Laser turbidity data for Mem2

turbidity and particle counting. The significance of this is very considerable, and is currently being investigated by us.

REFERENCES

1. Charles Liu, Masatake Fushijima, Jennifer Hayes, James Moy. Finding a Needle in a Haystack: Testing Integrity of Membrane Filters in Drinking Water Applications. *Proceedings of the WQTC*, November 1999.
2. A. K. Banerjee *et al.* Ultra Low Range Instrument Increases Turbidimetric Sensitivity by Two Orders of Magnitude. *Proceedings of the WQTC*, November 1999.

CHAPTER · 10

Assessing the Integrity of Reverse Osmosis Spiral-Wound Membrane Elements with Biological and Non-Biological Surrogate Indicators

Courtney L. Acker, Engineer
Carollo Engineers, Walnut Creek, CA

Christian K. Colvin, Process Engineer
CH2M Hill, Englewood, CO

Benito J. Mariñas, Professor
University of Illinois at Urbana-Champaign, Urbana, IL

James C. Lozier, Principal Technologist
CH2M Hill, Tempe, AZ

INTRODUCTION

The control of persistent pathogens, such as *Giardia lamblia* cysts, *Cryptosporidium parvum* oocysts, and enteric viruses, in drinking water has been receiving increased attention from regulatory authorities. Common methods of disinfection, both chemical and physical, are not always effective in inactivating these pathogens. Studies have shown that reverse osmosis (RO) membranes may provide an effective barrier against pathogens, as long as the integrity of the membranes is not compromised. If a membrane element is defective or fails during either installation or operation, integrity testing methods must be in place to detect potential breakthrough of microorganisms. This study investigated a new method of integrity assessment for RO membrane elements, consisting of challenge testing with polystyrene fluorescent-dyed microspheres as non-biological surrogate indicators of waterborne microorganisms. A biological surrogate indicator, MS2 *phage*, was used simultaneously with microspheres of similar size. The main focus of this study was to establish a correlation between the passages of these two surrogate indicators through the RO membranes.

EXPERIMENTAL DESIGN

Experimental Apparatus

Spiral-wound elements were tested with the closed loop system presented in Figure 1. Feedwater was drawn from a high-density polyethylene tank using a constant speed pump (CAT 341, CAT Pumps USA, Minneapolis, MN), and forced through the inlet of the membrane element vessel (Model S2.5, Advanced Structures, Inc., Escondido, CA). The vessel contained a single spiral-wound membrane element, 2.5 inches in diameter and 40 inches in length. A diversion line, part of the closed loop system, allowed a portion of the flow exiting the pump to bypass the membrane and flow directly back into the feed tank. Due to the high output of the pump, compared to the capacity of the membrane element, the diversion was necessary to restrict the amount of flow routed to the element. The amount of flow delivered to the membrane element was controlled with a needle valve (Whitey, Highland Heights, OH). A second needle valve was used in the concentrate return line to adjust the pressure in the element to achieve permeate flow rates within a target range. Permeate and concentrate flow rates from the element were measured with flow meters (Model RMC-141 and Model RMC-144, Dwyer Instruments, Inc., Michigan City, IN) prior to being returned to the feed water tank. The flow meters were calibrated volumetrically. Permeate and concentrate sample valves (Whitey, Highland Heights, OH) were located in the corresponding return lines for sample collection.

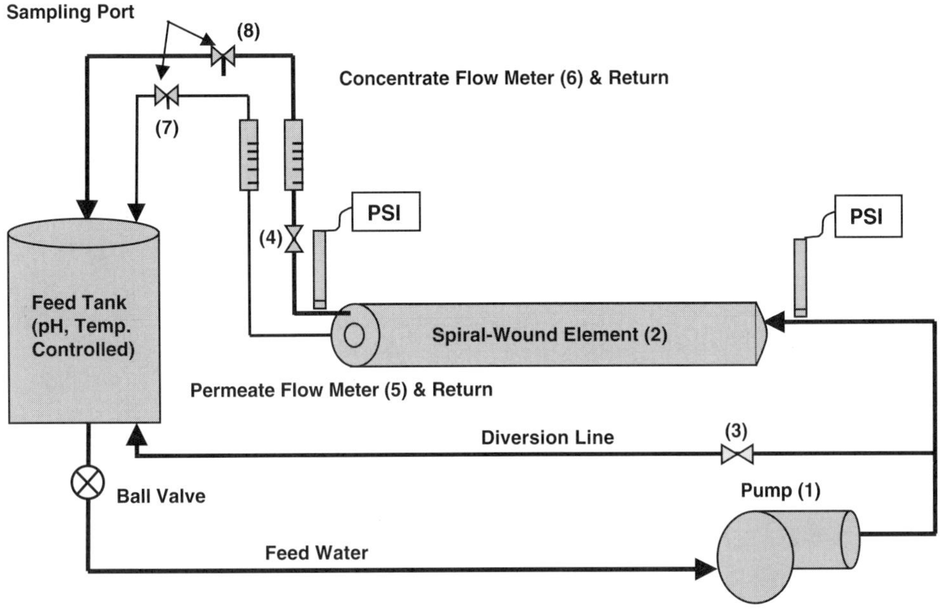

Figure 1 Schematic of closed loop system used for testing spiral-wound membrane elements

Experimental Matrix

Three membranes were investigated in the spiral-wound configuration. Manufacturer's specifications for the spiral-wound elements tested in this study are presented in Table 1. Summaries of the experimental conditions for all membrane testing performed are presented in Tables 2, 3, and 4 for membranes A, B, and C, respectively. Initially, each element was tested to characterize the membrane performance with 2,000 mg/L NaCl at pH 6.0 (±0.5) and 25°C (±1°C). The results of this characterization were compared to manufacturers' specifications.

After the preliminary characterization was complete, MS2 *phage* viruses and NaCl were added to the feed tank. It was suspected that viruses would not reach a steady-state concentration in the feedwater quickly. Thus, tests were performed to determine

Table 1 Manufacturer's specifications for spiral-wound RO membrane elements tested

	Membrane		
Parameter	A	B	C
Model	BW30HP(1)	ESPA1(2)	ROGA(3)
Nominal Diameter, in.	2.5	2.5	2.5
Membrane Active Area, ft^2	26.4	28	27
Target Feed Concentration, mg/L	2,000 (NaCl)	1,500 (NaCl)(4)	700 (TDS)(4)
Feed Pressure, psi	225	150	200
Feed pH	8	6.5–7.0	7.0
Temperature, °C	25	25	25
Water Productivity, (gal/d)	650	750	260
Product Water Recovery, %	15	10	15
Nominal Solute Rejection, %	99	–	95
Minimum Solute Rejection, %	98	98	–

(1) Manufactured by Dow Chemical, Minneapolis, MN.
(2) Manufactured by Hydranautics, Oceanside, CA.
(3) Manufactured by Koch Fluid Systems, San Diego, CA.
(4) This was lower than tested concentration.

the virus feed concentration over time (referred to as time tests). The feed tank was seeded with a stock solution of approximately 108 viruses/mL and samples were taken over the course of one hour. The time tests were performed at a constant flow rate to rule out differences in virus passage based on flow.

Once the dynamics of the virus feed concentration were determined for a given membrane, the passage of the MS2 *phage* at various flow rates was examined. If MS2 *phage* was observed in permeate samples, then further testing was performed after adding the non-biological surrogate indicator (0.02-μm fluorescent-dyed polystyrene microspheres) to the feed solution. These experiments would reveal the relationship between the passage of the non-

Table 2 Operating conditions and feed solution composition for experiments performed with FilmTec BW30HP spiral-wound element

Exp.	pH	Temp (°C)	NaCl (*mg/L*)	MS2 *Phage* Virus (*pfu/mL*)	Microspheres (*mg/L*)	Permeate Flux (*gfd*)
1A	6.04	25.1	1,990	–	–	5.3–25.2
2A	5.75	24.9	1,980	–	–	5.6–24.9
3A	6.13	25.1	1,960	–	–	5.4–27.4
4A(1)	6.04	25.3	1,850	4.55×10^7	–	8.1–36.7
5A(1)	6.05	25.2	1,810	–	–	7.8–36.7

(1) Experiments conducted after exposing the membrane element to free chlorine for approximately 1,000 ppm-h.

Table 3 Operating conditions and feed solution composition for experiments performed with Hydranautics ESPA1 spiral-wound element

Exp.	pH	Temp (°C)	NaCl (*mg/L*)	MS2 *Phage* Virus (*pfu/mL*)	Microspheres (*mg/L*)	Permeate Flux (*gfd*)
1B	5.58	25.2	1,810	–	–	6.8–28.7
2B	5.88	26.3	1,810	–	–	8.7–28.6
3B	6.23	24.5	N/A(1)	5.8×10^7	–	6.2–28.6
4B	6.14	24.5	N/A(1)	1.2×10^9	10.3	6.5–28.1
5B(2)	6.20	23.8	N/A(1)	2.2×10^8	7.1	6.5–26.1

(1) N/A: Solute testing was not completed for this experiment.
(2) Experiments conducted after compromising the o-ring sealing the permeate pipe connection to the RO vessel end-cap.

biological surrogate indicator and actual microorganisms. Tables 2–4 list the concentrations of the various constituents that were tested with the three membrane elements.

Testing was also conducted with membrane elements of compromised integrity. Chemical and physical approaches for

Table 4 Operating conditions and feed solution composition for experiments performed with Fluid Systems ROGA-HR spiral-wound element

Exp.	pH	Temp (°C)	NaCl (*mg/L*)	MS2 *phage* Virus (*pfu/mL*)	Microspheres (*mg/L*)	Permeate Flux (*gfd*)
1C	6.06	25.4	1,860	3.6×10^4	–	9.3–22.8
2C	5.94	24.9	1,710	1.3×10^5	–	6.7–18.2
3C	6.16	25.0	1,700	5.5×10^6	–	6.7–17.9
4C(1)	6.16	24.7	1,770	6.3×10^8	10.4	6.7–19.4
5C(1)	6.03	24.5	1,850	4.1×10^8	9.9	6.7–19.9
6C(2)	6.13	24.2	1,850	3.2×10^8	8.5	6.5–21.1

(1) Compromised element with a small pinhole.
(2) Compromised element with a larger perforation.

compromising the membrane integrity were used, as noted here. The FilmTec BW30HP membrane was exposed to 1,000 (mg/L)–hour of free chlorine, the upper limit specified by the manufacturer. The Hydranautics ESPA1 membrane element was tested with a faulty o-ring, the seal between the permeate tube and the pressure vessel end-cap. For this purpose, a small crack was induced in the end-cap o-ring and the element was installed and tested for virus and microsphere rejection. The integrity of the Fluid Systems ROGA membrane element was challenged by means of perforating the actual membrane. The tape-wrapped element was unwrapped to expose the active surface of the membrane. First, a small pinhole was created using a pin to prick the surface of the membrane, just barely passing through to the polysulfone layer. After this imperfection was created, the membrane was rewrapped and tested with feed solutions containing sodium chloride, phage and microspheres. Next, a larger perforation was induced onto the membrane surface and similar testing was conducted.

MATERIALS AND METHODS

Chemical Solutions

Various analytical and experimental solutions mentioned in subsequent sections were prepared according to the following procedures:

Solute Challenge Solution: A stoichiometric quantity of crystallized NaCl (>99.9% purity) (Fisher Scientific, Itasca, IL) was dissolved in 15–25 gallons of distilled-deionized (DDI) water to produce a feed solution with an NaCl concentration of 2,000 mg/L.

Tryptone Yeast Extract (TYE) Broth: 1.0 g Bacto tryptone, 0.1 g Bacto yeast extract, 0.1 g Bacto dextrose (DIFCO Laboratories, Detroit, MI), 0.8 g NaCl (>99.9% purity) (Fisher Scientific, Itasca, IL), and 0.022 g CaCl2 (>95% purity) (Mallinckrodt Baker Inc., Paris, KY) were added to 100 mL of deionized water. The resulting mixture was subsequently stirred until all constituents were completely dissolved (approximately 10 minutes).

Base Agar: 1.5 g Bacto agar was added to 100 mL of TYE broth and gently mixed. The mixture was then autoclaved for 20 minutes and placed in a 45°C water bath to cool. The agar was then dispensed in 15 mL volumes into 15 × 100 mm Petri plates and allowed to harden. Once hardened, the plates were inverted and stored at 4°C.

Soft Agar: 0.7 g Bacto agar was added to 100 mL of TYE broth and gently mixed. The mixture was then autoclaved for 20 minutes and placed in a 45°C water bath to cool. The soft agar mixture was used in the MS2 *phage* assay procedure described herein.

Diluent: 0.85 g NaCl and 0.022 g $CaCl_2$ were added to 100 mL of DDI water and the resulting mixture was stirred to ensure proper mixing. The diluent was then sterilized in an autoclave for 20 minutes and stored at 4°C until usage.

MS2 *Phage*

MS2 Phage *Stock Preparation*

A plaque assay method provided by the U.S. Environmental Protection Agency, Cincinnati, was used with some modifications to obtain an initial stock of MS2 *phage* virus. All tests were performed with 15597-B1 MS2 *phage* virus obtained from American Type Culture Collection (Manassas, VA), received as a freeze-dried pellet. Plates of relatively low-dilution virus suspension, which resulted in complete "plaque out" (i.e., more plaques than could be counted), were prepared as described in a subsequent section and used to produce large quantities of viruses. The top agar of these plates was harvested with the aid of diluent and placed in a tube to be compacted by centrifugation at 1,900 rpm for 15 minutes. Subsequently, the supernatant, containing the *phage*, was filtered through a 0.22 μm filter to remove any remaining residuals. The *phage* was then stored at 4°C until addition to the system. Typically, stock concentrations ranged from 10^{11} to 10^{13} plaque forming units (pfu)/mL.

E. coli Preparation

A freeze-dried pellet of *E. coli* (ATCC, 15597) was obtained and kept at 4°C until ready for use. Half of the pellet was dissolved in approximately 100 mL of TYE broth and incubated, under slight agitation, overnight at 37°C. Once the initial colony of *E. coli* was prepared, it was kept alive through the use of hard agar slants. The hard agar slants were prepared by adding 7 mL of hard agar into 15-mL centrifuge tubes. The agar was allowed to harden while the centrifuge tubes were tilted, thereby increasing the surface area of the hard agar. The hard agar slants were inoculated with *E. coli* using an inoculation loop that was flamed before and after each inoculation for sterilization. The slants were incubated overnight at 37°C and subsequently stored at 4°C for up to 7 days before being used to prepare *E. coli* broth. The *E. coli* broth was prepared by introducing *E. coli* from the hard agar slants (previously stored at 4°C) into approximately 100 mL of TYE broth and incubated, under slight agitation, overnight at 37°C. The cycle between *E. coli* broth and slants was repeated at a minimum of every seven days to keep the colony of *E. coli* alive.

MS2 Phage *Concentration Determination*

The plaque assay procedure was used to determine the titer of the prepared MS2 *phage* stock solutions as well as MS2 *phage* concentration in both feed and permeate samples. *E. coli* broth and soft agar solution were prepared before the assay was performed. Then, feed and permeate samples were diluted, if necessary, using TYE broth in order to decrease the concentration of MS2 *phage* under 200 pfu/mL. This concentration of MS2 *phage* could easily be counted on each plate with the naked eye. Next, a test tube rack containing 10 mL sterile glass vials with lids was introduced in the 45°C water bath. Approximately 4 mL of soft agar was dispensed into each glass vial with care not to contaminate. Once the soft agar was added, 0.1 mL of *E. coli* broth was added to each vial. Then, 1 mL of each MS2 *phage* sample was also added to individual vials. The contents of the vials were then emptied onto 15 × 100 mm Petri dishes, containing base agar, and allowed to harden. Once the soft agar mixture was firm, the plates were inverted and incubated overnight at 37°C for a minimum of 18 hours. Next, the plates were removed from the incubator and the plaques were counted. The number of plaques observed on each plate was assumed to correspond directly to the concentration of MS2 *phage* in pfu/mL.

Several *E. coli* control samples were included with the MS2 *phage* samples in order to ensure that contamination was not a factor. For the *E. coli* controls, 1 mL of *E. coli* broth (instead of 1 mL of sample) was added with the same pipette to the control vials. The absence of MS2 *phage* in the control vials ensured that the pipette was not contaminated during the assay procedure.

Microspheres

A stock solution of orange fluorescent carboxylate-modified FluoSpheres® with a concentration of 2% (20,000 mg/L) was obtained from Molecular Probes (Eugene, OR). The actual particle size was reported as 0.024 ± 0.005 µm. Maximum excitation/emission wavelengths were 540 nm/560 nm respectively.

Microsphere Concentration Determination

Microsphere fluorescence was determined using a Bowman Series 2 Luminescence Spectrometer (SLM Aminco, Rochester, NY). An excitation wavelength of 533±4 nm or 533±8 nm (depending on concentration) and an emission wavelength of 566±8 nm were used to analyze all microsphere samples. Although the maximum emission/excitation wavelengths were 540/560 nm, the wavelengths of 533/566 nm were used to decrease interference due to light scattering. A high-quality quartz spectrofluorometry cuvette was used in order to minimize inconsistencies caused by cuvette imperfections.

Microsphere samples were analyzed in three concentration ranges: high (>50 μg/L), intermediate (1–50 μg/L), and low (<1 μg/L). Prior to sample analyses, microsphere standards were prepared for each of these anticipated concentration ranges. When analyzed with the spectrofluorometer, high concentrations of microspheres produced very stable readings. Therefore, a single-point standardization was used to develop a calibration curve for samples in this concentration range. Intermediate and low concentrations of microspheres often produced readings with moderate variability. An average of 30 readings was used in the development of 5-point linear regression calibration curves for samples in these concentration ranges.

The calibration curves developed for the microsphere concentration ranges were then used to determine the microsphere concentrations in samples taken during membrane testing. A new calibration was performed prior to the analysis of each set of samples.

RESULTS AND DISCUSSION

Membrane A—BW30HP

Solute Experiments

Experimental results obtained for all five experiments performed with the BW30HP element are presented in Figure 2. As depicted in

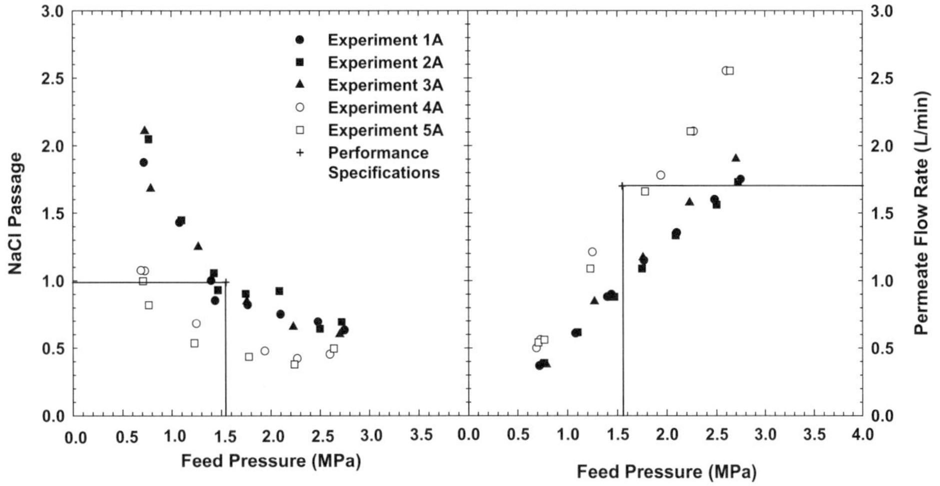

Figure 2 Performance of FilmTec BW30HP element before (Experiments 1A–3A) and after (Experiments 4A–5A) free chlorine exposure (see Table 2 for experimental conditions)

the plots for water and sodium chloride, good reproducibility was observed prior to chlorine treatment (i.e., Experiments 1A–3A). A comparison to manufacturer's performance specifications revealed that the permeate flow rate observed for Experiments 1A–3A in this study was approximately half of that expected and the NaCl rejection was in excess of the average of 99 percent for a new element. These findings are not very surprising because the element used was installed in the unit for nearly eight years after use in various other projects. Some level of membrane compaction and/or fouling would be expected for such an aged element. This element was tested in this study for the purpose of assessing if an older version of a membrane reported to serve as an excellent barrier for MS2 *phage* when new (Adham et al., 1998) would allow virus passage. Virus testing results are discussed in the following section.

EXPOSURE TO FREE CHLORINE. Exposure of the membrane element to free chlorine at 1,000 (mg/L)–hour resulted in an increase in water flux but a decrease in solute passage, as shown in Figure 2, Experiments

4A and 5A. These changes do not appear to be consistent with membrane damage as a result of chemical attack because in such a case the passage of both water and solute would be expected to increase. A possible explanation for the changes observed is that chlorine might have reacted with fouling material from prior use of the membrane element. A lower level of concentration polarization for the "cleaned" element would be consistent, at least in part, with the changes observed in membrane performance.

Virus Experiments

FEED CONCENTRATION DYNAMICS. Initial virus testing efforts with the FilmTec BW30HP element revealed that steady-state conditions were not being reached with respect to the virus feed concentration. Experimental results for testing performed to assess the dynamics in virus feed concentration are presented in Figure 3. In Seeding A, the feed tank was seeded with a stock solution volume that resulted in an initial concentration of approximately 106 viruses/mL. The concentration of viruses dropped within the first 20 minutes of operation to approximately 102 viruses/mL, or a 4-log reduction. For the remaining 40 minutes of the testing period, the concentration leveled off. After testing for one hour, the feed tank was reseeded (Seeding B) with stock solution at a higher initial concentration of approximately 107 viruses/mL. The first 20 minutes after Seeding B yielded a decrease in virus concentration by approximately 1 log. However, by the end of the second sixty-minute period, the concentration of viruses had decreased to roughly 5.0×10^4 viruses/mL, or an overall reduction in excess of 3 logs. Two subsequent tests were conducted (data not shown in Figure 3), with similar results to the second seeding event. Therefore, a steady state was not being reached within the system. A possible explanation for these observations is that the viruses might be adhering to the various surfaces of the testing apparatus components. In order to ensure that the virus concentration in the feed tank remained at fairly constant levels during the time of testing, additional virus stock was injected into the tank prior to taking a feed sample.

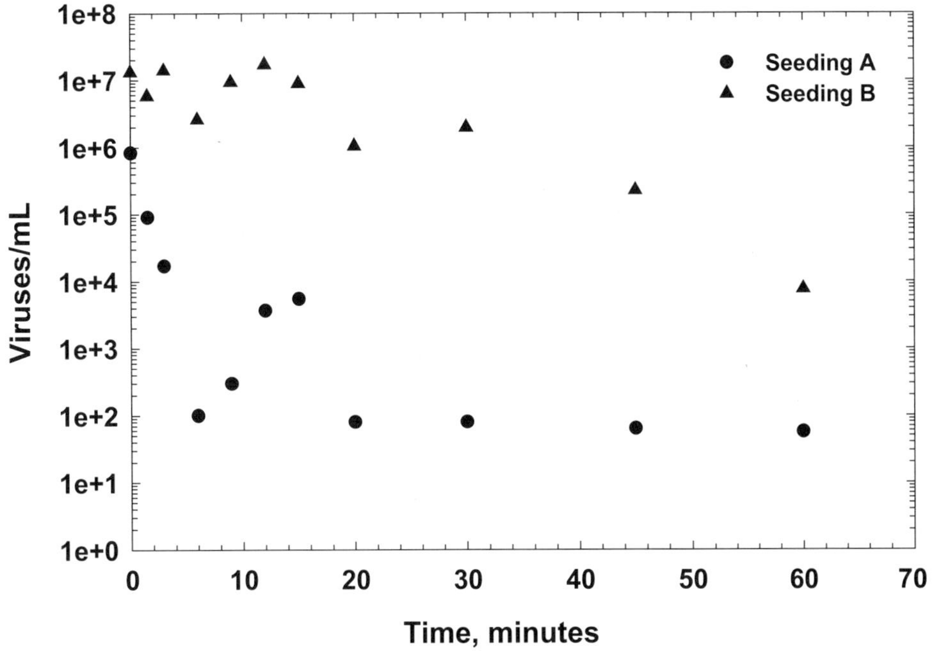

Figure 3 Results of testing on BW30HP to evaluate the decrease in the feed concentration of viruses

During the time test performed on the BW30HP membrane, permeate samples were also taken to assess virus passage through the membrane. No viruses were found in any of the permeate samples. Therefore, virus removal levels were as high as 7-log with the FilmTec BW30HP element prior to chlorine exposure.

ASSESSMENT OF VIRUS REMOVAL WITH FILMTEC BW30HP ELEMENT EXPOSED TO FREE CHLORINE. Experiment 4A was designed to assess the rejection of viruses by the FilmTec BW30HP element after exposure to 1,000 (mg/L)–hour of free chlorine. Experimental results for Experiment 4A are presented in Figure 4. Five data points were obtained as a function of the hydraulic pressures indicated in the figure, which corresponded to permeate fluxes of 5.4 to 27.4 gfd. The feed virus concentration ranged from 1.0×10^6

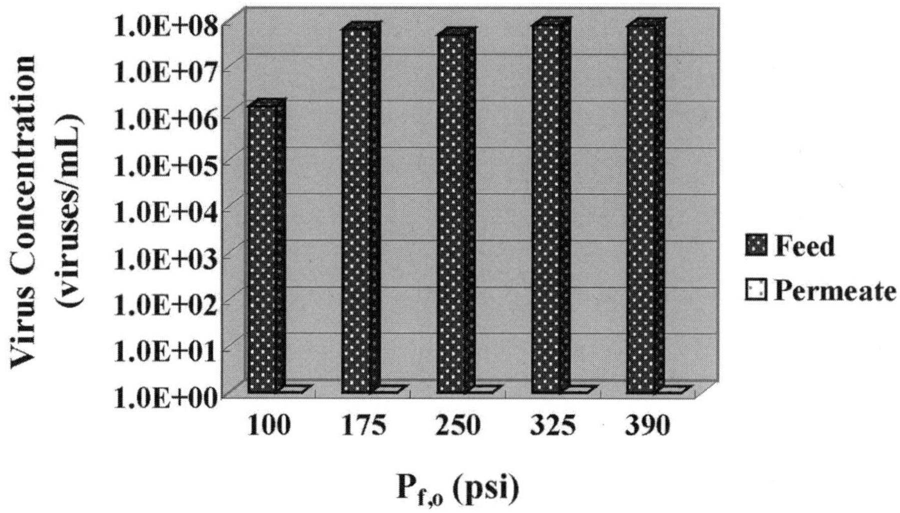

Figure 4 Virus removal observed for Experiment 4A performed with the FilmTec BW30HP element

to 8.0×10^7 viruses/mL. As depicted in the figure, the virus removal exhibited by the membrane was greater than 6 to 8 logs. The removal rates confirmed that exposure to free chlorine did not compromise the integrity of the membrane element.

Membrane B—ESPA1

Solute Experiments

Water flux and sodium chloride passage observed in Experiments 1B and 2B with the Hydranautics ESPA1 element are presented in Figure 5. In general, the experimental results obtained were consistent with performance specifications provided by the membrane manufacturer, which are also indicated in Figure 5. Notice that the specified solute passage corresponds to a minimum solute rejection value (see Table 3), and not to a nominal value. The slightly greater value for the water flux specifications is also consistent with the membrane performance being provided for a more dilute solution.

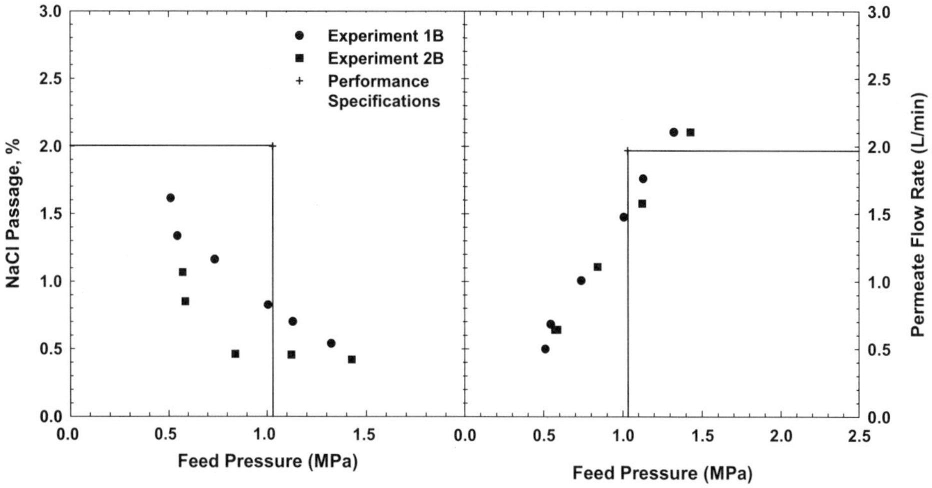

Figure 5 Performance of Hydranautics ESPA1 element prior to virus addition (see Table 3 for experimental conditions)

CONCENTRATION DYNAMICS. The stability of virus concentration in the feed tank after a seeding event was assessed with the Hydranautics ESPA1 element as part of Experiment 3B. Experimental results obtained are presented in Figure 6. As shown, the virus feed concentration decreased from an initial value of 1.2×10^8 to 1.4×10^2 viruses/mL within 45 minutes. These results are similar to those discussed with the FilmTec BW30HP membrane. However, in contrast to the BW30HP results, virus passage was observed during the first 6 minutes of testing with the ESPA1 element. The corresponding rejection was approximately 7-logs and therefore, once the virus feed concentration dropped below 107 viruses/mL, the permeate concentrations were found to be below detection.

EFFECT OF WATER FLUX. The effect of water flux on virus passage was assessed as part of Experiment 3B. Virus stock solution was added approximately four minutes before sampling at each pressure tested. Experimental results for these tests are presented in Figure 7 in terms of virus feed and permeate concentrations versus

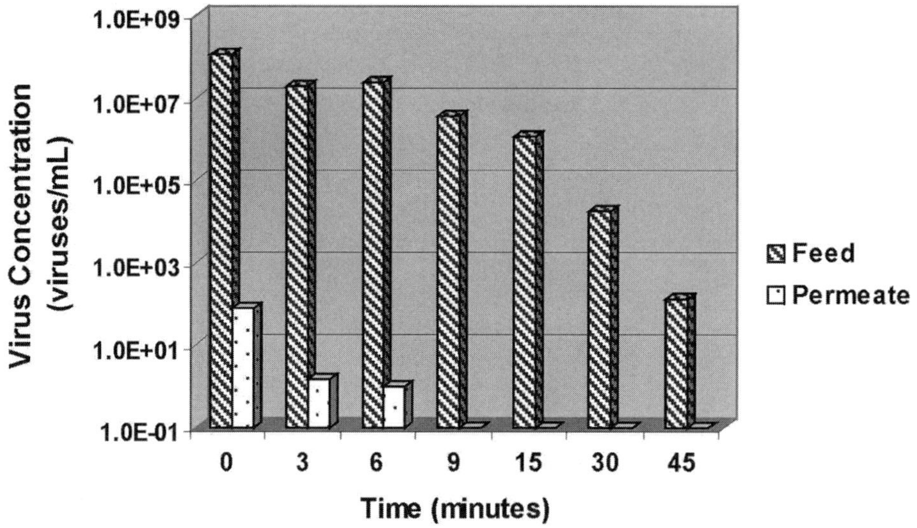

Figure 6 Virus removal observed during time test (Experiment 3B) with the Hydranautics ESPA1 element

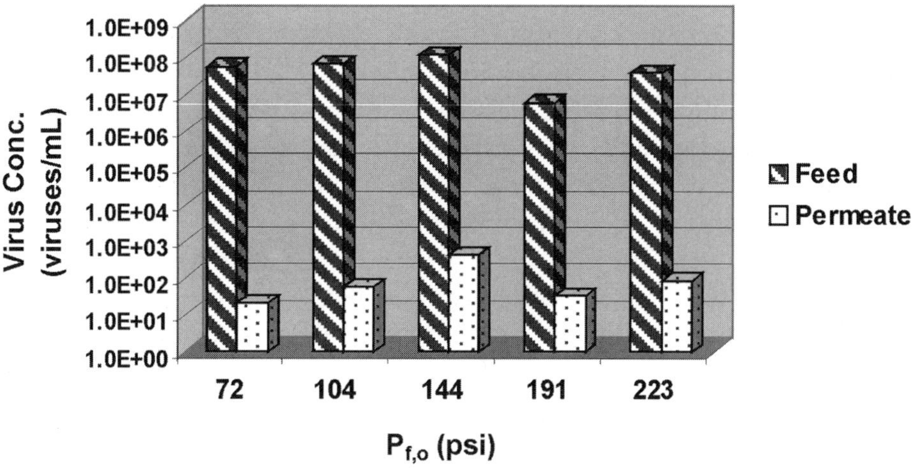

Figure 7 Virus passage observed for Experiment 3B with the Hydranautics ESPA1 element

feed pressure. The virus concentration achieved in the feed tank ranged from 6.1×10^6 to 1.2×10^8 viruses/mL, and the permeate concentrations corresponded to virus rejections of 5.3- to 6.4-logs. These rejections appeared to be independent of water flux, consistent with the occurrence of advective passage through membrane imperfections (i.e., small pinholes).

VIRUS/MICROSPHERE EXPERIMENTS. Because virus passage was observed in Experiment 3B, the objective of Experiment 4B was to correlate the passage of viruses and microspheres with the Hydranautics ESPA1 element. The initial addition of microsphere stock solution resulted in a feed concentration of approximately 10 mg/L. However, it was assumed that, similar to the phenomena observed for viruses, the concentration of microspheres in the feed tank would decrease over time. Hence, additional microsphere dosages of 10 mg/L were added before each sampling event. A stabilization period of only 1 hour was allowed between samples.

The virus results for Experiment 4B are shown in Figure 8. Virus feed concentrations ranged from 8.0×10^8 to 1.8×10^9 viruses/mL

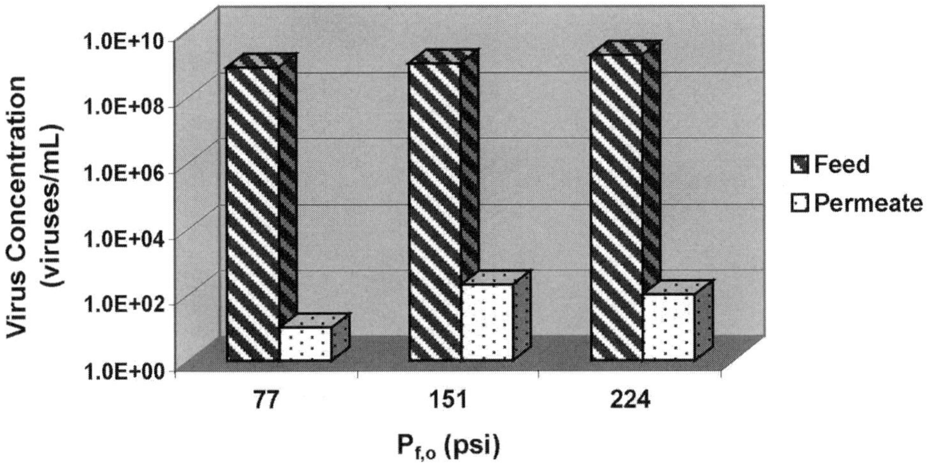

Figure 8 Virus passage observed for Experiment 4B performed with the Hydranautics ESPA1 element

and the corresponding rejections ranged from 6.7- to 7.9-logs. Again, as in Experiment 3B, no trend in virus removal with change in water flux was observed.

Experiment 4B results for microspheres are presented in Figure 9. The fluorescence of all permeate samples was found to be below the detection limit somewhere in the range of 0.01–0.1 μg/L of microspheres. Consequently, the removal of microspheres was in excess of 5- to 6-logs. Although these results are consistent with the observations for viruses, a direct correlation could not be established due to analytical detection limitations for the microspheres.

VIRUS/MICROSPHERE EXPERIMENTS WITH COMPROMISED O-RING. Experiment 5B was performed to assess if a faulty o-ring from the seal between the permeate tube and the vessel end-cap would affect the performance of the system. Experimental results for this test are presented in Figures 10 and 11 for viruses and microspheres, respectively. The virus concentration in the feed was 1.9–2.6 × 10^8 viruses/mL, and permeate concentrations corresponded to 6.2- to 7.0-logs removal, generally consistent with previous results obtained with this element.

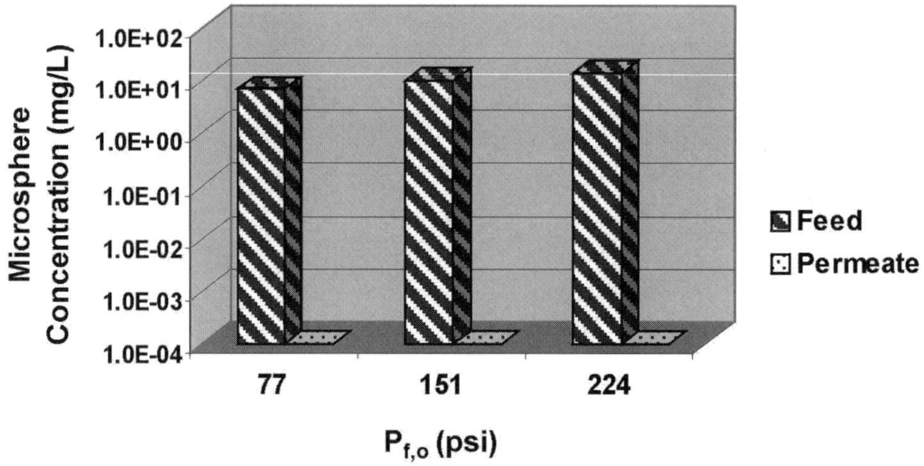

Figure 9 Microsphere passage observed for Experiment 4B performed with the Hydranautics ESPA1 element

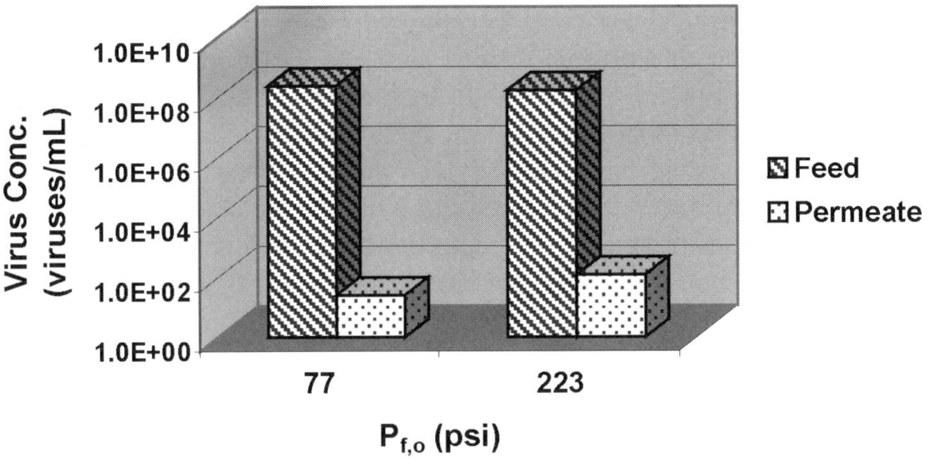

Figure 10 Virus passage observed for Experiment 5B performed with the Hydranautics ESPA1 element installed with a compromised o-ring

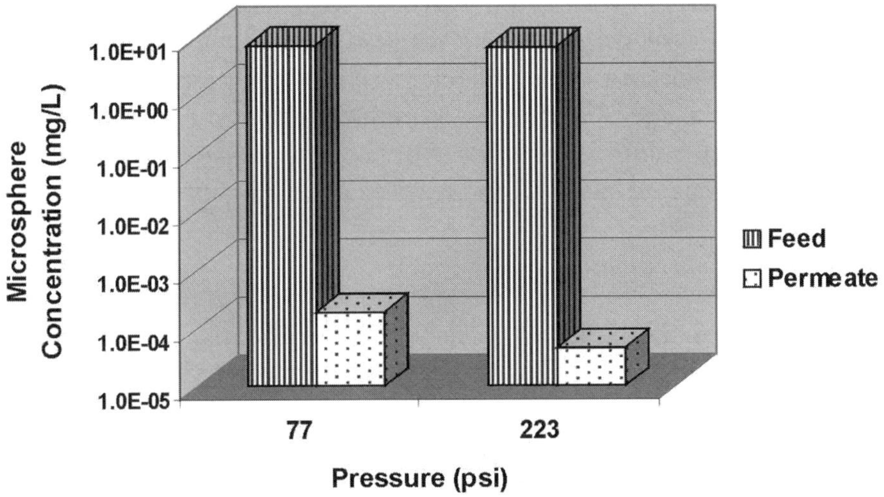

Figure 11 Microsphere passage observed for Experiment 5B performed with the Hydranautics ESPA1 element installed with a compromised o-ring

Microsphere removals in the order of 5- to 6-logs are presented in Figure 11 but the permeate concentrations were within the detection limit range of 0.1 to 0.01 µg/L. Consequently, the microsphere results should be interpreted as corresponding to rejection levels in excess of 5-logs. Because virus and microsphere rejections were approximately the same as those found with the same membrane element prior to inducing the crack in the o-ring, the cracked o-ring appeared to continue to serve as an effective barrier against the passage of particles greater than approximately 20 nm.

Membrane C—ROGA-HR

Solute Experiments

Water and solute passage results for all experiments (Experiments 1C–6C) performed with the Fluid Systems ROGA-HR membrane element are presented in Figure 12. Membrane performance before inducing the imperfections (Experiments 1C–3C) was the same for the three experiments and was consistent with performance specifications. The small pinhole resulted in a slight increase in solute passage without noticeable changes in water permeability (Experiments 4C and 5C). The larger pinhole (Experiment 6C) resulted in more than double the passage of sodium chloride consistent with a corresponding increase in water permeability of approximately 6–7%.

Virus Experiments

FEED CONCENTRATION DYNAMICS. Experiments 1C–2C included tests designed to assess the rate of virus concentration decline in the feedwater over time. Two separate virus seedings were done; however, they were not consecutive. Both were performed after the tank had been filled with a fresh batch of feedwater. Both seedings demonstrated consistent decline in the concentration from the initial feed concentration of approximately 4×10^5 to 3×10^6 viruses/mL. After roughly 20 minutes, the feed concentration had decreased

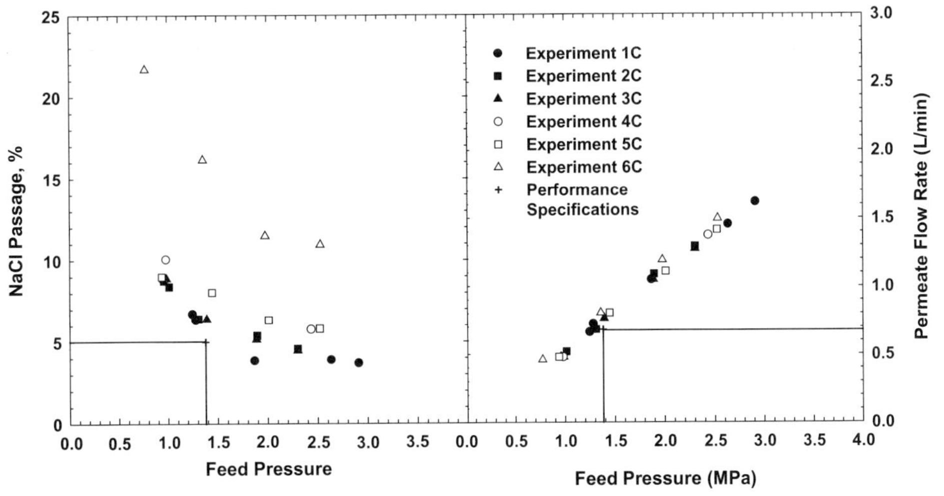

Figure 12 Performance of Fluid Systems ROGA-HR element before (Experiments 1C–3C) and after inducing relatively small (Experiments 4C–5C) and large (Experiment 6C) pinholes (see Table 4 for experimental conditions)

nearly 5-logs. This information was used in determining the sampling protocol for the virus testing at variable pressure, discussed in the next section. Permeate samples taken during the time tests resulted in virus concentrations below the detection limit of 1 virus/mL.

VIRUS PASSAGE EXPERIMENTS. Virus passage data obtained with the Fluid Systems ROGA-HR element before inducing the pinholes are presented in Table 5 for Experiments 1C–3C. Some difficulties were encountered in maintaining a constant feed concentration for Experiment 1C. Nevertheless, removals in excess of 4.8-logs were achieved at the higher pressures investigated. In Experiment 2C, virus feed concentrations were more constant at 1.0–1.6 × 10^5 viruses/mL. Once again, virus rejections greater than approximately 5-logs were observed. Experiment 3C was performed to assess if higher log removals could be achieved with the Fluid Systems

Table 5 Virus removals for Experiments 1C–3C for ROGA-HR

		Virus Concentration (viruses/mL)		
Exp.	Feed Pressure, *MPa*	Feed	Permeate	Log Removal
1C	1.2	0	0	–
–	1.9	7.0×10^2	0	2.8
–	2.6	7.3×10^4	0	4.9
–	2.9	6.9×10^4	0	4.8
2C	1.0	1.5×10^5	0	5.2
–	1.3	1.0×10^5	0	5.0
–	1.9	1.6×10^5	0	5.2
–	2.3	1.0×10^5	0	5.0
3C	1.4	7.5×10^6	0	6.9
–	1.9	3.0×10^6	0	6.5
–	2.3	6.0×10^6	0	6.8

ROGA-HR element. Even though the feed virus concentration ranged from 3.0×10^6 to 7.5×10^6 viruses/mL, no viruses were detected in permeate samples. Thus, removals achieved with the ROGA-HR element prior to inducing imperfections were in excess of 6.5-logs.

VIRUS/MICROSPHERE EXPERIMENTS WITH INDUCED PINHOLES. The small and large imperfections induced in the surface of the Fluid Systems ROGA-HR element for Experiments 4C to 6C, described subsequently, were characterized with an Olympus BX60 (Leco Corporation, Chicago, IL) microscope and a CH250 Photometric Charge Captured Device with a Kodak KAF 1400 chip (Photometrics, AZ) at a magnification of 40 times. The sizes of the small and large pinholes were estimated at approximately 10 and 100 μm, respectively.

The ROGA-HR element was first tested after inducing a relatively small pinhole in order to attempt the development of a correlation between viruses and microspheres at measurable microsphere rejection levels. Experiment 4C results for viruses and microspheres are presented in Figures 13 and 14, respectively. As

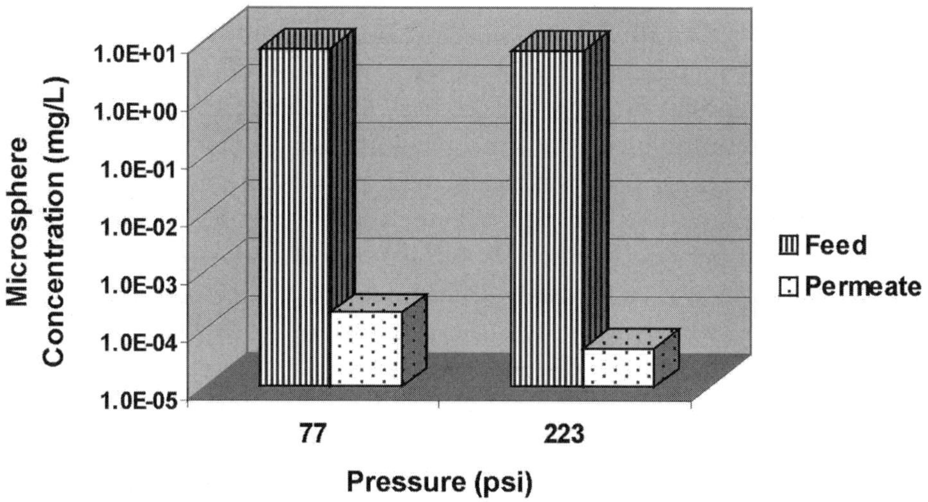

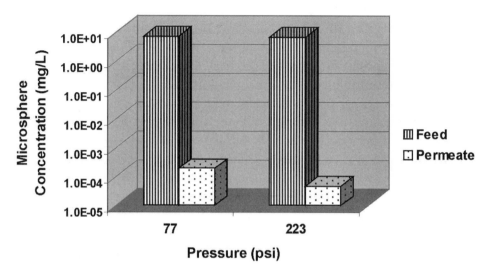

Figure 13 Virus passage observed for Experiment 4C performed with the Fluid Systems ROGA-HR element compromised with a small pin-sized hole

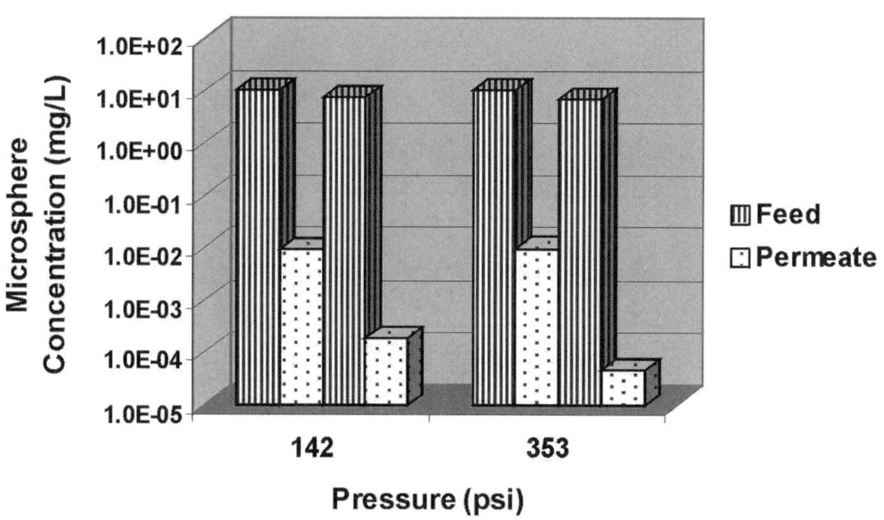

Figure 14 Microsphere passage observed for Experiment 4C performed with the Fluid Systems ROGA-HR element compromised with a small pin-sized hole

depicted in the figures, there was good agreement between removals for viruses, 2.4- to 2.8-logs, and microspheres, 3.0-logs. Furthermore, no dependence of particle passage on water flux was observed indicating that the passage is consistent with advection through the induced pinholes, as previously hypothesized.

Experiment 5C was performed to assess if passage of viruses and microspheres through the small pinhole would change within 24-hour period. The results for Experiment 5C are presented in Figures 15 and 16 for viruses and microspheres, respectively. As can be seen from the figures, the removal of both viruses, 6.6- to 7.6-logs, and microspheres, >4.7- to 5.1-logs, increased considerably compared to those in Experiment 4C. A possible explanation for the increase in removal is that the particles and perhaps organic matter from the virus stock were clogging the small imperfection, causing a healing effect to occur in the membrane.

Experiment 6C was performed with the same ROGA-HR element after inducing a second pinhole approximately ten times larger than the previous imperfection. Again, the element was tested with viruses and microspheres simultaneously. The experi-

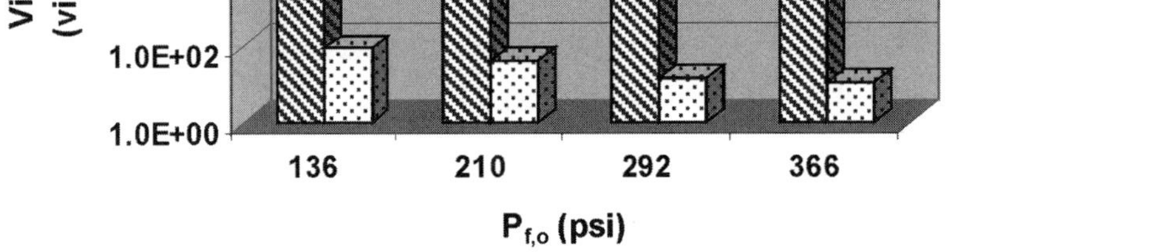

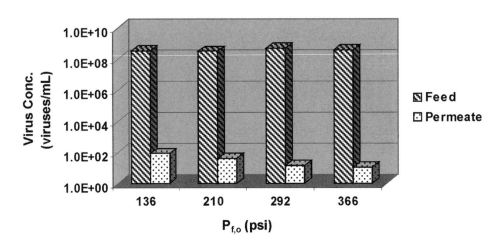

Figure 15 Virus passage observed for Experiment 5C performed with the Fluid Systems ROGA-HR element compromised with a small pin-size hole

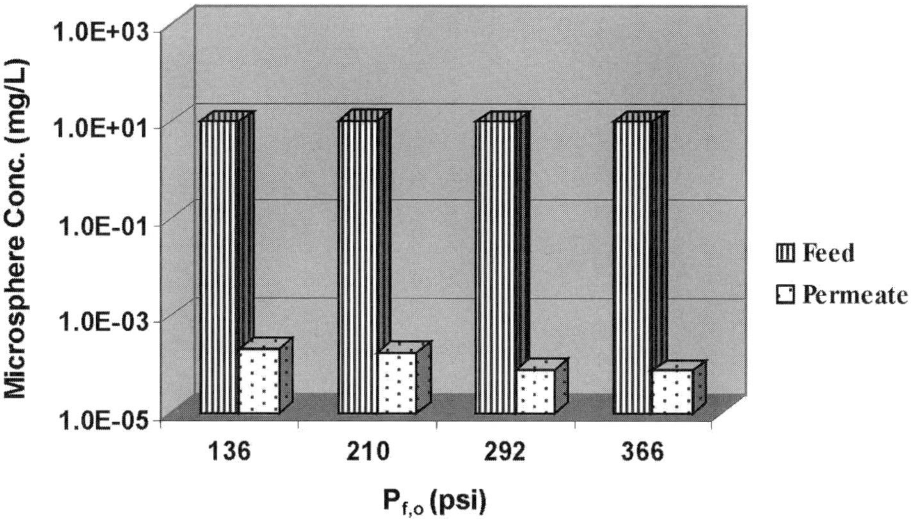

Figure 16 Microsphere passage observed for Experiment 5C performed with the Fluid Systems ROGA-HR element compromised with a small pin-size hole

mental results are presented in Figures 17 and 18 for viruses and microspheres, respectively. Once again, good agreement was found between removals for viruses, 1.0- to 1.6-logs, and microspheres, 0.9- to 1.2-logs.

CONCLUSIONS

All three membrane elements investigated exhibited at least 6-log removal of MS2 *phage* virus. The virus passage that was observed indicated that the virus permeate concentration was a function of the concentration of viruses in the feedwater and passage occurred primarily by advective transport through imperfections.

Exposure of the FilmTec BW30HP (Membrane A) element to chlorine at approximately 1,000 ppm-h did not result in performance deterioration with respect to virus passage. The uncompromised

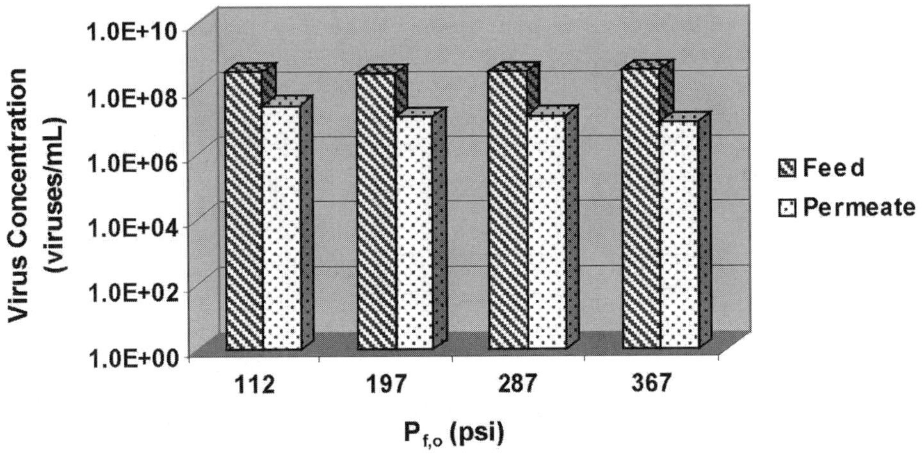

Figure 17 Virus passage observed for Experiment 6C with the Fluid Systems ROGA-HR element compromised with a large pinhole

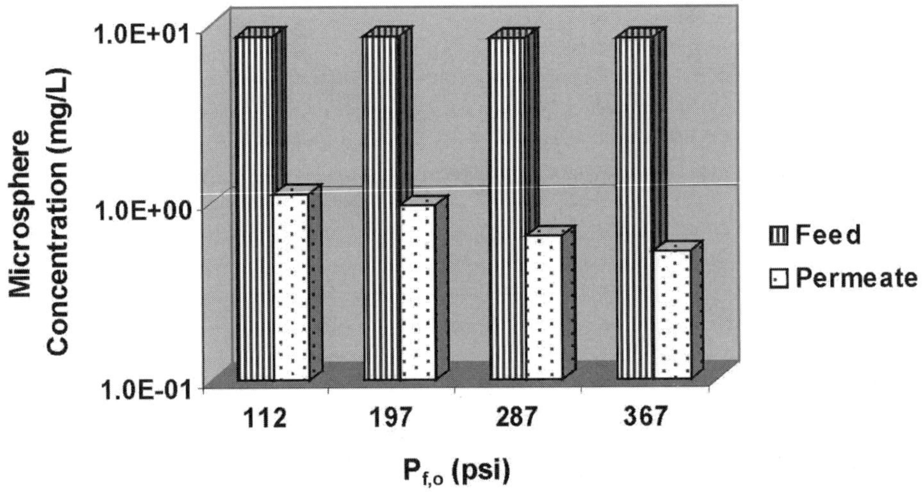

Figure 18 Microsphere passage for Experiment 6C performed with the Fluid Systems ROGA-HR element compromised with a large pinhole

Hydranautics ESPA-1 (Membrane B) membrane was installed with a cracked o-ring seal. No deterioration in the performance was observed.

Perforation of the surface of the Fluid Systems ROGA-HR (Membrane C) membrane element with a small pin-sized hole caused moderate deterioration in the performance of the membrane. The compromised ROGA-HR membrane element provided removals of approximately 3-logs for both viruses and microspheres. However, after approximately 1 day, the imperfection became clogged, increasing the removal of viruses and microspheres to nearly 6-logs once again. A perforation about ten times larger than the small pinhole was induced on the ROGA-HR membrane surface, causing greater performance deterioration. The removal of both viruses and microspheres was only 1-log. Based on the imperfection testing results, the non-biological surrogate indicator behaved very closely to the MS2 *phage* virus.

REFERENCE

1. Adham, S., P. Gagliardo, D. Smith, D. Ross, K. Gramith and R. Trussell. Monitoring of Reverse Osmosis for Virus Rejection. *Proceedings of the 1998 American Water Works Association Water Quality Technology Conference*, San Diego, CA, November 1–4 (1998).

CHAPTER · 11

Development of an Innovative Method to Monitor the Integrity of a Membrane Water Repurification System

Paul Gagliardo, Yelidiz Chambers, Rhodes Trussell
Metropolitan Wastewater Department, City of San Diego, CA

Samer Adham, Brian Gallagher
Montgomery Watson, Pasadena, CA

Mark Sobsey
School of Public Health, University of North Carolina, Chapel Hill, NC

OVERVIEW

The concept of recycling wastewater for potable reuse is becoming more attractive as sewage treatment regulations become more stringent and water supply options become more scarce. The chemical and microbial risks associated with potable reuse of wastewater must be identified. In developing a methodology to assess microbial-based risk to consumers from recycled water, a technique to enumerate pathogens should be employed. In drinking water applications the typical indicator organisms tested for are total and fecal coliform. While these are good indicators of contamination, they are relatively large in size (0.1–10 microns). The Information Collection Rule (ICR) has recommended the use of MS-2 bacteriophage as an indicator organism, as it is similar in size and shape to human enteric viruses, but is not a human pathogen. This chapter analyzes a membrane filtration method for the concentration and enumeration of indigenous coliphage as a way to monitor virus removal through low- and high-pressure membrane systems.

BACKGROUND

The City of San Diego has, since the 1970s, been analyzing the feasibility of using wastewater as a source commodity in the production of drinking water. The Total Resource Recovery Project (TRRP), jointly funded by state and federal grants as well as the City of San Diego, sought to demonstrate that recycled wastewater could be safely and reliably used as a supplement to the existing water supply. The TRRP has become affectionately known as the Aquaculture project due to the fact that aquaculture ponds with water hyacinths are used as the secondary treatment unit process. AQUA I and II (1) were constructed in the early 1980s in Mission Valley. AQUA II, a 0.3 mgd system, began operation in 1984 and ran until 1993. In 1993, the San Pasqual Water Reclamation Facility (AQUA III), a 1 mgd wastewater treatment system, began operation in the San Pasqual Valley, and is currently running. (2)

Health Effects Studies (1, 2) were performed on the effluent from both AQUA II and III and the risk associated with using this water as an alternative to raw imported water was analyzed. Based on chemical, toxicological and microbial risk assessments, the recycled water was shown to be less of a potential health risk than the imported water supply.

The City of San Diego has recently completed construction on the 30 mgd North City Water Reclamation Plant (North City). This facility produces Title-22 quality water that can be used for most purposes other than drinking, according to California Health and Safety Codes and DHS Water Reclamation Guidelines.

In an effort to beneficially reuse this commodity in a cost-effective manner, a reclaimed water distribution system was constructed to deliver the product. Based on an optimization study conducted by the City, the most cost-effective system size would deliver 8,700 AFY. This left 21,500 AFY of product water available at North City. Using the Aquaculture HES as a foundation, the DHS was approached as to the concept of repurifying the reclaimed water and supplementing a surface water reservoir with this effluent. DHS responded with cautious optimism in August 1993 and the project began to gain momentum.

The City and DHS partnered during all stages of the project. A 1994 Feasibility Study (3) was prepared and DHS requested that some issues be resolved. This request generated the creation of the Aqua 2000 Research Center (Aqua 2000) where pilot testing has been performed since 1995. As pilot experiments were developed and performed, and results analyzed, it became more and more certain that the system configuration, MF or UF, RO, IX, O3 and pipeline chlorination, could reliably and safely produce a drinking water–quality product. A critical question kept arising: how will the City monitor the performance of the system?

Much research has gone into the development of a viable, cost-effective and non-labor-intensive monitoring program that provides operators with real-time insight into the system functioning. An online low method detection TOC analyzer is being used to monitor the performance of the RO system. TOC results as low as 10 μg/l and removal efficiencies approaching three logs have correlated well with full-scale virus seeding experiments. This technique shall be used in the full-scale Water Repurification project to monitor unit process performance and serve as an indirect indication of pathogen removal. But the holy grail was still be to found. A way was needed to directly measure virus size pathogens, with high sensitivity, low cost, and to obtain these results quickly. That is the focus of this chapter.

Pathogen Monitoring

Bacteriophages have been used for many years as models for the behavior of viruses in water treatment processes. Bacteriophages are classified into six (6) morphological groups, A through F. (4) Morphological groups A, B and C have double-stranded DNA and have heads and tails of various sizes. Group D has single-stranded

DNA and Group E has single-stranded RNA. Both Groups D and E have cubical symmetry. Group F exhibits single-stranded DNA and is long filamentous phages.

Coliphages are bacteriophages that infect strains of *E. coli*. Male-(F-) specific RNA coliphages are the most appropriate group of phages to be utilized as indicator organisms because their physical structure most closely resembles human enteric viruses. Additionally, they are typically found in high numbers in sewage, are persistent in the environment, and do not appear to multiply in the environment.

A male specific coliphage, MS-2, has been used extensively and successfully in the TRRP HES as well as the Aqua 2000 pilot testing (5, 6, 7) to delineate the ability of the system to eliminate viruses from the influent waters.

Both the AQUA II and III systems were tested for virus removal/deactivation by seeding MS-2 phage in the influent stream. Effluent of the entire system was tested to determine, quantitatively, the ability of the system to provide disinfection efficacy. The results are shown in Table 1.

The 1995 pilot testing at Aqua 2000 (5) also utilized seeded MS-2 phage to determine system virus removal efficiency. In these and subsequent experiments, unit processes were tested individually. In the Aquaculture HES experiments the system was tested as a whole. A more extensive testing program was conducted in 1997 at Aqua 2000. (6) The 1995 and 1997 microbial challenge study results are presented in Table 2. These results have been accepted by DHS as a model for virus removal/deactivation through the proposed Water Repurification System. Additional seeding experiments were performed in 1998 on MF and RO that confirmed earlier results (Table 3).

The above-mentioned experiments were performed at various laboratories using various methodologies and host organisms for phage enumeration. As part of the 1997 testing, samples were split and analyzed by two laboratories to determine any difference in testing procedure. No significant difference was noted. Also as part of the 1997 work, indigenous coliphage was routinely monitored.

These data are presented in Figure 1. The data suggest that not only can seeded MS-2 phage be used to document system virus removal capability, but indigenous coliphage can be used to document actual system performance during full-scale operation, and membrane integrity.

Until recently, the two main methods for enumerating bacteriophages were the Double Agar Layer (DAL) method (4) and the Single Agar Layer (SAL) method (see Appendix A). The main drawback of these methods is the need for overnight incubation of the plates prior to reading of plaques and the need to plate several plates to increase the sensitivity of the detection limit. Both the SAL and DAL methods can utilize various host organisms. The work presented thus far in this chapter used *E. coli* C-3000 or Famp as a host organism. When MS-2 phage seeding studies were being performed, the results were consistent. The C-3000 host organism enumerates somatic as well as male-specific coliphage. The Famp only enumerates male-specific phage. When MS-2, in large concentrations, is seeded before unit processes, they overwhelm any other coliphages present. Thus, one can say with confidence that the coliphages being enumerated are exclusively MS-2 whether C-3000 or Famp is being used. This is not the case when indigenous phages are being monitored. The Famp counts will be significantly lower than the C-3000 counts due to the fact that C-3000 can be infected by somatic as well as male-specific coliphages.

An AWWA Research Foundation study (8) detailed a membrane filtration (MF) method for the enumeration of male-specific coliphage. This method is similar to the standard MFM for the enumeration of coliform bacteria. Test results using various host organisms showed good results as compared to the DAL and SAL methods. Typically, the MFM methodology calls for incubation overnight. But when the membrane is plated face-down on the host lawn, plaque formation can be detected in four to six hours. A series of experiments were conducted to determine if the MFM, using various host organisms, could produce statistically similar results to the DAL or SAL methods.

These experiments were performed utilizing indigenous coliphage in the wastewater, as well as seeded MS-2 coliphage. The experiments were also performed at various levels of treatment: secondary (pond) effluent, tertiary filter effluent, microfiltration (MF) and ultrafiltration (UF) effluent, and reverse osmosis (RO) effluent. Influent levels of phage were measured to all unit processes in order to calculate virus removal capabilities.

As previously noted during the 1997/1998 pilot testing, indigenous coliphage was monitored throughout the system. Significant amounts of phage were detected after secondary and tertiary treatment (Figure 1). The microfiltration unit was variable in its ability to remove coliphage, while the reverse osmosis was consistently able to remove 100% of the indigenous coliphage present.

It was further demonstrated in the 1997 work (microbial challenge study report) using MS-2 seeding data as well as indigenous phage measurements that the MF's ability to remove coliphage was related to the fouled condition of the membrane. Immediately after cleaning, or at low trans-membrane pressure (TMP), the phage log removal would drop below 1.0. At extremely fouled conditions the MF log virus would approach an asymptote of 3.2.

Figure 2 shows the results of MS-2 seeding experiments performed during 1997 and 1998. Data are plotted as log MS-2 removal on the y-axis versus specific resistance (psi/gfd @ 20°C) on the x-axis. The 1997 data using the single agar method (SAL) with Famp host organism were plotted originally generating the characteristic curve. MS-2 seeding experiments using the DAL and MFM (C-3000 and Famp hosts) were performed in 1998 and plotted over that original curve. The correlation was very good, showing that the four methods were consistent in their enumeration of seeded MS-2 phage.

Figure 3 shows the data set using only information from the DAL experiments. Figure 4 shows the data set using only information from the MFM-Famp experiments. Figure 5 shows the data set using only information from the MFM C-3000 experiments.

Figure 6 shows all the indigenous coliphage experimental data previously mentioned plotted together.

The microfiltration data are important because they show that all four coliphage enumeration methods predict the same performance for the MF unit process. All show a trend toward zero log virus removal at a specific resistance of 0.16–0.18. They also show a trend toward a maximum log removal of 3.2 at a specific resistance between 0.50 and 0.60 psi/gfd.

Table 4 shows data from all sample locations in the pilot plant for each of the three enumeration methods performed by the City lab. At high phage concentrations (pond, secondary effluent) the DAL consistently enumerates higher amounts of coliphage than either of the MFM methods. For tertiary treated wastewater effluent all three methods enumerated similar amounts of coliphage. At the microfiltered sample location (clean samples), where there is a relatively low concentration of coliphage, the MFM methods consistently enumerate larger numbers. All methods showed no coliphage in the RO effluent. This resultant high bias for DAL at high phage concentration and high bias for the MFM methods at low phage concentrations is graphically shown in Figure 7.

Tables 5, 6 and 7 show the individual data points (defined in Figure 8) during 1998 comparing the MFM C-3000 method and the DAL method. As can be seen by the mean number enumerated at high phage concentration, the MFM C-3000 has a 50–59% recovery compared to the DAL method for secondary effluent, an 88–100% recovery for tertiary effluent and a 250% recovery rate for MF effluent.

These data are further reduced in Figure 9 where both the MFM methods are charted against the DAL method. Not only does the previous correlation show up, but the Famp host has a more extreme reaction on both the high bias and low bias ends of the spectrum. These data suggest the MFM-Famp method may be the choice method for membrane system disinfection efficacy monitoring.

The raw data used for the compilation of Figure 6 are contained in Tables 8, 9 and 10. These show a recovery rate between the Famp and C-3000 MFM methods of 35–44% for secondary effluent, 39–47% for tertiary effluent and 140–162% for MF effluent, again showing the increased ability of the MFM-Famp to enumerate indigenous coliphage in cleaner sample waters.

In 1998 a series of MS-2 seeding studies were performed on the microfilter and reverse osmosis systems. In order to further show the correlation between the enumeration techniques, all of the data collection was performed by splitting samples and testing by all three methods.

Table 11 shows the results of testing the MS-2 seed stock. Mean concentrations for the four trials are very similar.

Table 12 shows the data for the three methods on MF influent and effluent. Mean values of the six trials show excellent correlation for both MF effluent and influent valves.

Table 13 shows the results for reverse osmosis influent and effluent flows. Once again the 3 methods showed very similar results on the RO influent. On the RO effluent the MFM-Famp had four times the counts of the DAL method. This is consistent with the other data collected, showing a high bias by the MFM-Famp method over the DAL for very clean water samples.

Thus far we have shown that results obtained for the DAL and MFM C-3000 methods are similar on both indigenous and seeded phage experiments. Additionally, it has been shown that the MFM C-3000 and MFM-Famp have similar results across the board. There is some bias of total counts, depending upon the phage concentration in the sample and the cleanliness of the water sample, but the results are similar.

During all of these experiments MFM plagues were read at 4, 6, 8 and 24 hours' reading time. But the goal of this work was to show that not only does this method compare well with the standard DAL method, it could also be a rapid tool for detection.

Table 14 shows the raw data related to phage readings for MFM C-3000 and MFM-Famp methods at various incubation periods. If one considers a positive reading at 24 hours a "true" positive, then one can calculate the ability of a method to have true positive readings at shorter incubation periods. Figure 10 shows a plot of both methods and the number of positive readings (compared to the 24-hour baseline) at each incubation period. As can be seen at the 8-hour reading, 96% of the samples showed a true positive test result. At the 6-hour reading, 64% showed a true positive test result and at 4 hours 46% showed a true positive.

Breaking these data down further for each method (Figure 11), the Famp showed a 100% true positive test result at 8 hours, an 86% true positive result at 6 hours, and a 57% true positive result at 4 hours. The MFM C-3000 method shows a 93% true positive result at 8 hours, a 43% true positive at 6 hours and a 36% true positive result at 4 hours. These are data for sample point Aqua #5, which is the MF effluent.

CONCLUSION

The data presented above clearly show that use of the MFM-Famp method is a viable rapid-result integrity monitor for RO effluents. It can be utilized as an effective presence/absence test for membrane systems using incubation times of 6–8 hours. Hence, this method is indeed a breakthrough day-to-day indicator of membrane performance in virus removal from reclaimed water.

Table 1 AQUA III pathogen challenge studies, San Pasqual Reclamation Facility, total system removal efficiency

Organism		Expected Concentration	Observed Concentration	Log_{10} Removal
Attenuated Poliovirus2	#1	5.7×10^7	<0.002	>10.5
(pfu/L)	#2	8.5×10^7	<0.003	>10.4
	#3	6.4×10^7	<0.003	>10.3
MS-2 Coliphage	#1	1.1×10^{11}	<0.03	>12.6
(pfu/L)	#2	1.3×10^{11}	220	8.8
	#3	9.3×10^{10}	<0.06	>12.1

Table 2 Aqua 2000 pathogen challenge study, 1995 and 1997 experiments, unit process coliphage (MS-2) removal efficiency

Unit Process	Average Influent Concentration *(pfu/ml)*	Average Effluent Concentration *(pfu/ml)*	Average Log_{10} Removal	# of Trials
Microfiltration (1995)	3×10^8	1×10^6	1.8	12
RO—CA (1995)	2×10^8	8×10^4	3.8	6
RO—TFC (1995)	2×10^8	2×10^5	3.3	6
Microfiltration (1997)	3×10^5	3×10^4	2.6	24
Ultrafiltration (1997)	3×10^5	95	5.2	18
RO—HR (1997)	4×10^5	3×10^2	3.0	18
RO—DOW (1997)	4×10^5	1.2	5.4	18
RO—ESPA (1997)	6×10^5	12.5	4.7	18
RO—ULP (1997)	8×10^5	2×10^2	3.4	9

Table 3 Aqua 2000 pathogen challenge study, 1998 experiments, unit process coliphage (MS-2) removal efficiency (DAL method)

Unit Process	Average Influent Concentration *(pfu/100 ml)*	Average Effluent Concentration *(pfu/100 ml)*	Average Log_{10} Removal	# of Trials
Microfiltration	30×10^5	18×10^4	1.2	6
RO—DOW	11.9×10^6	4.7×10^1	5.4	3
RO—ESPA	11.9×10^6	2×10^2	4.8	3
RO—DOW/LE	11.9×10^6	2.2×10^1	5.7	3
RO—HR/Modified	11.9×10^6	4.8×10^3	3.4	3

Table 4 Comparison of indigenous coliphage detection (MDL = 1 pfu/100 ml)

	Count *(nd)*	Mean	Median	Range
Double Agar Method				
Pond Effluent	10 (0)	3.4 E+4	3.8 E+4	1.0 E+2 – 7.2 E+4
MF/UF Influent	14 (0)	1.8 E+3	1.4 E+3	1.0 E+2 – 4.5 E+3
MF Effluent	14 (5)	5.7 E+0	3.0 E+0	nd – 3.2 E+1
RO—DOW Effluent	11 (11)	nd	nd	nd
RO—Hydranautics Effluent	11 (11)	nd	nd	nd
Membrane Filtration Method C3000 Host				
Pond Effluent	10 (0)	2.0 E+4	1.9 E+4	3.0 E+2 – 4.2 E+4
MF/UF Effluent	14 (0)	1.6 E+3	1.4 E+3	3.0 E+2 – 3.5 E+4
MF Effluent	14 (2)	1.4 E+1	7.5 E+0	nd – 7.0 E+1
RO—DOW Effluent	12 (11)	nd	nd	nd – 1.0 E+0
RO—Hydranautics Effluent	12 (12)	nd	nd	nd
Membrane Filtration Method Famp Host				
Pond Effluent	10 (0)	1.0 E+4	6.3 E+3	1.0 E+2 – 3.5 E+4
MF/UF Effluent	14 (0)	7.6 E+2	7.5 E+2	1.0 E+2 – 1.7 E+3
MF Effluent	14 (3)	2.4 E+1	9.5 E+0	nd – 9.5 E+1
RO—DOW Effluent	12 (12)	nd	nd	nd
RO—Hydranautics Effluent	12 (12)	nd	nd	nd

Table 5 Indigenous coliphage monitoring, percent recovery, MFM vs. DAL, C-3000 host, sample location Aqua #1 (pfu/ml)

Date	MFM	DAL
February 17	160	400
February 24	31	34
March 3	420	510
March 10	280	720
March 16	200	360
March 24	310	520
April 9	3	1
April 16	62	84
April 20	180	420
May 5	370	320
Median	190	380
Mean	201	337
% Recovery	50%–59%	

Table 6 Indigenous coliphage monitoring, percent recovery, MFM vs. DAL, C-3000 host, sample location Aqua #4 (pfu/ml)

Date	MFM	DAL
February 17	16	38
February 23	16	22
February 24	31	32
March 3	35	45
March 10	20	10
March 16 (am)	13	18
March 16 (pm)	12	22
March 24	12	7
April 9	3	1
April 16 (am)	12	12
April 16 (pm)	9	6
April 20 (am)	18	10
April 20 (pm)	9	12
May 5	15	16
Median	14	14
Mean	15.8	17.9
% Recovery	88%–100%	

Table 7 Indigenous coliphage monitoring, percent recovery, MFM vs. DAL, C-3000 host, sample location Aqua #5 (pfu/100 ml)

Date	MFM	DAL
February 17	14	3
February 23	16	3
February 24	10	4
March 3	4	4
March 10	4	<1
March 16 (am)	<1	1
March 16 (pm)	23	11
March 24	2	<1
April 9	1	<1
April 16 (am)	<1	<1
April 16 (pm)	39	19
April 20 (am)	5	<1
April 20 (pm)	70	32
May 5	15	3
Median	7.5	3
Mean	14.5	5.7
% Recovery	250%–254%	

Table 8 Indigenous coliphage monitoring, membrane filter method (MFM), Famp vs. C-3000 host, sample location Aqua #1 (pfu/ml)

Date	Famp	C-3000
February 10	9	13
February 17	140	160
February 24	10	31
March 3	59	420
March 10	220	280
March 16	51	200
March 24	3	310
April 9	1	3
April 16	67	62
April 20	150	180
May 5	350	370
May 20	23	64
May 26	23	27
June 2	21	61
June 16	11	340
June 23	13	61
July 7	1	15
July 14	31	110
July 21	4	6
July 28	1	4
Median	22	63
Mean	59.4	135.8
% Recovery	35%–44%	

Table 9 Indigenous coliphage monitoring, Famp vs. C-3000 host, sample location Aqua #1 (pfu/ml)

Date	Famp	C-3000
February 10	6	7
February 17	17	16
February 23	7	16
February 24	11	31
March 3	4	35
March 10	5	20
March 16 (am)	3	13
March 16 (pm)	1	12
March 24	1	12
April 9	1	3
April 16 (am)	12	12
April 16 (pm)	14	9
April 20 (am)	8	18
April 20 (pm)	8	9
May 5	14	15
May 19	9	20
May 26	1	1
June 2	1	7
June 16	1	5
June 23	2	8
July 7	7	19
July 14	3	3
Median	5.5	14
Mean	6.2	13.2
% Recovery	39%–47%	

Table 10 Indigenous coliphage monitoring, membrane filter method (MFM), Famp VS. C-3000 host, sample location Aqua #5 (pfu/100 ml)

Date	Famp	C-3000
February 10	63	42
February 17	37	14
February 23	16	16
February 24	15	10
March 3	2	4
March 10	2	4
March 16 (am)	3	1
March 16 (pm)	85	23
March 24	1	2
April 9	1	1
April 16 (am)	1	1
April 16 (pm)	60	39
April 20 (am)	4	5
April 20 (pm)	95	70
May 5	20	15
May 19	19	15
May 26	6	2
June 2	1	1
Median	10.5	7.5
Mean	23.9	14.7
% Recovery	140%–162%	

Table 11 Aqua 2000 pathogen challenge study; MS-2 seeding experiments—1998; comparison of MFM-Famp, MFM C-3000 and DAL methods (pfu/100 ml)

MS-2 Seed Stock			
Trial	Famp	C-3000	DAL
1	87×10^7	39×10^7	160×10^7
2	69×10^7	61×10^7	82×10^7
3	48×10^7	44×10^7	40×10^7
4	69×10^7	57×10^7	99×10^7
Mean	68×10^7	50×10^7	95×10^7

Table 12 Aqua 2000 pathogen challenge study; MS-2 seeding experiments—1998; comparison of MFM-Famp, MFM C-3000 and DAL methods (pfu/100 ml)

Microfiltration Influent			
Trial	Famp	C-3000	DAL
1	41×10^5	36×10^5	58×10^5
2	34×10^5	34×10^5	35×10^5
3	58×10^5	36×10^5	16×10^5
4	39×10^5	45×10^5	24×10^5
5	46×10^5	34×10^5	26×10^5
6	46×10^5	38×10^5	24×10^5
Mean	44×10^5	37×10^5	30×10^5

Microfiltration Influent			
Trial	Famp	C-3000	DAL
1	16×10^4	8.1×10^4	14×10^4
2	8.5×10^4	8.5×10^4	9.6×10^4
3	41×10^4	34×10^4	32×10^4
4	24×10^4	22×10^4	21×10^4
5	19×10^4	20×10^4	20×10^4
6	27×10^4	17×10^4	13×10^4
Mean	22×10^4	18×10^4	18×10^4

Table 13 Aqua 2000 pathogen challenge study; MS-2 seeding experiments—1998; comparison of MFM-Famp, MFM C-3000 and DAL methods (pfu/100 ml)

Reverse Osmosis Influent			
Trial	Famp	C-3000	DAL
1	73×10^5	60×10^5	110×10^5
2	82×10^5	39×10^5	160×10^5
3	79×10^5	79×10^5	88×10^5
Mean	78×10^5	59×10^5	119×10^5

Reverse Osmosis Effluent			
	Famp	C-3000	DAL
RO—DOW(1)	120	89	41
RO—DOW(2)	110	69	41
RO—DOW(3)	110	110	58
Mean	113	89	47
RO—DOW LE(1)	76	66	27
RO—DOW LE(2)	120	61	23
RO—DOW LE(3)	100	68	17
Mean	98	65	22

Table 14 Response time of MFM to indigenous coliphage, C-3000 host vs. Famp host, sample site Aqua #5, 100 ml sample volume

	C-3000 Host				Famp Host			
Date	4 hr	6 hr	8 hr	24 hr	4 hr	6 hr	8 hr	24 hr
February 10	0	0	7	42	0	25	47	63
February 17	0	5	11	14	5	24	26	37
February 24	2	2	5	10	3	5	8	15
March 3	0	0	1	4	0	0	1	2
March 10	0	0	1	2	0	0	0	1
March 16 (am)	0	0	0	0	0	1	1	3
March 16 (pm)	2	10	22	23	4	32	62	85
March 24	0	0	1	2	0	0	0	0
April 9	0	0	1	1	0	0	0	0
April 16 (am)	0	0	0	0	2	2	2	2
April 16 (pm)	1	25	37	39	29	40	51	60
April 20 (am)	0	0	4	5	0	2	4	4
April 20 (pm)	1	45	55	70	33	70	85	95
May 5	0	0	11	15	1	5	11	20
May 20	2	4	*	15	5	11	*	19
May 26	0	0	1	2	0	3	3	6
June 2	0	0	0	0	0	0	0	0

*8-hour sample note read

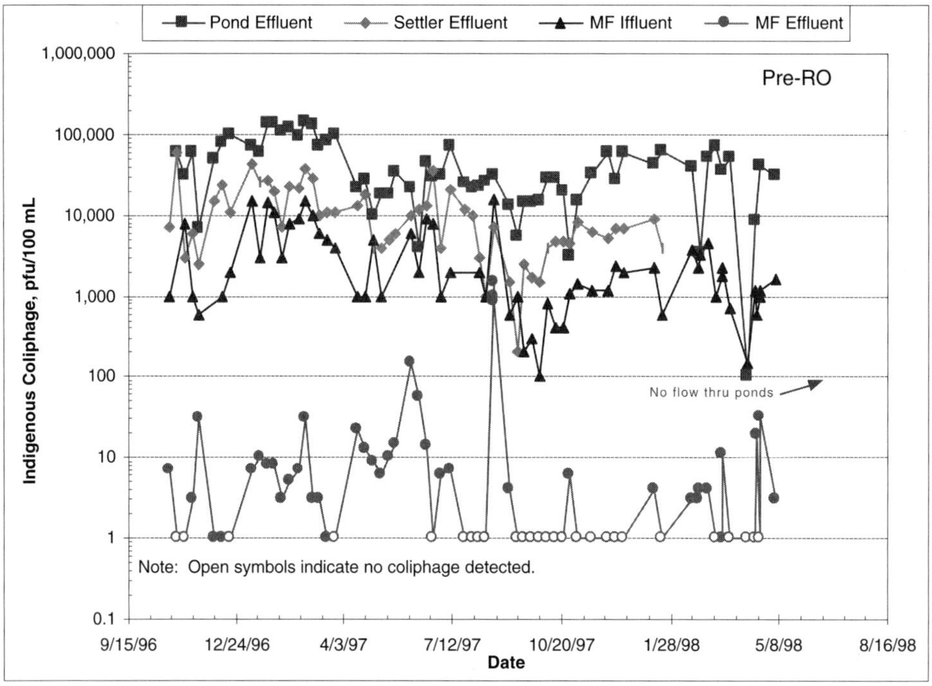

Figure 1 Indigenous phage removal (DAL)

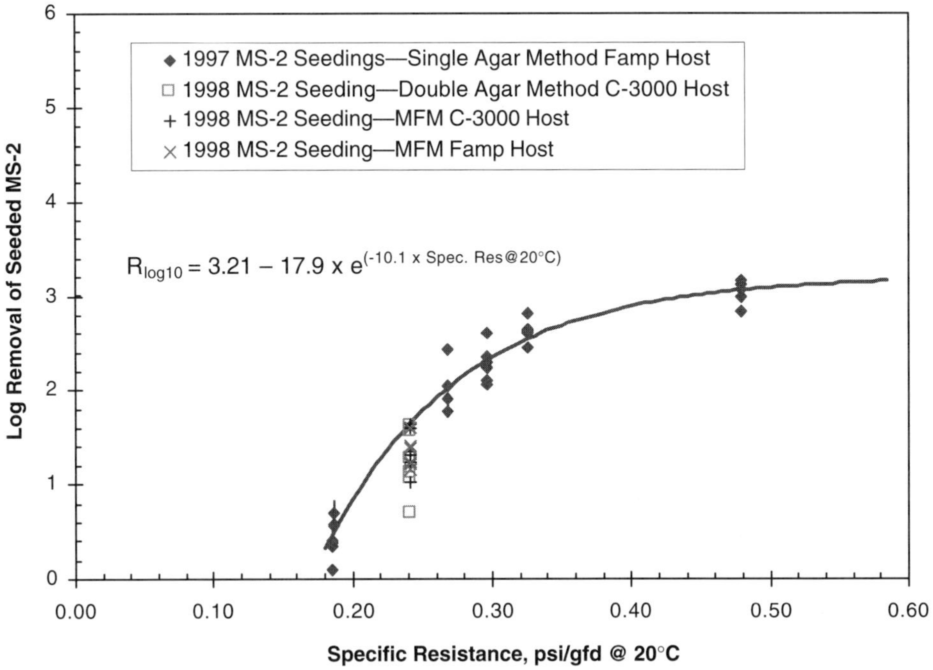

Figure 2 Log virus removal vs. specific resistance for microfiltration (seeded MS-2 phage)

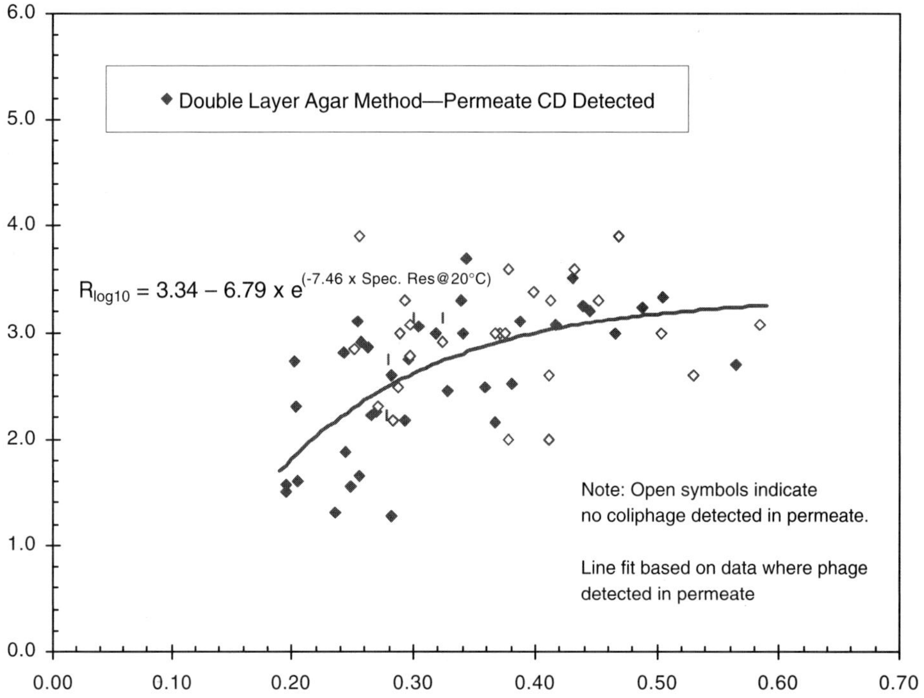

Figure 3 Log virus removal vs. specific resistance for microfiltration (indigenous phage [DAL] C-3000 host organism)

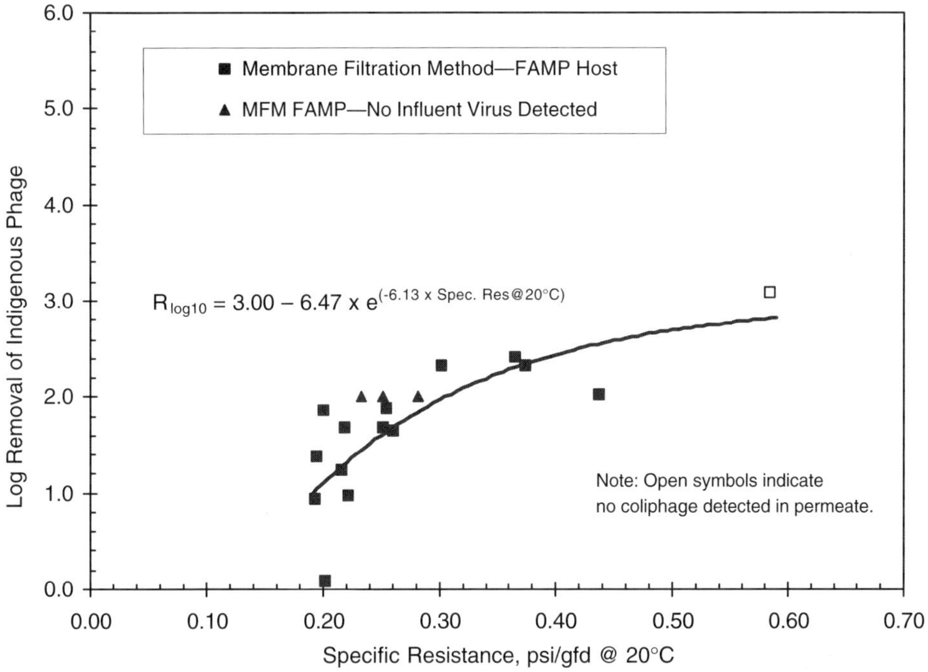

Figure 4 Log virus removal vs. specific resistance for microfiltration (indigenous phage [MFM] Famp host organism)

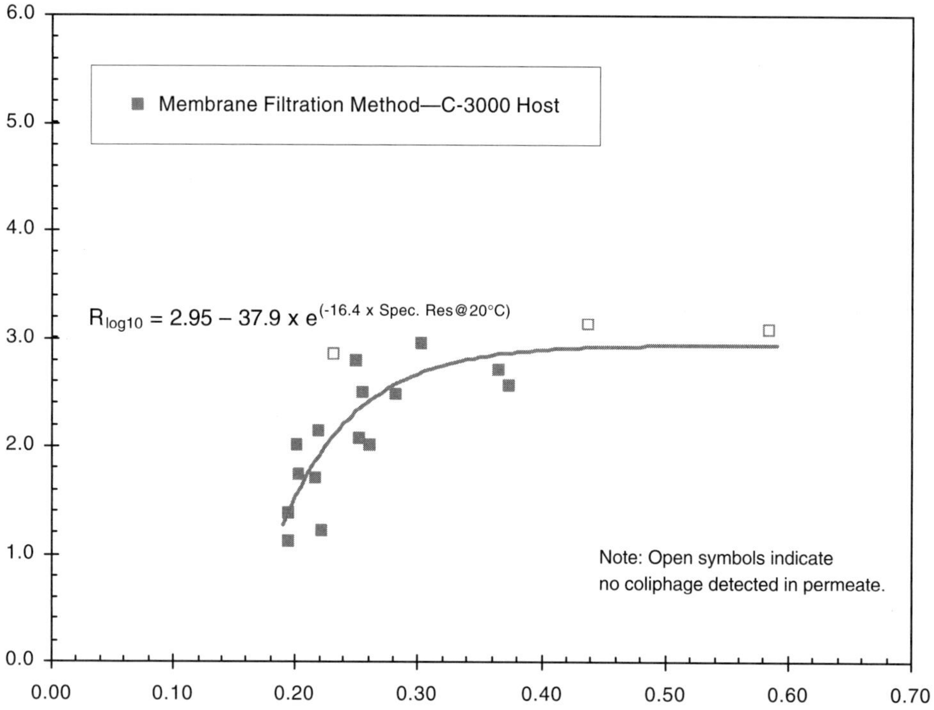

Figure 5 Log virus removal vs. specific resistance for microfiltration (indigenous phage [MFM] C-3000 host organism)

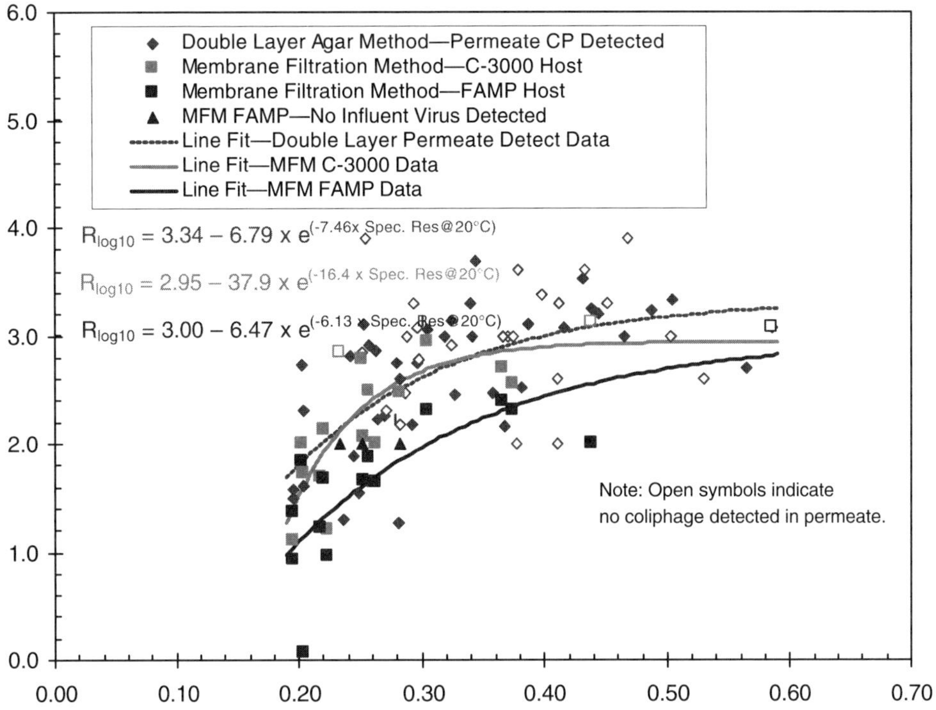

Figure 6 Log virus removal vs. specific resistance for microfiltration (indigenous phage)

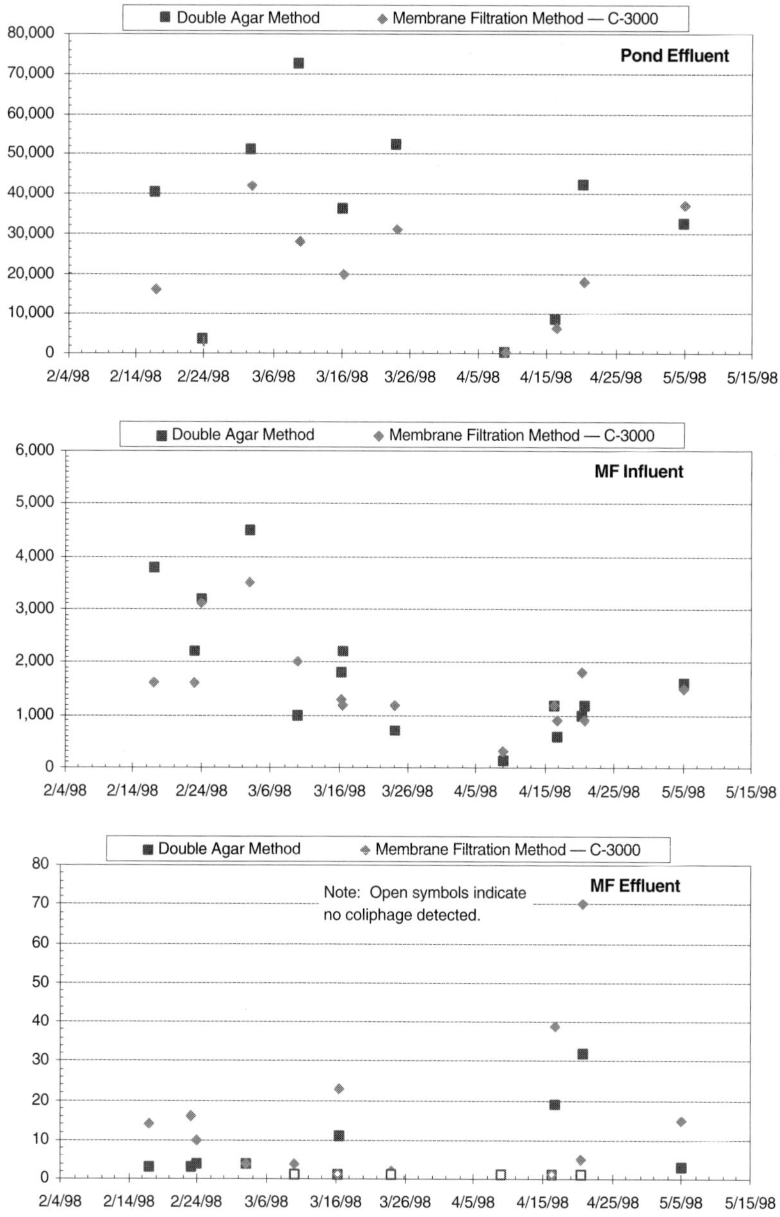

Figure 7 Comparison of indigenous coliphage enumeration by double layer agar method and membrane filtration method using C-3000 host organism

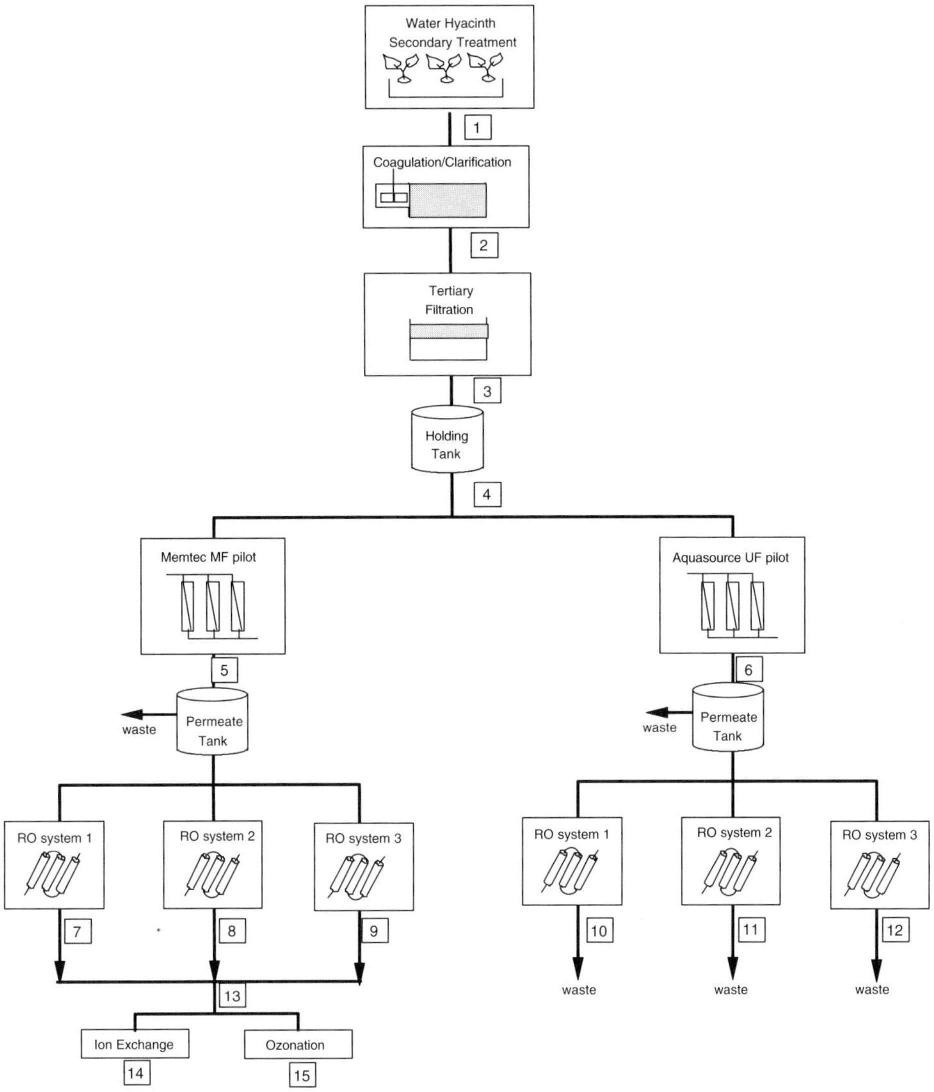

Figure 8 Sampling points location in the AWT pilot train

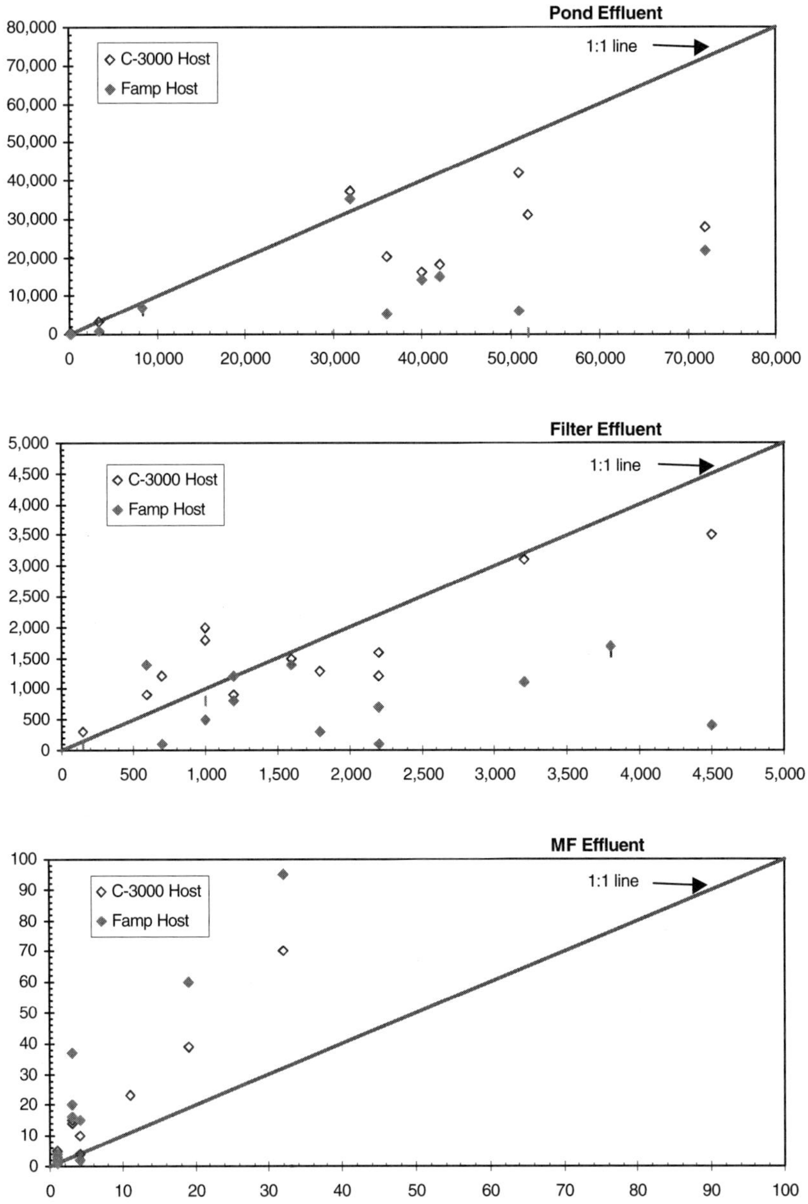

Figure 9 Comparison of phage enumeration by double agar layer and membrane filtration method for indigenous coliphage at pond, filter and MF effluents

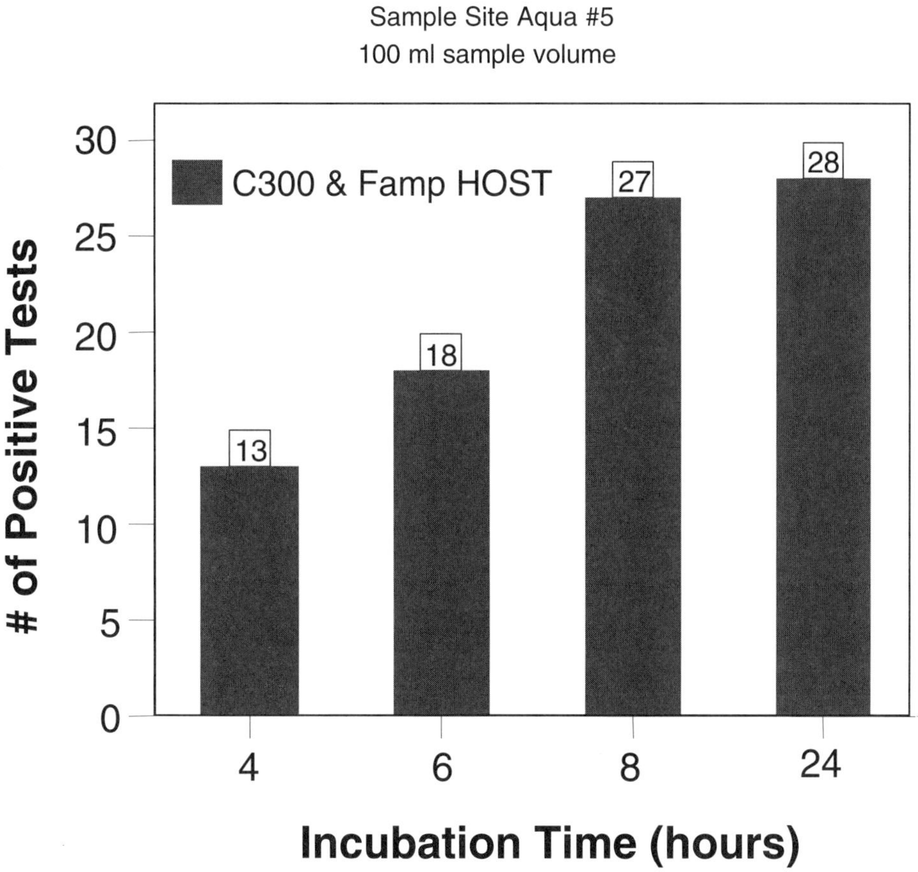

Figure 10 Response time of MFM to indigenous coliphage

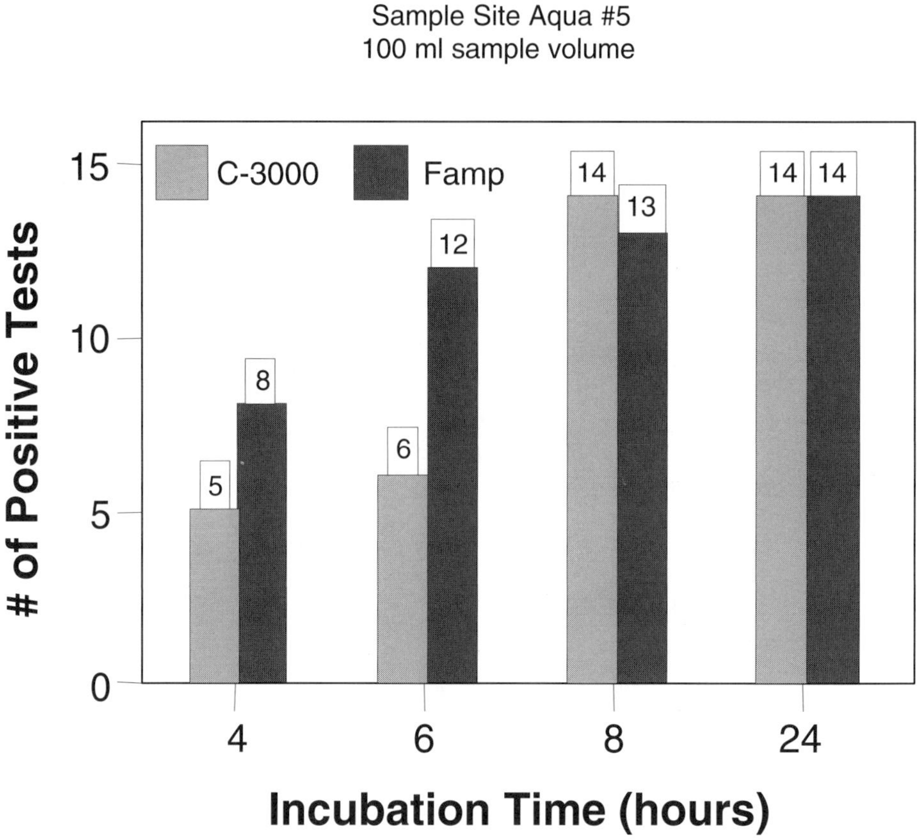

Figure 11 Response time of MFM to indigenous coliphage

REFERENCES

1. Western Consortium for Public Health, *The City of San Diego Total Resource Recovery Project, Health Effects Study*, September 1992.
2. Western Consortium for Public Health, *The City of San Diego Total Resource Recovery Project, Health Effects Study*, January 1997.
3. Montgomery Watson, *Water Repurification Feasibility Study.* Final Report prepared for San Diego County Water Authority. June 1994
4. Adams, M.H. 1959. *Bacteriophages.* New York: Interscience Publisher
5. Montgomery Watson, *Advanced Treatment Pilot Study.* Final Report prepared for the City of San Diego, December 1995.
6. Montgomery Watson, *Membrane Prequalification Pilot Study.* Final Report prepared for the City of San Diego, October 1997.
7. Soller, J.A., A.W. Olivieri, D.M. Eisenberg, J.N. Eisenberg, R.C. Cooper. *Microbial Challenge Studies at Aqua 2000 Research Center and Estimation of Process Train Performance with Respect to Microbial Agents*, prepared by the Public Health Institute, EOA, Inc., BioVir, and Montgomery Watson, for the City of San Diego, December 1997.
8. Sobsey, M.D., K.J. Schwab, T.R. Handzel. A simple membrane filter method to concentrate and enumerate male-specific RNA coliphages. *Jour. AWWA*, 82 (9), 52–59.

APPENDIX A: COLIPHAGE ASSAY METHODS

Double Agar Layer (DAL) Method

1. Prepare bottom agar using 40 g Tryptic Soy Agar (TSA) per one (1) litre nanopure water, autoclave for 20 minutes at 121°C cool to 50°C, dispense into sterile petri dishes (10–12 ml each). Let plates cool and store inverted at 4°C.
2. Prepare top agar using 30 g Tryptic Soy Broth (TSB) and 10 g Bacto Agar per one (1) litre nanopure water, boil and add 3.5 ml into each test tube, autoclave, cool and store at 4°C.
3. Melt top agar and keep in water bath at 47°C.
4. 1.0 ml of water sample is added to top agar (dirty water); for clean samples 5.0 ml of water sample is added to each plate.
5. Vortex and pour onto bottom agar plates, solidify and incubate inverted at 37°C overnight.
6. Count plaques.

Single Agar Layer (SAL) Method

1. Water sample (100 ml) placed in 37°C water bath for 10 minutes.
2. Five (5) ml of overnight, log phase host bacterium added to sample, mixed and returned to water bath for three (3) minutes.
3. Sample placed in 47°C water bath for three (3) minutes.
4. Sample added to equal volume of 2x agar medium, mixed gently, and poured into 8–150 mm diameter plates.

5. After hardening, plates inverted and incubated at 37°C for 18 hours.
6. Clear plaques counted as coliphage.

Membrane Filtration Method (MFM)

1. 0.1 ml of a 4-hour host culture is added to 3.0 ml of molten top agar medium and poured over bottom agar (60 nm plates).
2. pH of sample should be 6–9.
3. $MgCl_2$ added to water sample to achieve 0.05M final concentration.
4. Sample (volume 2–2000 ml) filtered through 47mM diameter filters with 0.45 μm pore size (Type HA, Millipore, Bedford, MA). Filtration rate at 7–10 minutes per 100 ml of sample.
5. Filter removed from holder and placed, face-down, on agar plates containing host cells.
6. Plates inverted and incubated at 37°C overnight (6–8 hours minimum).
7. Clear circular plaques counted against violet background.

CHAPTER · 12

Membrane Integrity Monitoring at the UF/RO Heemskerk Plant

Joop C. Kruithof
Peer C. Kamp
Henk C. Folmer

N.V. PWN Water Supply Company North Holland
Velserbroek, The Netherlands

BACKGROUND

At the Heemskerk water treatment plant of N.V. PWN Water Supply Company North Holland (PWN), 20 Mm^3/year (15 MGD) pretreated IJssel Lake water is treated by ultrafiltration (UF) and reverse osmosis (RO). Disinfection is a major objective of the direct surface water treatment plant, where no chemical disinfection is applied. The highest disinfection requirement was needed for virus removal and amounted to 3.6 log units. Integer UF membranes removed MS-2 phage spikes completely (>5.4 log); severely compromised membranes still removed 2.7 log. RO showed a high although always incomplete removal of MS-2 phages (up to 4.8 log). Therefore, membrane acceptance and integrity monitoring are critical issues.

As a membrane acceptance test, vacuum testing was applied with a criterion of 10 kPa/min. Many standard modules failed to pass the test. During operation UF was monitored by 1 μm particle counting with 0.05 μm particle counting as a quality control. A 5 log removal could be established. For RO conductivity measurement was applied enabling up to 2 log removal monitoring. In the short term, monitoring of conductivity will be replaced by sulfate, increasing the monitoring range with 1 log unit.

The total monitoring range of 7–8 log units easily satisfies the maximum disinfection requirement for the membrane steps for viruses of 3.6 log.

INTRODUCTION

Since the end of 1999, 20 Mm^3/year (15 MGD) pretreated IJssel Lake water has been subjected to an integrated membrane system (IMS) at water treatment plant Heemskerk. The direct surface water treatment plant based on the combined application of ultrafiltration and reverse osmosis has the following quality control objectives:

- removal of pathogenic microorganisms
- removal of organic micropollutants (e.g., pesticides)
- removal of inorganics (chloride, sodium, sulfate, hardness)
- biological stability

Reverse osmosis was chosen as the barrier against all pollutants. With the application of reverse osmosis, membrane fouling is a major concern. Therefore pretreatment has been a research topic for many years (1). Initially CSF treatment followed by an additional (direct) filtration step was pursued as pretreatment for cellulose acetate membranes. Besides colloidal fouling, biofouling resulted in high cleaning frequencies. Therefore alternative pretreatment options were pursued.

After a thorough research effort the combination of ultrafiltration with reverse osmosis with ultra-low-pressure composite membranes was selected for the posttreatment of CSF pretreated water. At the Andijk pilot facilities, research was conducted to determine the design criteria for the Heemskerk water treatment plant. For a number of ultrafiltration and reverse osmosis membranes system performance and fouling assessment were established. After the final membrane selection in cooperation with the selected manufacturer, a lot of attention has been paid to the development of a membrane acceptance test based on vacuum hold testing.

Because in the total treatment system no chemical disinfection is applied, removal of pathogenic microorganisms together with membrane integrity monitoring is investigated thoroughly. For UF, particle counting has turned out to be a suitable monitoring technique. For RO, conductivity and sulfate monitoring have been selected.

MEMBRANE ACCEPTANCE

For the Heemskerk water treatment plant the ultrafiltration and reverse osmosis membranes were tested on their performance before loading. The ultrafiltration membrane manufacturer X-flow has tested each element by permeability and vacuum hold testing. At Heemskerk a vacuum hold test was carried out onto each pressure vessel loaded with four 1.5 m elements in order to test both the elements, O-rings and connectors. The reverse osmosis membrane manufacturer Hydranautics has tested each module on chloride retention, feed-concentrate pressure drop, MTC and vacuum hold. During the pilot testing a vacuum hold test was developed with a vacuum decrease criterion of 10 kPa/min (1.5 psi/min). At Heemskerk the same testing was carried out for a random module selection. Some testing data are presented here.

Ultrafiltration Vacuum Hold Testing

Initially UF membrane acceptance was proposed based on a pressure hold test. PWN research showed that the vacuum hold testing applied for RO with a criterion of 10 kPa/min (1.5 psi/min) can be used for UF testing as well. Figure 1 shows an example of a vacuum hold test with a compromised module before sealing and with the same element after sealing the leaking fibers.

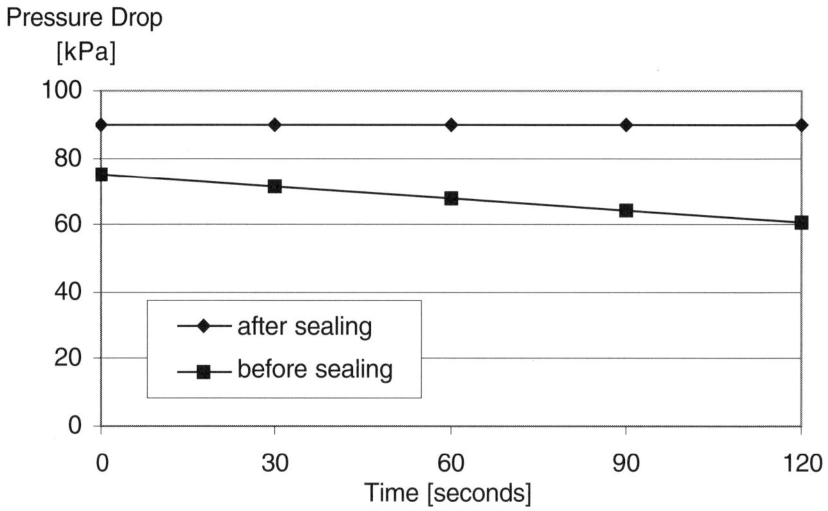

Figure 1 Vacuum hold testing results of UF module

Reverse Osmosis Vacuum Testing

For the first pilot study 43 membrane elements were subjected to vacuum hold testing (see Figure 2). Initially a pressure drop of 20 kPa/min (2.9 psi/min) was used as a selection criterion. Only one element failed this test. When experience with batches from various manufacturers showed that lower values are realistic, PWN sharpened the criterion to 10 kPa/min (1.5 psi/min). Based on this criterion, 36 out of 43 elements of the first batch failed the test. A second batch showed a significant improvement: only 7 out of 43 elements did not meet the new requirement. In cooperation with Hydranautics the 10 kPa/min criterion is set for the Heemskerk plant. A random selection out of the 1,344 modules of the full-scale installation all passed the test.

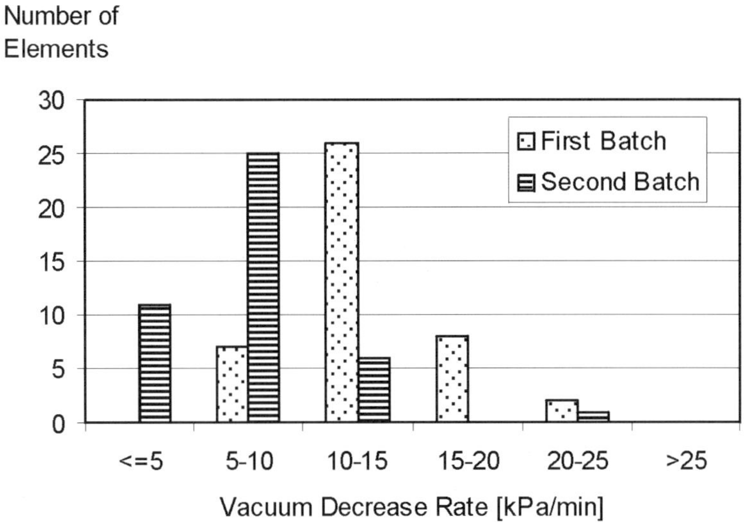

Figure 2 Vacuum hold testing results of two batches RO modules

DISINFECTION ASSESSMENT

Removal of Microorganisms

The disinfection requirements for the Heemskerk water treatment plant were based on an acceptable infection risk of 10^{-4} per person per year (2). Based on measurements in IJssel Lake water and the maximum allowable concentrations, the required log removal for *E. coli*, faecal *Streptococcae*, *Clostridia*, viruses, *Giardia* and *Cryptosporidium* were calculated (see Table 1). The remaining required log removal at Heemskerk is as low as 1.7 log for *Clostridia* and as high as 3.6 log for viruses.

Combined UF-ULPRO is a very robust disinfection barrier but always the disinfection capacity must be established. Since the concentrations of indicator organisms such as *E. coli* and *Clostridia* in the UF feed are already low, the required disinfection capacity for UF and RO cannot be established based on measurements of these organisms. Therefore MS-2 phages spiking experiments were carried out.

MS-2 Phage Challenge Studies

UF filtration showed a reduction varying from 2.7 to > 5.4 log units (see Table 2). The low 2.7 log reduction was caused by membrane defects. ULPRO showed a high although always incomplete removal. These results emphasize that integrity monitoring is essential if membrane filtration processes are used as disinfection barrier (1).

Table 1 Concentration of microorganisms, disinfection requirement and required log removal

Micro-organisms	Max. Content IJssel Lake (*n/L*)	Max. Allowable Content (*n/L*)	Total Required Inactivation (*log*)	Inactivation Capacity Pretreatment (*log*)	Additional Required Inactivation Capacity (*log*)
E. coli	6,000	10^{-2}	5.8	2.0	3.8
Faec. *Strept.*	3,000	10^{-1}	4.5	2.5	2.0
Clostridia	4,500	10^{-1}	4.7	3.0	1.7
Viruses	0.1	2.5×10^{-7}	5.6	2.0	3.6
Giardia	5.2	7.0×10^{-6}	5.9	2.5	3.4
Cryptosporidium	2.6	3.3×10^{-5}	4.9	2.0	2.9

Table 2 Calculated log reduction for MS-2 phage challenge studies

	Feedwater (*pfu/ml*)	Product Water (*pfu/ml*)	Log Reduction
UF	25,000	< 0.10	> 5.4
UF	18,000	0.25	4.9
UF (compromised)	34,000	6.7	2.7
RO	18,000–30,000	0.17–24	3.0–4.8
ULPRO	62,000	0.8	4.6

INTEGRITY MONITORING

Methodology Selection

A number of analytical methods are selected for integrity monitoring. The log removal capacity that can be monitored by these methods is summarized in Figure 3. For ultrafiltration, particle counting turned out to be an appropriate method. The number of log units that can be monitored is roughly the same for 0.05 and 1 μm particle counting. Therefore PWN has decided to use 1.0 μm particle counting for online monitoring and 0.05 μm particle counting for additional quality checks.

Because the number of particles > 1.0 μm in the RO feed is low, a reduction of 2–3 log units, for example, cannot be monitored by particle counting. Since retention of inorganics by RO is very high, this offers possibilities for integrity monitoring. Electric conductivity measurement is simple and cheap and able to monitor 1–2 log units removal. Initially this method was pursued and is applied at the full-scale plant at Heemskerk at this moment. However, the decrease of the sulfate concentration can be monitored by 3 log units (from 100 to 0.1 mg/l). This promising technique is selected for additional RO monitoring.

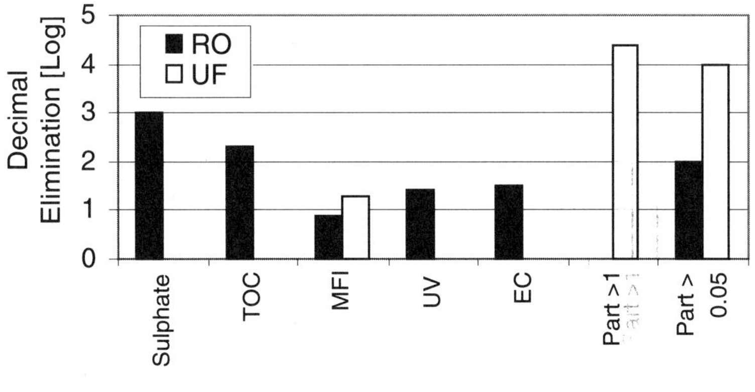

Figure 3 Log removal by UF and RO

Particle Counting for Ultrafiltration Integrity Monitoring

Originally PWN pursued 0.05 μm particle counting for UF integrity monitoring (see Figure 4).

The UF feedwater contained between 5 and 10 × 10^6 particles (0.05 μm) per milliliter, which is reduced to values between 0.2 and 7 × 10^4 particles per milliliter, corresponding with a 2–3.5 log reduction. The low value of 2 log was caused by membrane compromise. Under normal conditions 0.05 μm particle counting can be used for 3 log units monitoring in CSF pretreated IJssel Lake water. Comparable results were achieved by 1 μm particle counting. This method is applied for monitoring of the full-scale installation (see Figure 5).

After start-up a 5 log removal was achieved (Figure 5A). After a running time of about eight months, a decrease in log removal occurred (Figure 5B), indicating compromised fibers. This was confirmed by 0.05 μm particle counting (Figure 6).

Figure 6 shows a gradual increase of the 0.05 μm particle count in the permeate of UF unit 6. When the number of particles exceeded 1,500/ml additional measurements were carried out at both sides of the individual pressure vessels. The results of the measurements carried out September 15, 2000 are summarized in Table 3.

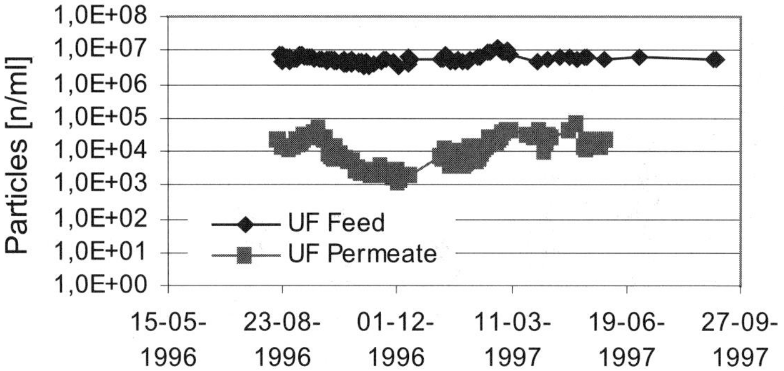

Figure 4 Particles 0.05 μm reduction for ultrafiltration integrity monitoring

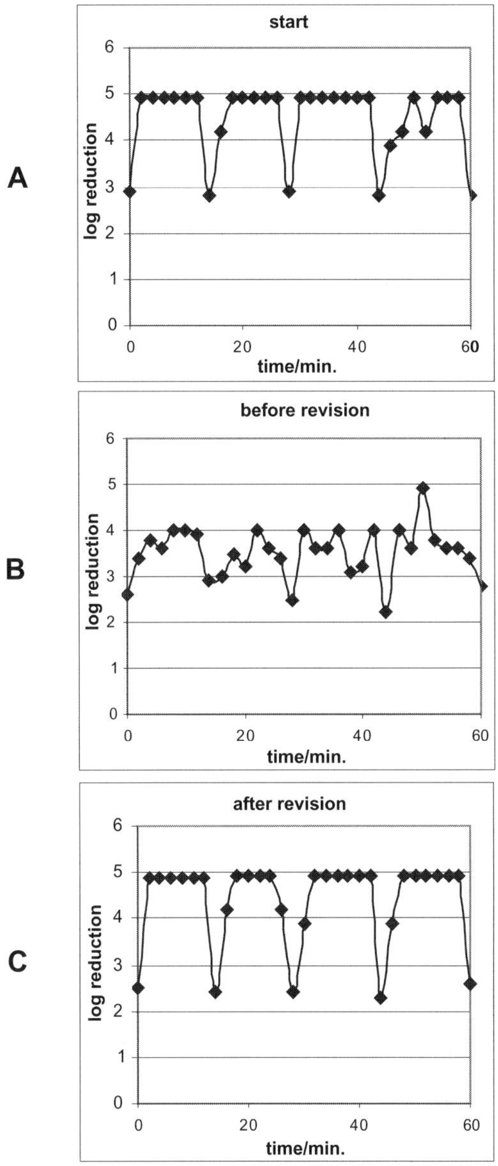

Figure 5 Integrity monitoring for ultrafiltration by 1 μm particle counting after start-up (A), with some compromised fibers (B), and after repairing the compromised fibers (C) of UF unit 6

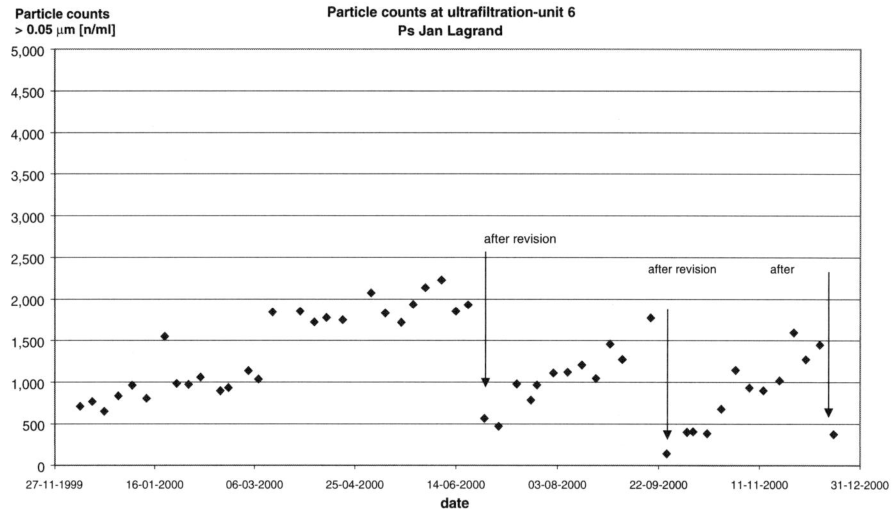

Figure 6 Integrity control for ultrafiltration by 0.05 μm particle counting in the permeate of UF unit 6

Table 3 Results of 0.05 μm particle counting in the permeate of the individual pressure vessels of UF unit 6. Date September 15, 2000

	Left Side					Right Side			
Row/ Position	A	B	C	D	Row/ Position	A	B	C	D
1	107	169	252	218	1	2,332	7,228	106	1,652
2	390	206	5,018	181	2	82	224	542	8,996
3	52	203	1,284	3,670	3	137	272	510	6,579
4	631	382	18,506	294	4	461	329	309	365
5	380	1,492	207	3,352	5	206	384	317	348
6	189	2,901	270	4,266	6	2,034	398	380	50

In total 14 particle counts were higher than 1,000/ml. After that measurement, filtration was stopped and the compromised modules were inspected and the broken fibers plugged off. Thereby the 0.05 μm particle count of the permeate of UF unit 6 dropped from about 2,000/ml to 500/ml after filtration was restarted (see Figure 6). After this revision the log removal monitored by 1 μm particle counting returned to 5 log units (see Figure 5C).

After filtration was restarted, the 0.05 μm particle count in the permeate gradually increased once again and reached 1,500 particles/ml at November 27, 2000. This time measurements in the permeate of the individual pressure vessels showed 12 particle counts higher than 1,000/ml. Once again the broken fibers were plugged off.

This strategy, based on 1.0 μm particle monitoring in combination with 0.05 μm particle counting for the permeate of the UF units and individual pressure vessels, adopted for the Heemskerk plant, gives very reliable results.

Electric Conductivity and Sulfate Measurement for Reverse Osmosis Integrity Monitoring

The electric conductivity of the UF permeate amounts to 98.7 mS/m. By RO, electric conductivity is lowered to 3.1 mS/m, enabling the monitoring of 1.7 log units removal. At this moment this measurement is applied at the Heemskerk plant. Since the sulfate concentration in the feedwater amounts to 140 mg/l and RO removes sulfate to a concentration of about 0.1 mg/l, sulfate measurement can be applied for integrity monitoring (see Figure 7).

For an operating time of six months a 3 log units removal could be established. Therefore a sulfate monitor based on ion chromatography is purchased to replace conductivity measurement for monitoring of the full-scale installation.

Integrity Monitoring of the Full Scale Installation

From Figures 5 and 7 it can be concluded that for UFs up to 3 log units removal can be monitored by 1 μm particle counting and for

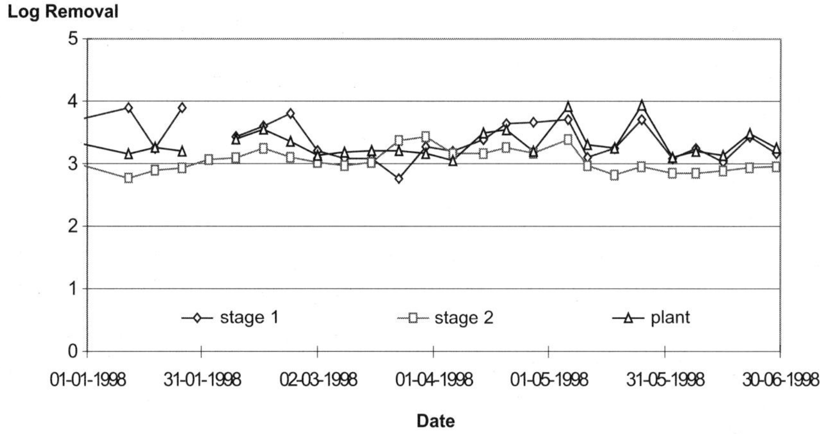

Figure 7 Sulfate measurement for reverse osmosis integrity monitoring

RO up to 3 log units removal by sulfate measurement. In total about 8 log units can be monitored, more than enough to guarantee the maximum required inactivation capacity of 3.6 log units for viruses.

EVALUATION

At the Heemskerk water treatment plant PWN has realised the first full-scale integrated membrane system in the Netherlands. The ultrafiltration (UF)/reverse osmosis (RO) plant servicing pre-treated IJssel Lake water has been in operation since November 1999. Disinfection is one of the major objectives of the direct surface water treatment plant. The disinfection requirements are based on an acceptable infection risk of 10^{-4} per person per year. Based on the presence of microorganisms in IJssel Lake, the minimum disinfection requirement was calculated for viruses, *Giardia* cysts and *Cryptosporidium* oöcysts. The highest disinfection requirement was needed for viruses and amounted to 3.6 log units for the combination of ultrafiltration and reverse osmosis.

Virus removal was investigated by spiking of MS-2 phages. Integer UF-elements showed a complete (>5.4 log) removal. For a compromised UF-element with a leakage of 0.001%, the elimination dropped to 2.7 log units. RO showed a 3.0–4.8 log removal. Elimination by RO was always incomplete, indicating small leakages in elements and/or interconnections.

Although MS-2 phage removal always was higher than 5.7 log units, membrane integrity proved to be a critical issue. Therefore a strategy for membrane acceptance and integrity monitoring was developed. Before installation both UF and RO elements have to pass a vacuum test with a criterion of 10 kPa/min. Pilot research has shown that many standard RO-elements failed to pass this test. In cooperation with the manufacturers the criterion of 10 kPa/min is set for membrane acceptance at the full-scale Heemskerk plant (768 UF and 2,016 RO modules).

Online monitoring is essential to control the disinfection potential. For a number of parameters, the elimination capacity by UF and RO was determined. For UF, particle counting proved to be an appropriate method. Initially particles of the size of viruses (~0.05 μ) were counted and showed a removal of more than 3 log units. However, up to 5 log units can be monitored by 1 μm particle counting. This last method is selected for online monitoring of UF with 0.05 μm particle counting as a quality check for the permeate of the UF units and the individual pressure vessels.

Because of the low number of particles, 1 μm particle counting cannot be used for RO monitoring. Electric conductivity (EC) measurement was selected, enabling monitoring up to 2 log units removal. In the short term this method will be replaced by online sulphate monitoring showing a removal capacity of up to 3 log units. For leakage assessment in situ EC measurement and 0.05 μm particle counting in the RO product pipe will be used. The developed strategy enables monitoring up to 8 log units removal capacity from pretreated IJssel Lake water, easily establishing the disinfection requirement of 3.6 log units for viruses.

REFERENCES

1. P.C. Kamp: *Proceedings AWWA Membrane Technology Conference*, Reno, pp. 31–38, 1995.
2. S. Regli, J.B. Rose, C.N. Haas, C.P. Gerda: *JAWWA* 83, 76–84, 1991.

CHAPTER · 13

Development of a New Online Membrane Integrity Testing System

S. C. J. M. van Hoof
NORIT Membrane Technology
Hengelo, The Netherlands

Joop C. Kruithof
N.V. PWN Water Supply Company North Holland
Velserbroek, The Netherlands

Peer C. Kamp
N.V. PWN Water Supply Company of North Holland
Velserbroek, The Netherlands

BACKGROUND

As membrane filtration systems are more commonly used in the water treatment industry, the call for a reliable, fast and online integrity testing system becomes louder. Especially where membrane filtration is used in potable water production for the removal of pathogenic microorganisms, the integrity of such a system is of the utmost importance. Membrane integrity testing can be performed in a number of ways, amongst which the pressure-hold or vacuum-hold test and the diffusive airflow test are well-known. Although relatively widely applied in membrane plants worldwide, all of these methods exhibit major drawbacks, like the fact that the plant has to be taken off-line and that there is no direct relationship between the measured data and the removal efficiency (log removal). At best an empirical relation has been established on the basis of a number of integrity tests and microbiological analyses. To overcome the above-mentioned drawbacks of conventional integrity testing systems, PWN Water Supply Company of North Holland and NORIT Membrane Technology have developed a new kind of integrity test, the Spiked Integrity Monitoring System or SIM®-System.

The newly developed method is based upon establishing the number of particles on the feed side as well as the permeate side of the membrane. The log removal can be directly calculated from the measurements. Therefore, there is a direct relationship between the removal efficiency and the measured feed and filtrate particle counts. In order to be able to measure the removal efficiency adequately, the number of particles in the feed cannot be too low. In relatively clean feedwater, i.e., from potable water production, this will pose a problem since the number of particles in the feed will be too low to demonstrate a high log removal. This problem has been solved by developing a dedicated system for dosing a limited amount of inert PAC particles for a short period of time at the feed side of the membrane plant during filtration. This renders a dramatic increase in the measured log removal potential to well above 6. Moreover, PAC-particle spiking increases the drop in log removal caused by a compromised fiber. One compromised fiber in a pilot system caused the log removal to drop 0.8 with PAC spiking. Without spiking the drop was only 0.2–0.3, proving the sensitivity of the new system. The particle size of the PAC is on the same order of magnitude or smaller than the microorganisms to be retained by the membrane barrier, *Cryptosporidium* (typically 2–7 μm) and *Giardia* (typically 4–12 μm).

INTRODUCTION

With an ever growing worldwide demand for water, and a decreasing availability, emerging technologies, like ultrafiltration (UF), hold the key to future water treatment. UF is typically used for the removal of suspended solids and microorganisms. It is applied in a number of fields, amongst which are the pretreatment to reverse osmosis or the reuse of wastewater treatment plant effluent. One of the rapidly growing applications for UF is as a last unit operation in a potable water treatment scheme, specifically for the removal of pathogenic microorganisms, like *Cryptosporidium* oocysts and *Giardia*. Both can pose a significant health threat since they cannot be killed by means of chlorinating. Recent outbreaks in the USA, Australia and the UK have significantly increased awareness of the risk of microbial contamination of the potable water. This has led the British government to put into effect new legislation, basically forcing water companies to either continuously monitor the water they put into supply or provide a "sufficient treatment plant capable of continuously removing or retaining particles greater than one micron diameter." This may be done by means of installing a "membrane or other filtration system that has been approved under regulation 25 of the Water Supply Regulations 1989" [1]. As a result of these rules, a number of UK water companies are installing, or planning to install, such systems.

MEMBRANE INTEGRITY

Membrane systems can play a crucial role in retaining the aforementioned microorganisms. A large number of investigations have clearly shown ultrafiltration membranes to be able to achieve a log removal efficiency higher than 5.8 [2], higher than 6 [3] and well over 8 [4]. An ultrafiltration system will surely provide sufficient treatment in removing or holding back one-micron particles, but any system will have to have some means of checking and maintaining its integrity. Practically all membrane systems used for the final treatment of potable water use membrane elements filled with capillary fibers. An integrated part of any potable water treatment plant should be a test to establish the integrity of these fibers. This does not necessarily mean that each and every fiber has to be intact. If the requirement for a plant is to provide a certain log removal for *Cryptosporidium*-sized particles, and a method were available to actually measure this parameter, an end-user might decide to keep a plant running as long as the required removal rate is met, instead of shutting it down at the first sign of a single fiber breakage. This would enable him to act upon such fiber breaks when the need arises and when taking part of the plant off-line would be convenient, while still being able to uphold the high standards in water quality and to satisfy the proper authorities.

INTEGRITY TESTING

Membrane integrity testing can be performed in a number of ways, which are in common practice since membrane filtration became an accepted technology.

Vacuum Test

The vacuum test is probably one of the oldest tests to establish the integrity of a membrane. It has been used to test reverse osmosis elements for a long time and can also be used to test ultrafiltration elements. The test applies a vacuum to one side of the membrane, after which this side is sealed. The other side of the membrane is kept at atmospheric pressure. The pressure increase at the vacuum side of the membrane is an indication of the integrity of the membrane. Even an intact membrane will show a certain pressure increase due to diffusion of air through fluid in the membrane wall.

Pressure-Hold Test

The pressure-hold test or pressure decay test is basically the opposite of a vacuum test. Air pressure is applied to one side of the membrane. After sealing this side, the decline rate of the pressure is monitored. Again, any element will show a certain, low decay rate due to diffusion phenomena [6].

Airflow Test

The airflow test or diffusive airflow test is a variation on the pressure-hold test. However, it does not measure the pressure decline rate. One side of the membrane is drained and pressurized, while the other side is filled with water. Air will be transported to the other side of the membrane, either by diffusion (in case of an intact system) or by a leaking fiber. The water that is displaced by air entering the other side of the membrane is measured and the flow is an indication of the integrity of the tested membrane [7].

Challenge Test

The ultimate integrity test by far is the challenge test with tracer microorganisms, which is the only one that measures the log removal performance of any system directly. It is, however, a relatively expensive test, very labor-intensive and not preferred as a test that is to be carried out on a production facility on a regular basis. On top of that, it takes a relatively long time for the results to be known, since the test requires that samples be cultivated.

Although relatively widely applied in membrane plants worldwide, all of these methods exhibit major drawbacks:

- Testing has to be performed off-line. This means the plant has to be stopped, drained, tested and restarted. The downtime of the plant will increase as the testing frequency increases and thus the net capacity will decrease. This can only be compensated by installing more membrane area.
- There is no direct relationship between the measured data and the removal efficiency (log removal). At best an empirical or mathematical relation has been established e.g., on the basis of a number of integrity tests and microbiological analyses [7], [8].

To overcome the above-mentioned drawbacks of conventional integrity testing systems and exploit the advantages of the challenge test, NORIT Membrane Technology and PWN Water Supply Company North Holland have developed a new kind of integrity test, the Spiked Integrity Monitoring System or SIM®-System.

SPIKED INTEGRITY MONITORING SYSTEM (SIM®-SYSTEM)

The newly developed method is based upon establishing the number of particles on the feed side as well as the permeate side of the membrane. The log removal can be directly calculated from the measurements. Therefore, there is a direct relationship between the removal efficiency and the number of particles at the feed and filtrate side of the membrane.

In order to be able to measure the removal efficiency adequately, the number of particles in the feed cannot be too low. In relatively clean feedwater, i.e., from potable water production, this will pose a problem since the number of particles in the feed will be too low to demonstrate a high log removal. Filtrate counts from intact membrane modules in pilot trials are typically on the order of 0.03 to 0.08 particles per ml (for particles > 1 μm). Feed counts are typically on the order of 300 to 800 particles per ml. Therefore, without further enhancement of the measuring technique, a log reduction of 3.6 to 4.4 can be measured at best.

This problem has been solved by developing a dedicated system for dosing a limited amount of inert Powdered Activated Carbon (PAC) particles for a short period of time at the feed side of the membrane plant during filtration. This renders a dramatic increase in the measured log removal potential to well above 6. Moreover, PAC-particle spiking increases the drop in log removal caused by a compromised fiber. One compromised fiber in a pilot system caused the log removal to drop 0.8 with PAC spiking. Without spiking the drop was only 0.2–0.3, proving the sensitivity of the new system. The particle size of the PAC is on the same order of magnitude or smaller than the microorganisms to be retained by the membrane barrier, *Cryptosporidium* (typically 2–7 μm) and *Giardia* (typically 4–12 μm).

Pilot Study

To increase the range of log removal measurement, a dedicated system for online dosing of a limited amount of inert particles to the feed of the UF system during a short period of time has been developed. A dedicated type of NORIT Powdered Activated Carbon (PAC) is used. This carbon type has no interaction with the membrane and is approved for use in water treatment (DWI, Kiwa ATA). The applied PAC has a particle size distribution as presented in Figure 1. Almost 70% of the particles have a diameter of less than 1.7 μm and therefore are on the same order of magnitude or smaller than microorganisms such as *Giardia* (4–12 μm) and *Cryptosporidium* (2–7 μm), to be rejected by the UF system.

A dosing station was constructed by means of a mixing vessel for storing the PAC suspension. A quantity of 7.81 g PAC was suspended in 85 liters of water. At 1.1×10^8 particles/mg PAC, the suspension contained 1.0×10^7 particles/ml. To avoid sedimentation of PAC, the vessel was stirred continuously. A dosing pump (prominent 04120 NP) fed the suspension into the UF feed at a rate

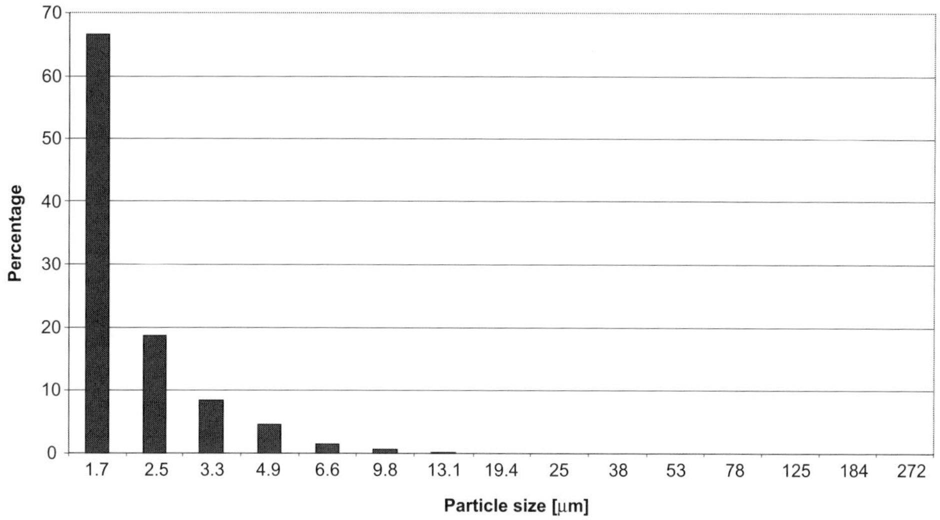

Figure 1 Particle size distribution of PAC used for integrity testing

of 60 l/h at a fixed period after a backwash cycle. Dosing pump operation in relation to UF plant operation is presented in Figure 2.

The UF pilot plant is equipped with two Met-One PCX 1 μm particle counters with a measuring range of 1–150 μm connected to the feed and permeate side. Particles are measured in samples of 40 ml automatically. Therefore the numbers of particles that can be measured are 0/40 ml (0/ml), 1/40 ml (0.025/ml), 2/40 ml (0.05/ml), etc. For log removal calculations 0 particles/40 ml will be considered to be less than 0.5/40 ml (< 0.0125/ml).

Pilot Results

For interpretation of the data, two measuring windows have been defined. The first window covers the data during PAC-spiking, the second window covers a reference measurement without PAC-spiking. A typical plot for a pressure vessel without compromised fibers is shown in Figure 3.

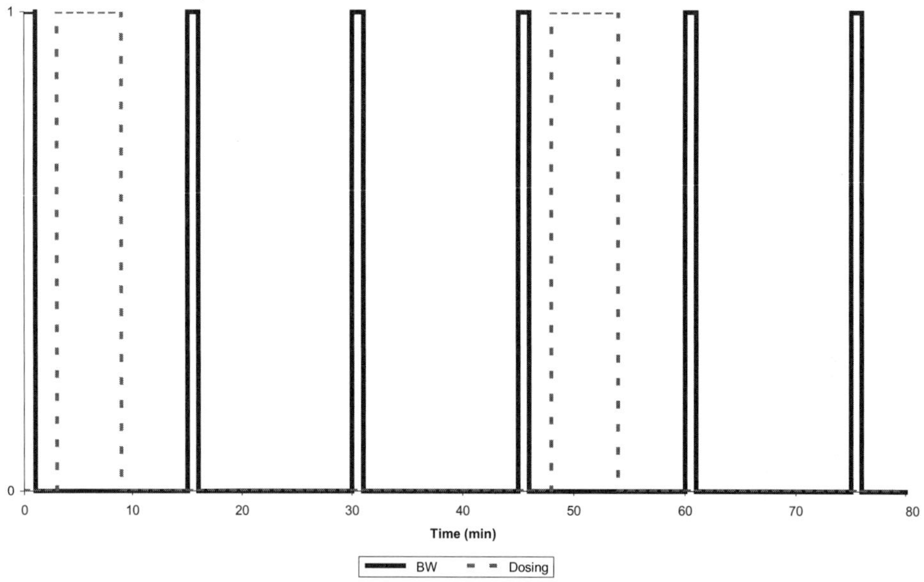

Figure 2 Operation PAC dosage in relation to UF backwash

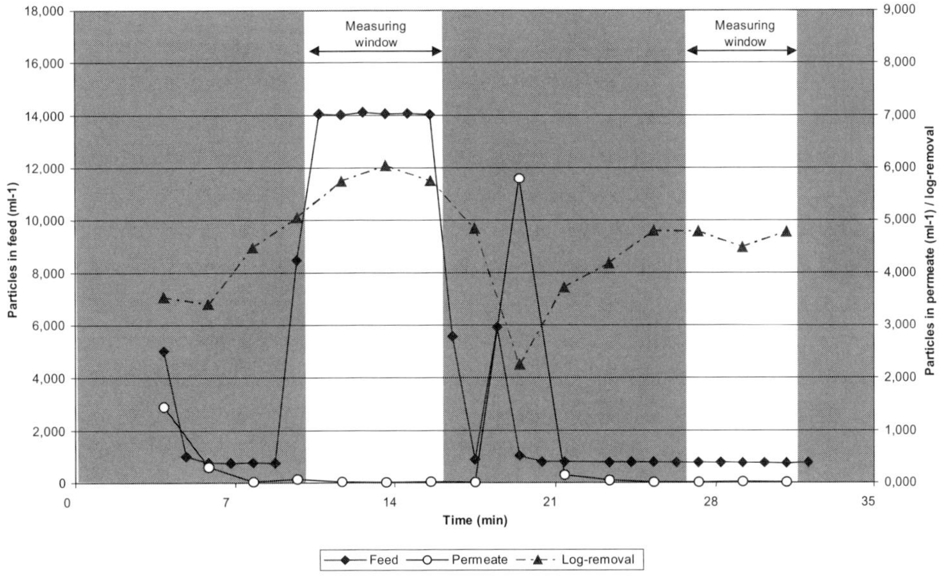

Figure 3 Typical plot for a non-compromised pressure vessel

The measurement started after a backwash. The elevated number of particles in both feed and permeate is caused by the backwash. Valve action releases some particles at both sides of the membrane. Moreover the backwash pump introduces particles at the permeate side. These phenomena yield a short peak on the feed and permeate of the membrane but are not related to a compromised module. The feed and the permeate particle count stabilized at 700–800 particles/ml and 0–0.08 particles/ml respectively.

After PAC-spiking the UF feed signal rose to a level of around 14,000 particles/ml, which took approximately two minutes due to the residence time in the particle counter. After the feed concentration reached its maximum value, the measuring window was started, lasting six data points. The permeate measurements did not react at all to the particle increase at the feed side. The calculated log removal amounted to 5.83 log units. After six particle counts PAC-spiking was stopped, resulting in a decrease of the feed side particle counts.

Then a second backwash was started lasting approximately 45 seconds. Once again the backwash is visible in the particle count on both the feed and the permeate side. After the backwash both feed and permeate measurements stabilized at 700–800 particles/ml and 0–0.08 particles/ml respectively. Another measuring window was started, once again lasting six data points. Without PAC-spiking the measured log removal amounts to 4.65 log units. A typical plot for a pressure vessel with an element with one compromised fiber is shown in Figure 4.

Again two measuring windows were taken into account. After PAC-spiking up to a particle count of 14,000 particles/ml, the increase at the feed side is followed by a small increase at the permeate side as well. The calculated log removal amounted to 5.07 compared to 5.83 log units, a signal that membrane integrity has been breached.

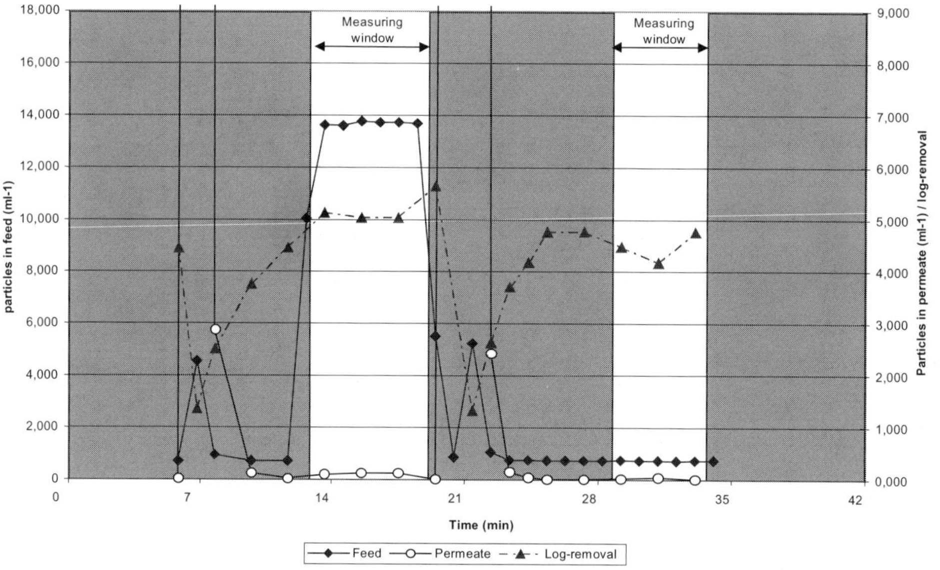

Figure 4 Typical plot for a pressure vessel with one compromised fiber

In addition, results from the nonspiked measuring window did not show a pronounced decrease of log removal for fiber compromising: 4.65 log and 4.40 log removal respectively. The results of both tests are summarized in Table 1.

By increasing the amount of particles in the UF feed the potential log removal can be increased to even higher log levels, thereby showing the potential of PAC dosing during in-line integrity testing.

Full-Scale Evaluation

Yorkshire Water Services, one of the large private water companies in the UK, awarded Earth Tech Engineering the contract to design and construct an ultrafiltration plant at the Keldgate Water Treatment Works, one of Yorkshire Water's principal treatment facilities. The design and manufacturing of the membrane racks and supply of the membrane elements were subcontracted to NORIT Membrane Technology. This project was chosen as the first full-scale membrane facility to be fitted with a NORIT SIM®-System [8]. Work on-site for the new treatment works began in January 2000. Commissioning of the plant was finished in November of 2000 and the plant was subsequently taken into supply. It treats water from four wells near the city of Hull, UK. The plant has a capacity of 90 MLD or 24 MGD and comprises 11 racks of 24 membrane housings, filled with four X-Flow S-225 UFC membrane

Table 1 SIM®-System results for non-compromised and compromised UF membrane systems

SIM®-System	UF System	Particle Conc. Feed (*n/ml*)	Particle Conc. Permeate (*n/ml*)	Log Removal (*Log*)
PAC	Non-compromised	14,055	0.021	5.83
	Compromised	13,689	0.140	5.07
no PAC	Non-compromised	768	0.017	4.65
	Compromised	733	0.029	4.40

elements each. The membranes themselves consist of hydrophilic PES/PVP fibers, with an internal diameter of 0.8 mm. Each element contains 35 m^2 or 377 ft^2. The plant has an installed membrane area of 36,960 m^2 or almost 400,000 ft^2. Since this plant is used as a last stage in the potable water production for the region, a removal of at least 4 log for *Cryptosporidium*-sized particles was one of the specifications in the design. To check and maintain the required log removal, the SIM®-System was installed on the plant. The system is capable of monitoring each of the eleven racks individually. It consists of a mixing and dosing section and a particle counting section.

Mixing and Dosing

The mixing and dosing section is used to makeup the PAC slurry and keep it in suspension. It contains a dosing pump section and a transport pump section. Since the dosing unit is physically a long distance from the individual dosing points on each of the eleven racks, the transport pumps are used to dilute the PAC suspension and transport it to the rack to be tested. After each test the transport piping is flushed with potable water to avoid sedimentation of PAC in the piping. The amount of carbon to be dosed is fixed. This means that the total amount of carbon particles in the feed stream to the UF unit is variable, as the feed flow to the unit will vary with demand of the plant.

Since a high log reduction was to be achieved by the measurement, it was decided to spike the feed to the membrane racks with $1.5*10^5$ particles per ml at the maximum capacity of the rack. The particle concentration may be higher as the feed flow is lower. This is taken into account in the log removal calculation.

Particle Counting

It is impossible to measure these high particle concentrations with any readily available particle counter. Therefore, the dosing section is calibrated to dose $1.5*10^5$ particles per ml at the maximum flow in the feed to the membrane rack of 440 m^3h^{-1}. This can be achieved by making up the PAC suspension at a predefined concentration and achieving a fixed and calibrated dosing rate. The number of particles generated by one gram of PAC was derived from the pilot experiments and was found to be $1.0*10^8$ particles per mg.

The particle counting section is equipped with one Met-One PCT-1 μm particle counter per three racks. An intricate system of sampling tubing and valves makes sure there is a constant sampling flow on all racks. The sampling tubing itself is made of Hytrell with a Teflon coating on the inside, in order to reduce disturbances of the measurement by the tubing itself, either by particle introduction by the tubing or by accumulating and subsequent shedding of particles from the permeate side of the plant. Valves in the sampling system are made of Teflon and are as smooth as possible on the inside. At any one time, only one rack is connected to the particle counter and actually measured. Regardless of which rack is being measured, the sampling flows of all racks are kept at the same rate at all times. Great care has been taken not to disturb the sampling flow during switching of the valves. Disturbances, like flow variations or shocks in the tubing, may generate particles, disturbing the measurements.

Despite all the precautions that have been taken in the design of the SIM®-System, commissioning of the plant will inevitably introduce a lot of particles, even in the permeate side (clean side) of the membrane. Before loading of the membrane elements, the whole plant is being flushed extensively. Flushing will introduce huge numbers of particles, which will only be removed after installing the membrane elements. Tests on a pilot plant have shown the number of particles to reach values as low as 0.03 to 0.08 per ml. This could only be achieved after weeks of operation and the use of a small backwash tank with no direct contact to outside air. Even then, stabilization of the particle counts was only achieved after weeks of running the plant, with particle measurements coming down gradually as the system was operated and all particles introduced at the permeate side of the membrane during commissioning were flushed out of the system.

Full-Scale Results

Testing the SIM®-System was one of the requirements in the take-over procedure of the Keldgate plant. The tests have been conducted automatically. In the control system, SIM®-tests are scheduled on a periodic basis, typically one test per rack per day. During the take-over tests the SIM®-tests were scheduled more frequently, to establish the correct working of the system. During each test, the PAC suspension was dosed during a period of six minutes. Because of the time lag between the dosing station of the PAC suspension and the PAC actually reaching the racks, a measuring window was defined for each rack. The measuring window had been established during commissioning, by dosing a high-salinity solution to each of the racks in exactly the same way as the PAC suspension was to be dosed, and measuring the salinity of the permeate as a function of time at each of the particle counter sampling points.

For one of the take-over tests, prior to the time the plant was put into operation, five SIM®-tests were conducted on each of the eleven racks. The results are displayed in Table 2.

A log 4 removal was required to pass the test, and the system showed very good performance, exceeding the pass criteria significantly on each measurement. Since the take-over test was performed relatively fast after the membrane elements were loaded, it is expected that the particle counts at the permeate side of the membrane will decrease and the measured log removal subsequently will increase.

Table 2 Results of take-over test 1, Keldgate WTW

Rack	1001	1002	1003	2001	2002	2003	3001	3002	3003	4001	4002
Test 1	5.0	5.2	5.3	5.4	5.3	5.4	4.2	5.7	6.6	4.5	5.0
Test 2	5.9	5.8	5.5	5.6	5.3	5.6	5.8	6.1	6.6	5.7	5.2
Test 3	5.3	5.6	5.0	5.2	4.7	5.6	4.5	6.1	5.9	5.2	5.6
Test 4	5.9	5.7	5.8	5.5	5.7	5.8	4.6	5.4	5.3	4.8	5.2
Test 5	5.5	5.9	5.9	5.1	6.3	6.1	5.2	6.6	5.1	5.2	5.2

CONCLUSIONS

As membrane filtration becomes increasingly more important in the production of potable water, it is evident that membrane integrity checks are desirable, if not mandatory. Most of the systems for membrane integrity testing that are supplied today are not capable of online testing. Online test equipment that is used (turbidity measurement or particle counters) does not have the sensitivity required of such tests. None of the existing systems can directly measure the removal efficiency of a membrane system. The SIM®-System solves all of these drawbacks. It is, in essence, an online challenge test, which renders results in seconds and is fully automated. The system has been extensively tested on pilot scale and has proven to be very effective in measuring the removal efficiency and spotting a defect in the membranes. Upon designing and installing the system on a 90 MLD plant, it has shown its capability of establishing the removal efficiency of any single rack, thus providing the security needed to safely operate this plant as a final barrier to pathogenic microorganisms.

ACKNOWLEDGEMENTS

The pilot study in this work has been performed in close cooperation with PWN Water Supply Company North Holland. The efforts of Nico Luten, Henk Folmer, Peer Kamp and Joop Kruithof are greatly appreciated. Brenda Franklin, Norman Johnson and John Lever of Earth Tech Engineering and Rosemary Smith and Derek Wilson of Yorkshire Water Services are thanked for their support and open-mindedness in introducing this system.

REFERENCES

1. Drinking Water Inspectorate Information Letter 16/99.
2. Kamp P.C., Combined use of ultrafiltration and reverse osmosis for direct treatment of surface water, proceedings 20th congress IWSA Durban (1995).
3. Panglish S., W. Dautzenberg, O. Kiepke, R. Gimbel, J. Gebel, A. Kirsch and M. Exner, Ultra- and microfiltration pilot plant investigations to treat reservoir water, *Desalination* 119 (1998) 277–288.
4. Khow J., A. Donn, J. Slade, M. Bauer, A. Rachwal, Performance of a novel hollow fibre UF membrane system: clarification and disinfection of a high turbidity surface water source, *blad*, nummer (jaar).
5. Adham S.S., J.G. Jacangelo and J.M. Laine, Low pressure membranes: assessing integrity, *American Water Works Association Journal*, 3 (1995) 62–75.
6. Johnson W.T., Predicting log removal performance of membrane systems using in-situ integrity testing, proceedings annual *Conference AWWA* (1997) 411–419.
7. Panglish S., U. Deinert, W. Dautzenberg, O. Kiepke, R. Gimbel, Monitoring the integrity of capillary membranes by particle counters, *Desalination* 119 (1998) 65–72.
8. Franklin B., F. Knops and R. Smith, The construction and commissioning of a 24 MGD ultrafiltration plant with on-line integrity testing to monitor and maintain the barrier to *cryptosporidium* sized particles. *AWWA conference* (March 2001).

Part 4 Membrane Filtration Applications and Operations

Startup and First Year of Operation of a 7 mgd Microfiltration Plant

Microfiltration Treatment of Filter Backwash Recycle Water from a Drinking Water Treatment Facility

Experiences with Planning, Construction, and Startup of a 14.5 MGD Microfiltration Facility

The Use of Microfiltration for Backwash Water Treatment

Implementation of a 25 mgd Immersed Membrane Filtration Plant: The Olivenhain Municipal Water District Experience

Demonstrating the Integrity of a Full-Scale Microfiltration Plant Using a Bacillus Spore Challenge Test

CHAPTER · 14

Startup and First Year of Operation of a 7 mgd Microfiltration Plant

Roger A. Olson, Superintendent of Water & Wastewater
City of Marquette, Michigan

INTRODUCTION

From the very inception of the idea that Marquette should forego the construction of a direct filtration plant and build a microfiltration plant, interest has been very high. Had the microfiltration technology progressed to a point where it was cost competitive both from a capital cost and an operation and maintenance standpoint? When the project was complete, would the cost of construction of the facility show the savings indicated in the preliminary study? Would the microfiltration equipment and the complex computer controls necessary for its operation prove reliable? Would water temperatures as cold as 1°C prove to be a problem in the full-scale project?

This chapter will address those issues and look at how reliable all of the new plant equipment has been. The quality of the water produced will be looked at from a turbidity and particle count point of view. Operating data, including backwash and chemical cleaning frequencies, maintenance requirements, and operational costs will also be discussed.

HISTORY

Located on the south shore of Lake Superior, the city of Marquette, Michigan, has taken advantage of this high-quality surface water source for over 100 years without substantial treatment required.

The untreated lake water has an average turbidity of 0.22 NTUs, no color, no taste or odor problems, total hardness of 45–50 mg/L, alkalinity of 45 mg/L, and TDS of 50 mg/L. The Marquette water treatment plant experienced an all-time-high turbidity of slightly over 5.0 NTU for two days in April 1985 when unseasonably warm weather and rain combined with a high snowmelt rate caused some localized flooding.

In August 1991, following a 12-year legal battle, Marquette County Circuit Court ordered the Michigan Department of Public Health (MDPH; now known as the Michigan Department of Environmental Quality) to grant the city a seven-year variance from Michigan's complete treatment requirements. In January 1992, the MDPH issued the variance, including requirements leading to compliance with the Safe Drinking Water Act (SDWA) rules by the time the variance expired on August 7, 1998.

A preliminary water study completed for the City of Marquette in 1993 recommended the construction of a 7.0 MGD direct filtration plant at an estimated cost of $9 million (1996 dollars). Operation and maintenance (O&M) costs were estimated to be an additional $160,000 per year. In June 1994, the city retained the services of Fishbeck, Thompson, Carr, & Huber, Inc. (FTC&H) of Ada, Michigan, to complete a microfiltration feasibility study. The feasibility study estimated a cost of $7.5 million to construct a 7.0 MGD microfiltration plant. O&M costs of the microfiltration plant were estimated at an additional $175,000 per year.

In August 1995, following a successful pilot study, the city accepted bids from Memtec America Corporation, Memcor Division (now USFilter/Memcor), and General Filter for the purchase of a 7.0 MGD microfiltration system. Memtec America, with the Memcor Continuous Micro Filtration (CMF) system, was the low bidder at $3,489,408.

MEMCOR EQUIPMENT SCOPE OF SUPPLY

- Eight Memcor microfiltration units Model 90 M10C, each comprising 90 M10C membrane modules
- CIP system, including circulation pump, spare circulation pump, heating system and controls, and 55 gallons Memclean C solution
- CIP transfer system including two air diaphragm pumps
- Two redundant operator interface consoles and a third operating computer
- One set redundant PLC master controller
- Air compressor system
- Interconnecting pipework
- Commissioning
- Training
- O&M Manuals
- Set of recommended spare parts
- Backwash energy dissipation tank
- Engineering design review
- Automatic backwashing strainers
- Manual CMF frame isolation valves
- Extended membrane warranty
- Sonic analyzer
- Integrity test fixture
- Performance and payment bond
- Module removal tool
- A guaranteed-not-to-exceed module replacement price

With the city accepting the bid, FTC&H could now begin final design of the facility. In February 1996, the city accepted the bids for the construction of the facility. Yalmer Mattila Construction of Houghton, Michigan, was the low bidder at $3,960,000. Construction of the plant was started in May 1996, and on September 24, 1997, though the facility was not completed, the City of Marquette began full-time operation of its microfiltration plant.

CONSTRUCTION COSTS

Feasibility study estimated construction costs: $7,500,000

Actual Costs

Memtec Bid:	$3,489,408.00	Yalmer Mattila Bid:	$3,960,000.00
Change Orders:	$70,564.00	Change Orders:	$20,652.94
TOTAL:	$3,559,672.00	TOTAL:	$3,980,652.94

Total microfiltration plant construction cost: $7,540,324.94

The Memtec change order was for supplying and programming instrumentation and controls. It became apparent during the design that the majority of the work should be done by Memtec and should be included in their contract.

Along with the installation of the purchased microfiltration equipment, the city's general contractor constructed a 11,800 sq. ft. addition (3,000 below grade) that included an administrative area and conference room, replaced high lift and low lift pumps, installed bulk chemical feed systems, and replaced its gas chlorine system with hypochlorite generators. Although many of these items were not included in the estimate from the feasibility study, the total cost of the project was very close to the feasibility study estimate. The total amount of the change orders by the general contractor was only 0.52% of their original bid.

STARTUP

On September 24, 1997, the city started the microfiltration equipment. Discussions with other facilities with complex SCADA systems and computer controls had plant operators prepared for weeks, if not months, of troubleshooting and working the "bugs" out of the system. This did not occur as precommissioning work by USFilter/Memcor had almost totally eliminated the problems. About all that was left was final customizing of computer screens and preparing automatic reports to be printed at the end of the day. Even from startup, equipment reliability has been very good. Problems we have seen have been minor in nature and for the most part caused the shutdown of one CMF unit until repairs could be made. The few problems we have encountered are the following.

Equipment Reliability

Compressors

The Memcor CMF system microfiltration arrays (CMF units) backwash using compressed air at preset intervals. Three 40 HP Atlas Copco compressors provide air for the backwash. When started up, all three compressors were programmed to load and unload at the same pressures. This caused the compressors to short-cycle and to load and unload every few seconds. It took two visits from our local Atlas Copco reps and a call to Atlas Copco's national headquarters to diagnose the problem. The problem was corrected by staggering the load and unload pressures by 5 lbs on each compressor.

AV-13

When the CMF units go through the backwash and membrane rewet cycles, Automatic Valve-13 (AV-13) opens to allow the high-pressure air to get to the membranes. When the water temperature dropped to about 5°C in December, the valves on five of the CMF units began to fail to open. The valves were removed and lubricated and the problem was alleviated. One of the valves had to be replaced as it would not open, and the valve stem snapped off when a Memcor service technician tried to install a larger valve actuator.

Flowmeter

Each of the CMF units contains its own magnetic flowmeter. If the computer fails to see an adequate flow from the unit, a flow control valve continues to open. If the flow control valve is open more than 95% for more than three minutes, the CMF unit will shut down. CMF unit #2 was shutting down for this reason. We were able to diagnose the problem ourselves because we could see an increased combined filtrate flow. The meter was replaced.

PLC Programming

Each of the CMF units contains a Programmable Logic Controller (PLC) that communicates with the master control panel PLC. For unknown reasons, possibly a power surge or brownout, CMF unit #7 lost its program. A Memcor service technician downloaded a program from another CMF unit onto his laptop computer and installed it into CMF #7. We have purchased a laptop, and are now able to do this ourselves if necessary.

Tri-Manifold

The air used in the backwash and rewet cycles goes from the air receivers through a "tri-manifold" to the CMF units. If the air pressure downstream of the tri-manifold is too low (i.e., backwash initiated), the three valves open in a staggered succession to allow a larger volume of air to reach the units during a backwash. In August one of the regulating valves stuck and allowed large volumes of air to escape. The escaping air caused two of the three compressors to overload and shut down, and the air pressure in the air receivers to drop below 120 psi. When the computer sees an air pressure in the receivers less than 120 psi, it will shut the CMF units down as they come up for backwashes. Seven of the eight CMF units had shut down before someone had the idea of closing the valve upstream of the tri-manifold to build up air pressure in the receivers.

MEMBRANE INTEGRITY

What the facility has not seen in the first year of operation is a failure of any of the membranes or modules in the CMF units. A Pressure Decay (membrane integrity) Test is conducted every 24 hours of operation as required by the Michigan Department of Environmental Quality.

TYPICAL OPERATING PROCEDURES

In the past year of operation, the community's water needs have averaged 3.03 MGD, with a low flow of 2.1 MGD and a maximum flow of over 5.5 MGD. Typically, all eight CMF units are used in the operation to maintain an adequate water level (20 ft) in the 500,000-gallon chlorine contact tank. As water is pumped from the tank into the distribution system, the eight CMF units will run at rates of as low as 200 gpm (.148 gpm/sq.M or 19.8 gpd/sf) and as high as 550 gpm (.407 gpm/sq.M or 54.5 gpd/sf). In the summer months when water demands are the highest, the CMF units would routinely operate at rates of 650 gpm (.481 gpm/sq.M or 64.4 gpd/sf). On the occasions that the plant operated at rates as high as 725 gpm (.538 gpm/sq.M or 71.9 gpd/sf) no problems were encountered. The maximum rate approved by the Michigan Department of Environmental Quality is .553 gpm/sq.M.

COLD WATER TEMPERATURES

As during the pilot study, it was found that the temperature of the water did not limit the ability of the CMF units to operate at their designed capacity. An increased TMP (trans-membrane pressure) of 3 psi was all that was necessary to push the water through at temperatures down to 2°C.

An interesting incident occurred when the water temperature reached 4–5°C, the point at which water is the most viscous. The CMF units had 700 to 800 hours since their last clean, TMPs were at an acceptable level (10–12 psi), but with 60-minute backwash intervals and high filter rates, the flow modulating valves wanted to open to 100%. This would occur at 45 minutes after backwash and started to shut the units down. A chemical clean of one of the units relieved the problem for the unit. The event only happened for a short period, and was corrected by going to 45-minute backwash intervals or operating the filters at a slightly lower rate.

BACKWASH INTERVALS AND PLANT EFFICIENCY

Backwash intervals are typically set at 60 minutes. At this setting, operators can run the plant even at its highest rate without triggering an "increased TMP" backwash. (The CMF unit will backwash if it sees a TMP increase of 1.0 psi between intervals.) Backwash intervals can be set as high as 120 minutes at low to moderate flows with <0.4 psi TMP increase. Plant efficiency has been between 89.4% and 95.7% with an average of 92.7%.

CLEANING FREQUENCIES

The computers controlling the microfiltration system are set up to call for a CIP (Clean-in-Place) of the membranes if the TMP rises to 17 psi, or when the total hours from the last CIP reach 1,000. The facility has not experienced TMPs higher than 12–13 psi. CIPs have been done when there are 1,000 to 1,200 hours on the CMF units, which typically takes slightly less than two months. TMPs during the pilot study ranged from 8 to 20 psi, and cleaning intervals averaged 28 days. The reason for the shorter intervals in the pilot study was operating at an average rate of 6 MGD, which is

Table 1

CMF Unit	2/26/98	2/28/98	4/20/98	4/21/98	6/12/98	6/14/98	8/9/98	8/10/98	9/29/98	9/30/98	gpm/	gpd/
Total Flow	CMF 6	CMF 6	CMF 6	CMF 6	CMF 6	CMF 6	CMF 6	CMF 6	CMF 6	CMF 6	sq.M	sq.ft.
GPM	TMP	TMP	TMP	TMP	TMP	TMP	TMP	TMP	TMP	TMP		
200	3.7	2.8	3.1	2.3	2.7	2.1	2.5	1.8	2.6	1.8	0.148	19.8
225	4.2	3.2	3.5	2.7	3.1	2.4		2.0	3.0	2.2	0.167	22.3
250		3.6	4.0		3.5	2.8				2.5	0.185	24.8
275					3.9	3.0				2.7	0.204	27.3
300	5.3			3.6	4.1	3.3	4.2		4.0	3.0	0.222	29.7
325	6.1	4.4	5.3		4.8	3.6	4.6	3.0		3.2	0.241	32.2
350	6.3				5.0	3.7		3.4	4.7	3.6	0.259	34.7
375		5.4			5.4	4.4			5.1		0.278	37.2
400									5.7		0.296	39.6
425					6.1	4.9				4.4	0.315	42.1
450	8.2				6.9	5.1		4.3	6.4	4.8	0.333	44.6
475					7.5	5.3			6.7		0.352	47.1
500		7.4							7.1		0.370	49.5
525								5.2			0.389	52.0
550	10.7				7.9	6.2	8.3	5.3	8.1	6.7	0.407	54.5
575	11.0				8.7	6.9	8.6				0.426	57.0
600	11.4				9.0	7.2		6.0	8.6		0.444	59.5
625						7.8					0.463	61.9
650	12.6		11.4	9.4	9.9	8.4	9.8	6.4	9.7	7.3	0.481	64.4
725				10.3							0.538	71.9
Temp °C:	3.5	3.5	5	5	10	10	18	18	18	18		
Hrs/CIP	1,200	20	1,000+	clean	1,054	20	1,200	clean	1,032	clean		

twice the facility's average flow. The CIPs are done using Memclean, which is a diluted sodium hydroxide solution with an added surfactant. The 2% solution is heated to 32° to 34°C and pumped into the CMF unit. The computer controls the dosing of the CIP solution with sodium hydroxide and Memclean to reach a proper strength. The process takes about two hours, and the only manual labor involved with the CIPs is the opening and closing of two valves on the CFM unit at the beginning and end of the cleaning cycle. The CIPs have been very effective in recovering TMPs felt to be at or very close to what the membranes were when new. Table 1 shows the effect of water temperature, flow and clean membranes on TMP, and shows membrane recovery after the CIP. Comparing clean and dirty membranes at temperatures of 3°C and 18°C, the TMP required at the warmer temperatures is about 3 psi lower. You can see that a clean with citric acid may soon be required as TMP did not fully recover with the last clean. Figure 1 is a chart that shows the TMP at .481 gpm/sq.M or 64.4 gpd/sf.

TURBIDITIES

Turbidities of the finished water continue to be in the 0.04 to 0.1 NTU range. Figure 2 is a chart of turbidities from a one-week period in September 1998. Turbidity for the past year remains consistent regardless of the raw water quality. Maximum raw water turbidity over the last year has been 1.8 NTU.

PARTICLE COUNTS

Particle counters were installed on raw water, each of the CMF units, and the combined filtrate. Figure 3 shows the filtrate particle counts for the same period as Figure 2.

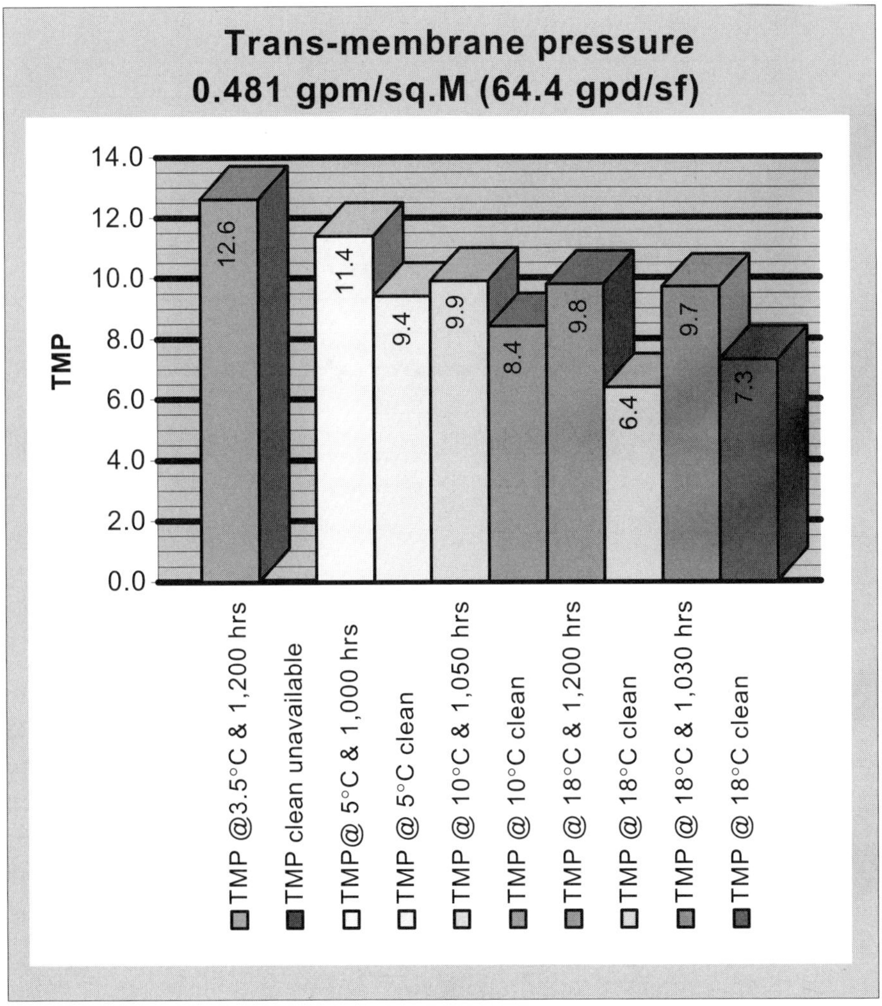
Trans-membrane pressure
0.481 gpm/sq.M (64.4 gpd/sf)
TMP
14.0
12.0
10.0
8.0
6.0
4.0
2.0
0.0
12.6
11.4
9.4
9.9
8.4
9.8
6.4
9.7
7.3
TMP @3.5°C & 1,200 hrs
TMP clean unavailable
TMP@ 5°C & 1,000 hrs
TMP @ 5°C clean
TMP @ 10°C & 1,050 hrs
TMP @ 10°C clean
TMP @ 18°C & 1,200 hrs
TMP @ 18°C clean
TMP @ 18°C & 1,030 hrs
TMP @ 18°C clean

Figure 1

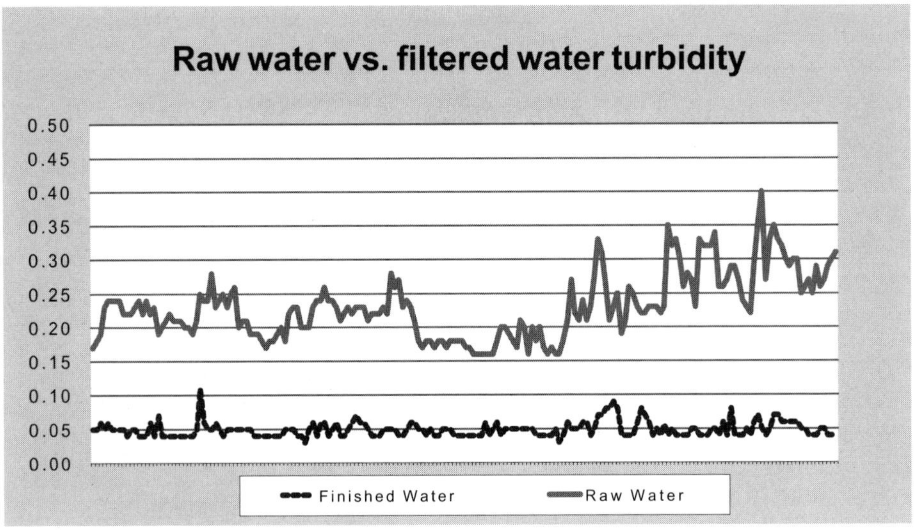

Figure 2

OPERATION AND MAINTENANCE COST

Figure 4 shows the annual treatment and pumping costs before microfiltration. Keep in mind that no filtration of the water was provided, and the only treatment was the addition of chlorine and fluoride. Figure 5 shows those costs with the microfiltration system. Each graph is based on an annual pumpage of 1,105,480,000 gallons or 3.03 MGD. The increased cost of $154,600 is less than the anticipated increased O&M cost of $175,000. The increases are a result of power costs per million gallons increasing from $88.67 to $128.43, increased maintenance costs, and establishing an annual sinking fund for membrane module replacement.

MORE OPERATIONAL COST INFORMATION

Figure 6 shows the microfiltration costs including actual MF-related labor. Discussions with staff indicate that two to three hours

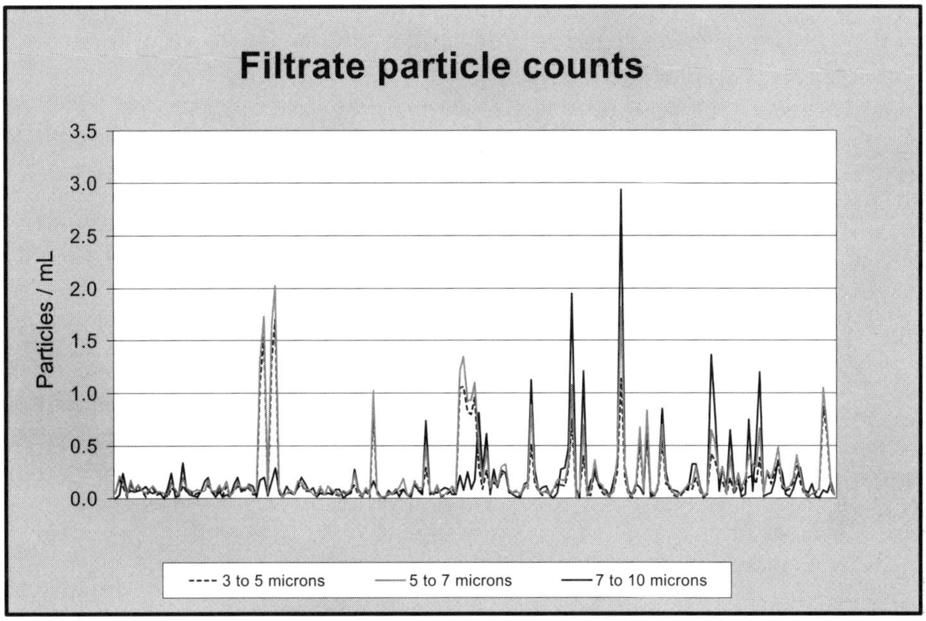

Figure 3

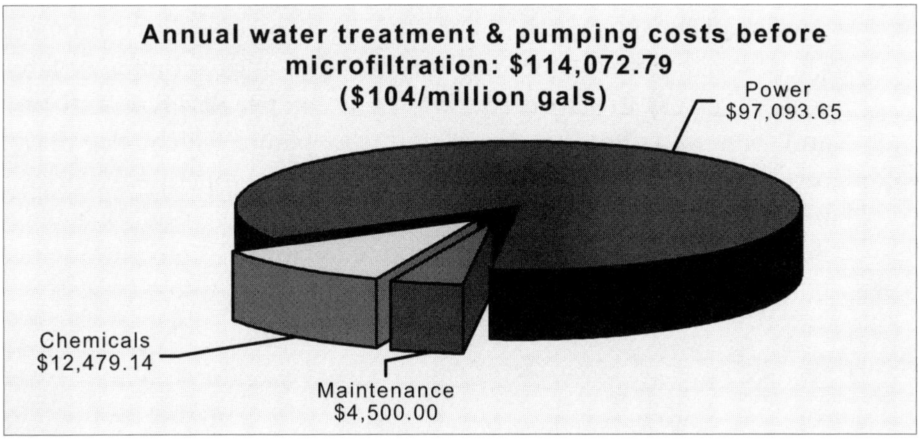

Figure 4

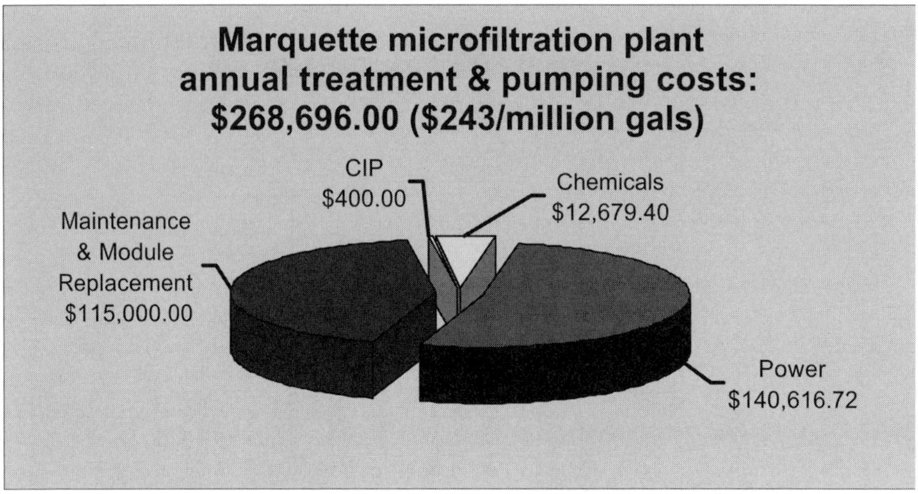

Figure 5

are spent on an average day dealing with the CMF system. This time includes monitoring chemical dosages, confirming computer readouts, and performing CIPs.

Figure 7 includes all facility operating labor. The chart was developed considering the Michigan Department of Environmental Quality requirement for full-time staffing, and contains all labor costs necessary to staff the facility 24 hours per day. Four full-time and one part-time operator staff the plant, which also has a certified bacteriological lab. The part-time operator works a minimum of three days per week and is there to assist on scheduled maintenance requiring more than one person.

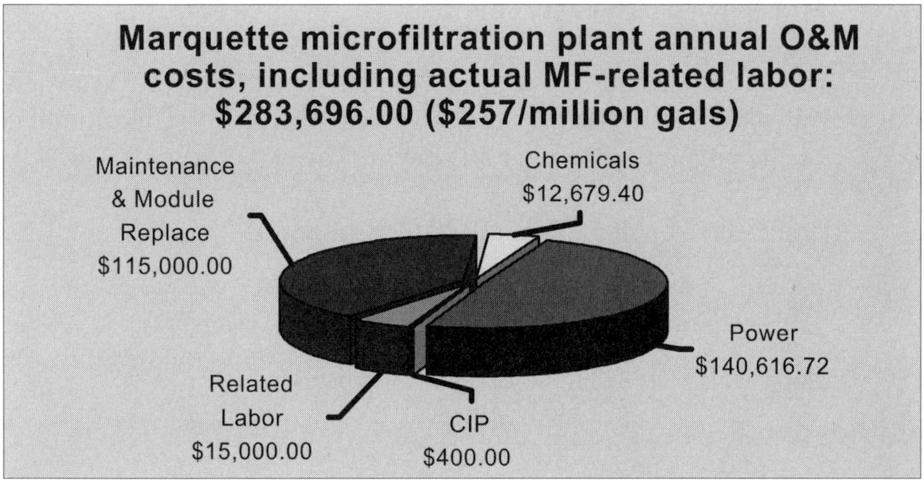

Figure 6

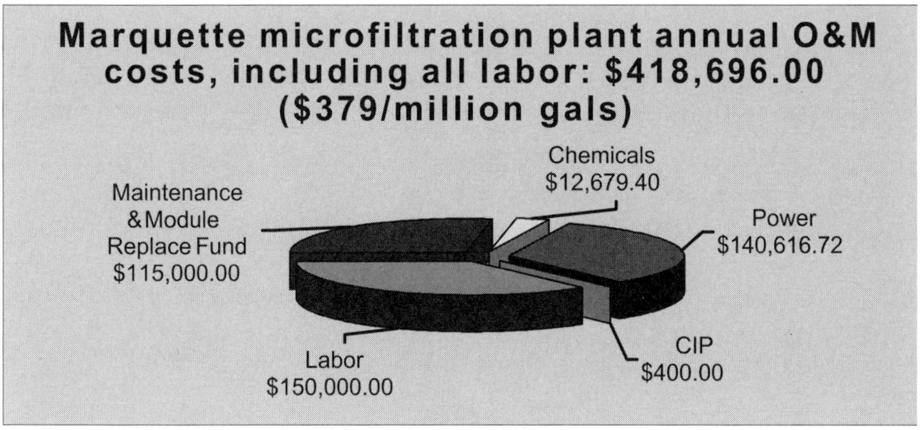

Figure 7

ACKNOWLEDGEMENTS

The author would like to acknowledge the staff of the Marquette Water Filtration Plant for the assistance provided in collecting data for the preparation of this document. They are:

James MacDonald, Filtration Plant Supervisor/Operator
Cynthia Zoll, Filtration Plant Operator
Roger Ohman, Filtration Plant Operator
Mark Spanton, Filtration Plant Operator
Mary Maki, Relief Filtration Plant Operator

CHAPTER · 15

Microfiltration Treatment of Filter Backwash Recycle Water from a Drinking Water Treatment Facility

David Y. Parker, Jr.
Michael J. Leonard
Phillip Barber
Gloria Bonic
Waymon Jones
Kenneth L. Leavell

INTRODUCTION

Concern over the threat of waterborne disease outbreaks has led to many treatment advances in the drinking water field. Physical treatment and disinfection in the late 19th and early 20th centuries led to the control of major waterborne diseases, such as typhoid and cholera [1]. The Surface Water Treatment Rule, promulgated by the United States Environmental Protection Agency (USEPA) in 1987, was designed to provide public health protection from waterborne protozoa (specifically *Giardia*) and viruses [2]. However, the significant Cryptosporidiosis outbreaks of the late 1980s and early 1990's raised serious doubts about the adequacy of existing drinking water treatment practices and regulatory standards.

Investigations surrounding the possible causes of waterborne disease outbreaks have led to concerns over the possibility of waterborne pathogen concentration and reintroduction into water plants through the reuse and recycling of filter backwash water in the drinking water treatment process. Furthermore, questions have arisen over the potential impacts of filter backwash recycle water on turbidity and organics (measured as disinfection by-product precursors) reintroduction into the overall drinking water treatment process, and the potential related drinking water regulatory compliance impacts.

These doubts led Congress to include requirements for USEPA to set forth regulations for the control of *Cryptosporidium* and to develop a filter backwash regulation as part of the Safe Drinking Water Act (SDWA) amendments of 1996.

The Interim Enhanced Surface Water Treatment Rule was promulgated by USEPA on December 16, 1998 [2]. It requires surface water treatment facilities and groundwater systems under the direct influence of surface water that serve ≥10,000 people to comply with lower turbidity standards and a treatment technique requirement for a 2-log (99%) inactivation of *Cryptosporidium.* EPA is currently developing similar requirements in the Long Term Enhanced Surface Water Treatment Rule for systems that serve ≤10,000 people, which will be promulgated in November 2000.

USEPA is currently developing a Filter Backwash Rule that should set forth minimum treatment requirements for recycled filter backwash flows in surface water treatment plants. This rule is scheduled for promulgation in August 2000.

In order to comply with these and other new drinking water regulations, some drinking water treatment plants may choose to install treatment for filter backwash recycle flows in order to reduce the public health risks from waterborne pathogens and in order to minimize turbidity and organic contaminant (specifically disinfection by-products precursors) reintroduction and concentration in the drinking water treatment process.

Previous research and full-scale applications of microfiltration for treatment of surface waters and for the treatment of secondary wastewater effluent have been well documented [3, 4, 5, 6, 7, 8, 9]. These studies focused on treatment effectiveness of microfiltration for particulate matter, microbial contaminants, disinfection by-product precursors, and as a tertiary wastewater treatment technology. In order to demonstrate the practicality, reliability, and viability of microfiltration as a process to treat filter backwash water, a U.S. Filter–Memcor CMF-S (submerged continuous microfiltration) pilot plant was installed and operated continuously for a period of four months at the Atlanta–Fulton County Water Treatment Facility in Alpharetta, Georgia.

The objectives of this study were to provide information on the removal of physical, chemical, and biological constituents in filter backwash water by microfiltration under "in-use" conditions. The specific aims for the project are to evaluate the reduction and/or removal of natural populations of total plate count bacteria, turbidity, particles, total organic carbon, total trihalomethanes, and haloacetic acids. Additionally, a separate study was conducted to determine the ability of the microfiltration system to remove protozoan cysts, viruses, coliform, and heterotrophic bacteria.

PILOT PLANT INSTALLATION AND OPERATION

The Atlanta–Fulton County Water Treatment Facility treats up to 90 MGD of potable water and utilizes the Chattahoochee River (upstream of Atlanta, Ga.) as the source water. The plant consists of a 400 million–gallon on-site presedimentation reservoir, rapid mix, four-stage flocculation, sedimentation with plate settlers, and dual-media filtration. Sludge is removed from the flocculation basins and sedimentation basins, and mixed with the filter backwash and rewash water in settling basins. Supernatant from the solids settling basins is returned to the head of the plant, while the solids are conditioned and dewatered using a plate and frame press.

The microfiltration system utilizes hollow fiber membranes with a nominal pore size of 0.2 micron. The microporous membranes were suspended in solution in bundles containing 13 square meters of surface area. Feed to the pilot plant was configured in a conventional "direct flow" or "dead end" mode.

The microfiltration pilot plant was located in an enclosed loading dock near the solids settling basins. A diagram of the pilot plant layout is shown in Figure 1. A small submersible feed pump (25 GPM) was suspended approximately 4 feet below the surface in solids settling clarification basin #2. This feedwater was dechlorinated based on demand using sodium metabisulfite and then entered a

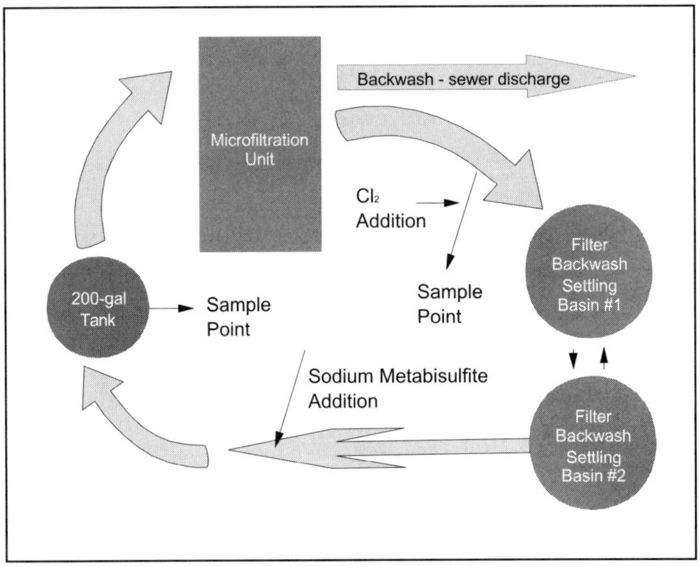

Figure 1 Pilot plant layout

200-gallon feed tank that continuously overflowed at approximately 5 GPM. The microfiltration pilot plant was operated at a flux of 31 GFD. Microfiltration effluent was returned to solids settling clarification basin #1 for recycle to the head of the water treatment plant. Microfiltration backwash water was directed to the sewer.

Samples were collected daily during the course of the study. The samples were taken from the influent and effluent. Daily sample parameters consisted of filtrate flow rate, trans-membrane pressure (TMP), feed and filtrate turbidity, feed and filtrate pH, feed chlorine residual, feed and filtrate temperature, feed and filtrate particle count, and particle count percent removal.

In addition to the routine daily sampling, intermittent sampling (four separate events) was conducted for Total Organic Carbon (TOC), Total Trihalomethanes (TTHM), Haloacetic Acids (HAA5), and Total Plate Count Organisms. Special challenges of dilute primary treated sewage and highly concentrated *Cryptosporidium parvum* cysts were dosed into the microfiltration system in order to evaluate its ability to remove waterborne pathogens that can be found in filter backwash recycle flows. All sampling was conducted and analyzed in accordance with *Standard Methods for the Examination of Water and Wastewater* [10] and/or approved USEPA methods.

RESULTS

Operating Pressure and Flow

The microfiltration pilot plant operating parameters of pressure and flow were recorded on a daily basis by plant operators at the Atlanta–Fulton County Water Treatment Facility. Figure 2 shows the flow and trans-membrane pressure (TMP) during the course of the study. The unit was operated at a constant flow rate of 12 GPM. As the membrane surface fouls, the TMP increases. Chemical cleaning of the microfiltration system was conducted once every three to four weeks using a combination of weak acids and bases. Chemical cleaning of the microfiltration membranes was required when the trans-membrane pressure (TMP) climbed over 12 psi.

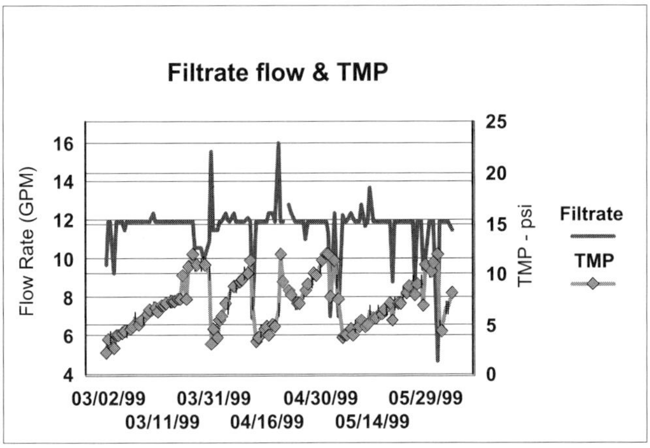

Figure 2

Feed Temperature and pH

Feed temperature and pH were monitored on a daily basis and are reported in Figure 3. Raw water temperatures ranged from 7.05 to 28 degrees Celsius due to seasonal variations. Feed pH ranged widely from 5.5 to 10.1 pH units. The microfiltration unit was nominally affected by the increasing water temperature and major pH changes.

Turbidity

The turbidity before and after the microfiltration pilot plant is shown in Figure 4. Raw water (from the backwash recycle sedimentation tank) turbidity ranged from 12.4 to 88 NTU. Finished water turbidity ranged from 0.013 to 0.55 NTU. Finished water turbidity levels fluctuated up to levels as high as 0.55 NTU during the first two months of the pilot study. These levels were much higher than expected based on previous performance. Investigation revealed incorrect sampling and analysis procedures to be the cause of the problem. A continuous-flow finished water turbidimeter was installed at the beginning of May 1999. This resulted in finished water turbidities reliably below 0.1 NTU

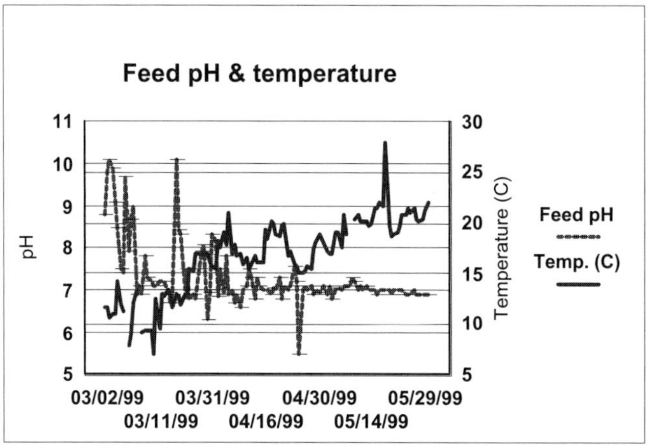

Figure 3

(0.0716 NTU average turbidity) during the last two months of the study.

Particle Counts

A particle counter was used to characterize the raw and finished particle counts for a size range of 3–15 μM particulate matter. Fluctuations in filtrate particle count levels ranged between non-detectable and 597 particles per 100 mL. A comparison of feed and filtrate particle counts is shown in Figure 5. Overall particle removal efficiency averaged 99.46% during the course of the study and is shown in Figure 6. Biological regrowth occurred in the finished water particle count sampling line, which caused finished water particle count levels to increase up to 597 particles per 100 mL during late March and early April 1999 (corresponding spikes can be seen in Figures 5 and 6). This problem was solved by feeding a low dose of chlorine into the finished water particle count sampling line. Elevated finished water particle count data also allowed the researchers to discover a minor problem with the microfiltration unit in late May 1999. The elevated finished water particle counts were caused by a broken fitting on the filtrate side that allowed a small amount of feedwater to pass into the filtrate

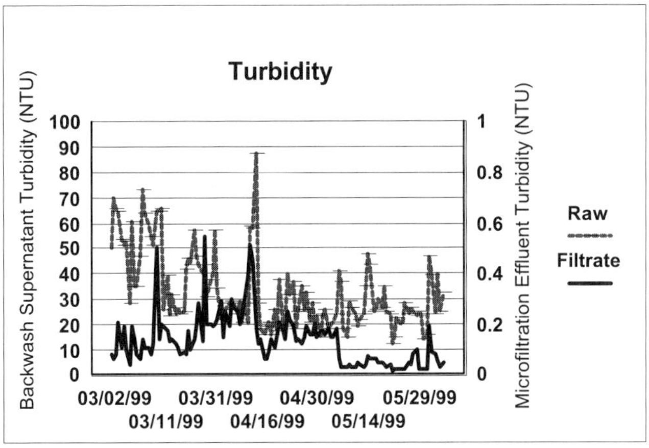

Figure 4

flow. The broken fitting was then isolated and replaced, thereby allowing the finished water particle counts to return to normal levels.

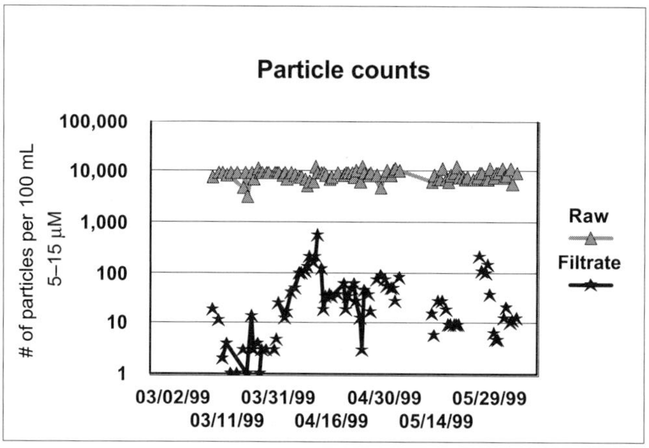

Figure 5

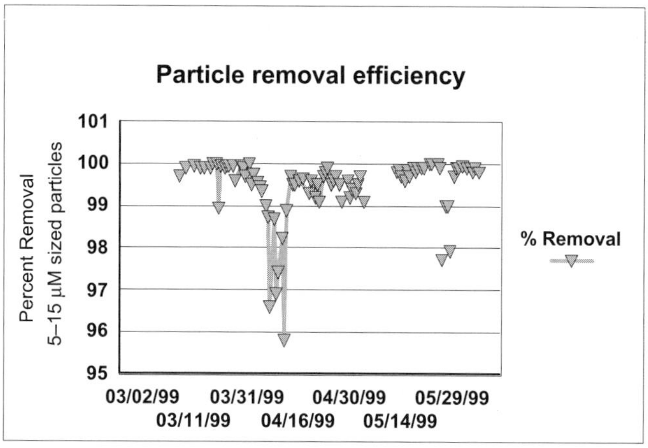

Figure 6

Total Plate Count Bacteria

Total Plate Count Bacteria were monitored four times during the operation of the pilot plant. The individual sampling results of this monitoring are illustrated in Figure 7, while Figure 8 shows the average removal (91.3%) of Total Plate Count Bacteria during these

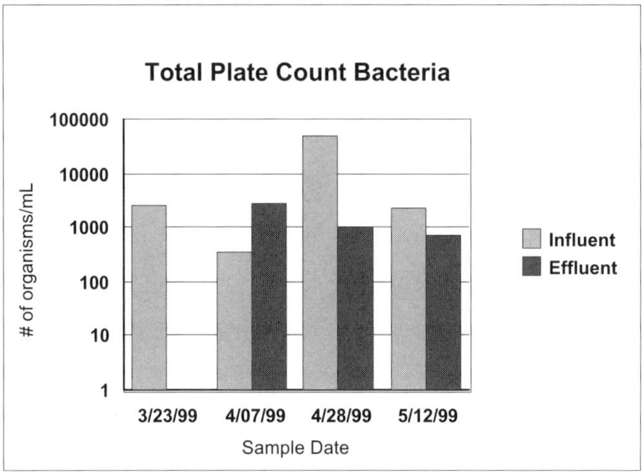

Figure 7

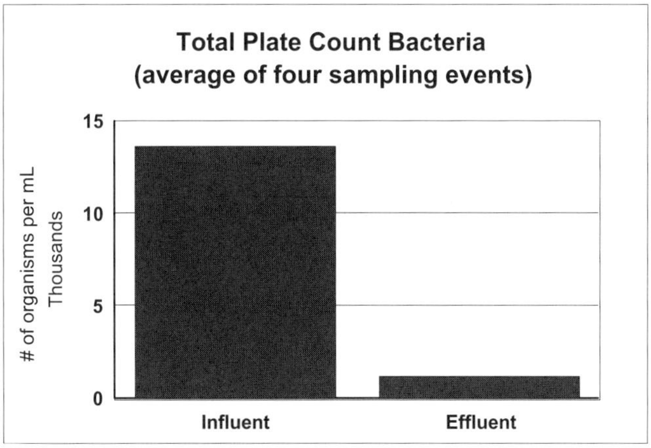

Figure 8

four sampling episodes. This monitoring and physical observation revealed that bacterial regrowth was occurring in the pilot plant effluent piping.

Total Organic Carbon Removal

The microfiltration system removed an average of 13.6% of the Total Organic Carbon (TOC) from the feedwater (Figure 9). This average was calculated based on four separate sampling events during the operation of the microfiltration pilot plant. This level of removal is consistent with the findings of previous microfiltration research projects [5].

Disinfection By-Product Formation

Average concentrations of Total Trihalomethanes (TTHMs) and Haloacetic Acids (HAA5) were measured in four separate sampling events during the operation of the microfiltration pilot plant. Although a reduction of 12.6% was seen in TTHM concentrations (Figure 10), the HAA5 levels increased on an average of 27.4% in the microfiltration effluent (Figure 11).

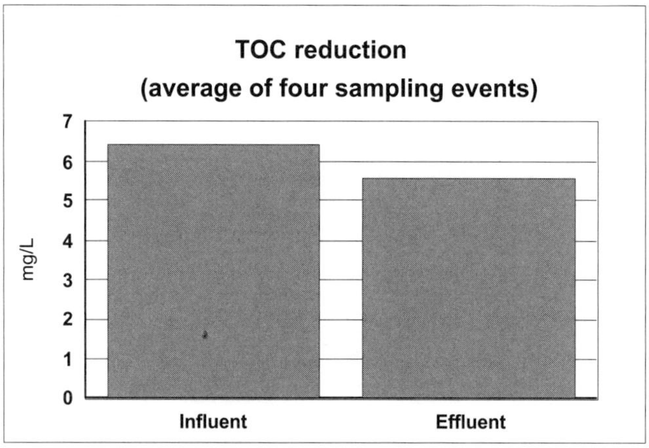

Figure 9

A possible explanation for the increase of HAA5 levels in the microfiltration effluent is the reaction between small amounts of chlorine and dissolved organic matter (resulting from bacterial regrowth) in the effluent stream of the pilot plant. This theory

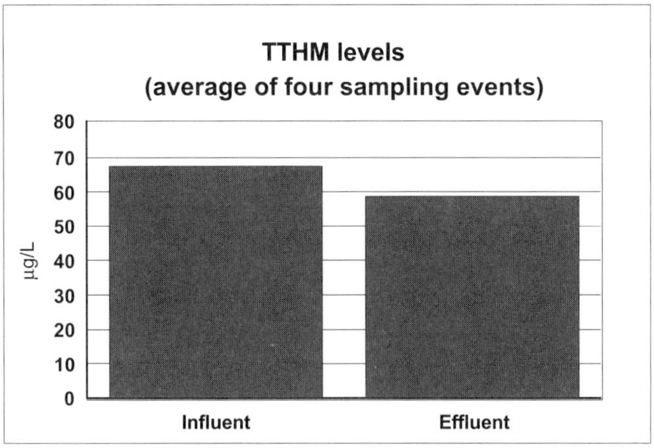

Figure 10

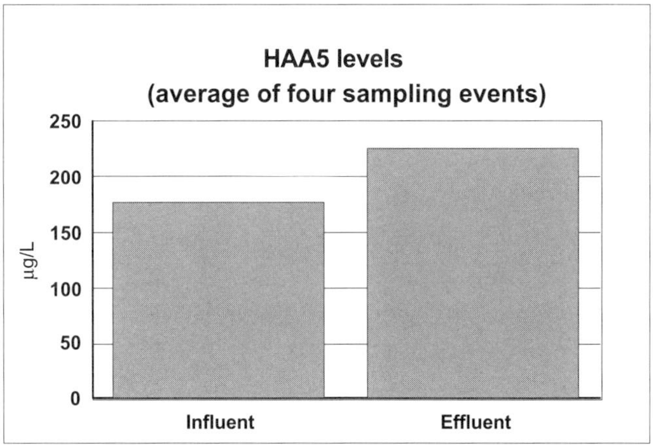

Figure 11

could explain the relatively smaller decreases and increased concentrations that appear in the latter three individual sampling events for TTHMs and HAA5 (Figures 12 and 13). However, further work should be done in this area to determine the actual cause of the increasing DPB formation.

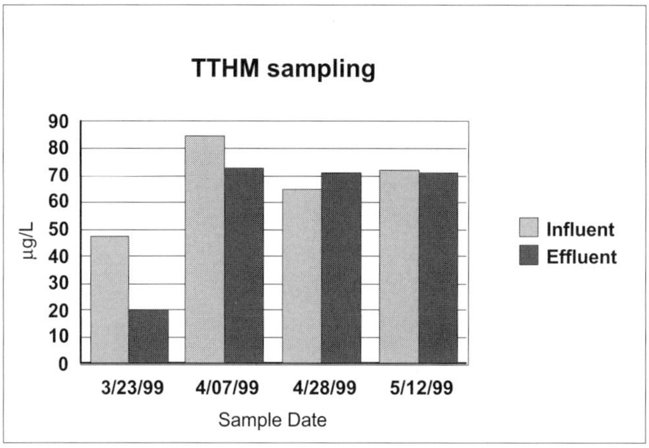

Figure 12

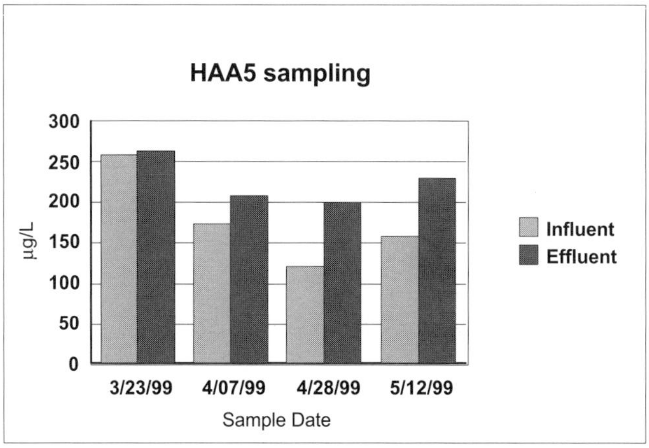

Figure 13

Special Microbial Sampling

In order to determine the ability of the microfiltration treatment unit to remove microbial pathogens (protozoa, bacteria, and virus), seeded challenges were performed. Special challenges of dilute primary treated sewage and highly concentrated *Cryptosporidium parvum* oocysts were prepared and dosed into the microfiltration system in order to evaluate its ability to remove waterborne pathogens that can be found in filter backwash recycle flows. (Note: the effluent and backwash water from these challenges were discharged to the sanitary sewer system and not recycled within the drinking water treatment plant.)

Protozoan Challenges

Inactivated *Cryptosporidium parvum* oocysts (obtained from Environmental Associates, Ltd. and shipped in a formaldehyde solution) were dosed into the microfiltration unit at a concentration of approximately 200,000 oocysts per 100 liters. Effluent samples results were <1 oocyst per 100 liters (5.3 log reduction). The influent and effluent samples were analyzed using the USEPA Information Collection Rule (ICR) methodology and USEPA method 1623. No significant difference was observed between the results of the two methods, which are listed in Table 1.

Table 1 Protozoan removal by microfiltration

Contaminant	Lab Method	Influent Concentration	Effluent Concentration	Log Removal
Cryptosporidium parvum	ICR	191,333 per 100 liters	<1 per 100 liters	5.3
Cryptosporidium parvum	1623	187,667 per 100 liters	<1 per 100 liters	5.3
Protozoan Surrogate (algae)	MPA	450,000,000	188	6.4

Protozoan surrogate removal by microfiltration was measured by Microscopic Particulate Analysis (MPA), which was conducted on influent and effluent samples during a challenge of dilute primary treated sewage effluent. The results of this sampling are also listed in Table 1. The sampling resulted in a detection of 4.5×10^8 algal cells in the influent sample, while only 188 algal cells were found in the microfiltration effluent (log reduction of 6.4).

Bacterial Challenges

Total coliform bacteria, heterotrophic plate count (HPC) bacteria, and aerobic spore concentrations were measured in influent and effluent samples during a challenge of dilute primary treated sewage effluent (Table 2). Over 18,800 total coliform organisms per 100 mL were found in the raw water, while less than one total coliform organism per 100 mL was found in the microfiltration effluent (>4.3 log reduction). The influent HPC concentration was 1.67×10^6 HPC organisms per mL, while the effluent was found to contain only 770 HPC organisms per mL (3.3 log reduction). Influent aerobic spore concentrations of 140 per mL were measured, while effluent concentration was found to be <0.04 per mL (>3.5 log reduction).

Viral Challenges

Removal of viral contaminants by microfiltration was measured in influent and effluent samples taken during a challenge of dilute primary treated sewage effluent (Table 3). Total culturable virus (TCV) was sampled using the USEPA ICR methodology. This

Table 2 Bacterial removal by microfiltration

Contaminant	Influent Concentration	Effluent Concentration	Log Removal
Total Coliform Bacteria	18800 per 100 mL	<1 per 100 mL	>4.3
HPC Bacteria	1670000 per mL	770 per mL	3.3
Aerobic Spores	140 per mL	<0.04 per mL	>3.5

Table 3 Viral removal by microfiltration

Contaminant	Influent Concentration	Effluent Concentration	Log Removal
Total Culturable Virus	83,405 per 100 L	122.7 per 100 L	2.8
Male-Specific Coliphage	35,200 per 100 mL	6.5 per 100 mL	3.7

resulted in a raw water TCV concentration of 8,3405 organisms per 100 liters, and a microfiltration effluent TCV concentration of 122.7 organisms per 100 liters (2.8 log removal). Male-specific coliphage samples taken at the same time yielded a raw water concentration of 35,200 organisms per 100 mL and an effluent concentration of 6.5 organisms per 100 mL (3.7 log removal). These findings were consistent with earlier research on viral removal by microfiltration [11].

CONCLUSIONS

The Memcor microfiltration technology performed admirably in the long-term continuous treatment of filter backwash supernatant. Although raw water fluctuations in turbidity and pH were quite severe, the microfiltration treatment unit was able to produce consistently low finished water turbidities and particle counts. The microfiltration technology proved capable of removing high levels of waterborne disease-causing microbial contaminants (protozoa, bacteria, and viruses) that can be found in filter backwash recycle applications. The microfiltration treatment unit also provided 13.6% removal of TOC from the filter backwash supernatant source water (indicating that a minor level of DBP reduction could be obtained in the drinking water treatment plant).

REFERENCES

1. Wolman, A., A. E. Gorman (1931). *The Significance of Waterborne Typhoid Fever Outbreaks 1920–1930.*
2. 40 C.F.R. Parts 141 and 142: Drinking Water. National Primary Drinking Water Regulations.
3. Olivieri, V.P., D.Y. Parker, G.A. Willinghan, J.C. Vickers. *Continuous Microfiltration of Surface Water.* AWWA Membrane Processes Conference. Orlando, Fla., March 10–13, 1991.
4. Olivieri, V.P., D.Y. Parker, D.F. Schrott, G.A. Willinghan, J.C. Vickers. *Continuous Microfiltration for the Treatment of Surface Water.* American Filtration Society Conference, Atlanta, Ga., October 1991.
5. Parker, D.Y. (1991). *Removal of Trihalomethane Precursors by Microfiltration.* Master's thesis, Johns Hopkins University.
6. Olivieri, V.P., G.A. Willingham, J.C. Vickers, C.L. McGahey. *Continuous Microfiltration of Secondary Wastewater Effluent.* AWWA Membrane Processes Conference, Orlando, Fla., March 10–13, 1991.
7. Yoo, R.S., D.R. Brown, R.J. Pardini, G.D. Bentson (March 1995). Microfiltration: A Case Study. *Journal AWWA*, 87(3), 38.
8. Kelley, W.A., R.A (June 1999). Olson Selecting MF to Satisfy Regulations. *Journal AWWA*, 91(6), 52.
9. Karimi, A.A., J.C. Vickers, R.F. Harasick (June 1999). Microfiltration Goes Hollywood: The Los Angeles experience. *Journal AWWA*, 91(6), 90.
10. *Standard Methods for the Examination of Water and Wastewater.* (1998). APHA, AWWA, and WPCF, Washington, D.C.
11. McGahey, C.L. (1994). *Mechanisms of Virus Capture from Aqueous Suspension by a Polypropylene, Microporous, Hollow Fiber Membrane Filter.* Doctoral thesis, Johns Hopkins University.

CHAPTER · 16

Experiences with Planning, Construction, and Startup of a 14.5 MGD Microfiltration Facility

Lance Schideman
Camp Dresser & McKee, Chicago Illinois

Nilaksh Kothari
Manitowoc Public Utilities, Manitowoc, Wisconsin

Dr. James Taylor
University of Central Florida, Orlando, Florida

OVERVIEW

In the wake of the *Cryptosporidium* outbreak in nearby Milwaukee, Manitowoc Public Utilities (MPU) implemented a new microfiltration water plant to provide upgraded protection against microbial challenges. Two of the more notable features of the project were the low cost and short implementation schedule. The total construction cost of the new microfiltration plant was $0.50 per gallon of capacity, and it was designed and constructed in 18 months. This chapter reviews the planning, construction, and first year of operations of the microfiltration facilities and highlights the key factors that contributed to the cost-effectiveness and compressed schedule of the project.

PLANNING PHASE

Manitowoc, Wisconsin, is located on the western shores of Lake Michigan, approximately 90 miles north of Milwaukee. For the most part, the raw water quality of Lake Michigan at Manitowoc is very good, but the *Cryptosporidium* outbreak in Milwaukee demonstrated the potential for failure of conventional treatment techniques to provide suitable water quality. Thus, MPU began a process to investigate upgrading their treatment facilities to ensure protection of their customers against similar microbial outbreaks.

The primary objective of the upgraded treatment facilities was to provide an additional and highly effective barrier against *Cryptosporidium*. Ozone and membrane filtration were determined to be the most proven alternatives for providing this barrier. Since membrane filtration had only been used at a handful of municipal water plants, the first step MPU took toward upgrading their facilities was to pilot a microfiltration (MF) and an ultrafiltration (UF) system to establish the performance characteristics with Manitowoc's raw water. MPU staff carried out the pilot study with technical guidance from Dr. James Taylor at the University of Central Florida. A summary of the membrane characteristics and pilot study findings is provided in Table 1.

Table 1 Summary of membrane filtration pilot study results

	Microfiltration System	Ultrafiltration System
Nominal Pore Size	0.2 micron	100,000 Dalton
Operational Mode	Dead-end, outside-in	Cross-flow, inside-out
Free Chlorine Tolerance	No	Yes
Maximum Transmembrane Pressure (TMP)	18 psi	30 psi
Optimum Flux Rates	92 gpd/sf	Flux rates were not sustained
Chemical Cleaning Interval	7–14 days	1–3 days
Minimum *Cryptosporidium* Sized Particle Removal	4-log	4-log

The MF system had much better performance, as demonstrated by a significantly longer interval between chemical cleanings. The UF system was not successful in maintaining flux rates for a sufficient amount of time to be reliable and cost-effective. The nominal pore size of the MF membrane (0.2 micron) is well below the size of *Cryptosporidium* cysts (2–5 microns). In comparing the two piloted membrane systems, the only downside for the MF system was that it was not tolerant of chlorine. However, that limitation could be accommodated with design and operational adjustments. At its conclusion, the pilot study clearly demonstrated that MF could be used to remove *Cryptosporidium* and meet the turbidity and particle removal requirements of existing and proposed future regulations. The MF system used in the pilot study, and later used in the full-scale plant, was the US Filter–MEMCOR CMF system.

Next, MPU commissioned Camp Dresser and McKee (CDM) to develop a conceptual design and cost estimates for comparing an ozone system upgrade with a microfiltration system. The design criteria and the estimated costs for each system are summarized in Table 2.

As shown in Table 2, the new microfiltration plant was expected to have lower capital and operating costs than the conventional plant upgraded with ozone. The cost-effectiveness of a new microfiltration system in comparison to ozone was aided by the following factors, some of which are atypical circumstances that are specific to MPU's situation.

Factors Contributing to the Cost-Effectiveness of a New Microfiltration System at Manitowoc Public Utilities

- Lower Staffing Requirements

  The state regulatory agency, Wisconsin Department of Natural Resources (WDNR), would have required 24-hour/day staffing of an ozone system to continually monitor the system operation. Typical microfiltration systems are highly automated and have a simpler treatment scheme for

Table 2 Design criteria and cost estimates of process alternatives

	Ozone Upgrade to Conventional System	New Microfiltration System
Design Flow of Process Comparison Study	11 MGD	11 MGD
Minimum Water Temperature	5°C	5°C
Minimum *Cryptosporidium* Inactivation/ Removal	3-log Inactivation	4-log Removal
Main System Operating Parameters	3 mg/l ozone dose 30 minutes contact time	90 gpd/sf at 62°F
Capital Cost	\$6,680,000	\$6,250,000
Operating Cost–Present Worth	\$2,770,000	\$1,940,000
TOTAL PRESENT WORTH	\$9,450,000	\$8,190,000

which unmanned operation is allowed by WDNR. This significantly reduced the labor costs for MF operations.

- Lack of Dissolved Contaminants

  Microfiltration is very effective at removing particulate matter, but it is not effective at removing dissolved contaminants from the raw water. However, typical dissolved constituents such as hardness, TOC, and taste and odor compounds are at low levels in Manitowoc's raw water and typically do not need to be removed. Thus, the pre- and post-treatment requirements for microfiltration were fairly minimal. Note that the MPU microfiltration system was designed with facilities to add powdered activated carbon and a coagulant for unusual situations when dissolved contaminants need to be removed.

- Availability of Heated Water

  MPU also operates a power plant adjacent to the water plant, which supplies heated raw water during the winter. Heated water raises the flux rates of the microfiltration process so that the equipment is more efficient during cold weather months. MPU typically maintains raw water temperatures above 5 degrees Celsius, even when the Lake Michigan temperature is near 0 degrees Celsius.

- No Need to Renovate Existing Facilities

  The ozone system cost estimate included approximately $1.2 million of capital improvements to the existing facilities to update facilities that did not meet current design standards or were in poor condition.

- Ability to Return Backwash Water to the Source

  The pilot testing demonstrated that the MF backwash water quality was similar to that of the raw water. This is primarily due to the fact that the microfiltration plant usually operates without any chemical pretreatment, so the backwash water does not contain any foreign impurities. Thus, MPU was able to obtain a permit for return of the backwash water to the source, which results in a much lower residuals processing cost. Note that the system was designed with facilities to discharge backwash to the sanitary sewer if pretreatment chemicals were to be added to the water.

- Reuse of Existing Facilities

  Certain facilities at the existing plant could be reused for ancillary facilities in the new microfiltration system to lower the overall system costs. The new microfiltration system reused the existing intake, clearwell, chemical systems, and backwash holding tank. The high and low lift pumping stations were reused, except all but one of the existing pumps were replaced. The project also included the conversion of the sedimentation tanks to additional clearwell storage.

In addition to a lower estimated cost, the following advantages of a microfiltration system were identified and ultimately led MPU to pursue a new microfiltration water treatment plant.

Additional Advantages of a New Microfiltration System over an Ozone Upgrade

- Greater process simplicity
- Smaller footprint of treatment facilities
- Less chemical addition
- Less residuals production
- Less interference with existing plant operations during construction
- Some existing facilities would become available for other uses

CONSTRUCTION PHASE

The actual capital cost was $7.3 million, with net change orders of less than 1%. Table 3 presents a comparison of estimated costs during the planning phase and the actual construction costs. The actual costs were higher than the planning phase estimates due primarily to an increase in the design capacity from 11 MGD to the 14.5 MGD that was made during the detailed design. In addition, MPU decided to build additional space in the building for future microfiltration units. The actual cost per gallon of capacity ($0.50) decreased from the planning phase estimates.

The project was designed in six months and the bid/award of the construction contract took three months. The construction was substantially complete and the plant was producing potable water

Table 3 Comparison of actual construction costs and planning phase estimates

	Planning Phase Estimated Costs	Actual Construction Phase Costs	Comments
Building	$ 590,000	$ 830,000	Added building space for future equipment
Process Piping and Equipment	$ 4,980,000	$ 5,950,000	Added plant capacity. (12 microfiltration units actually installed versus 10 planned)
Electrical and Instrumentation	$ 240,000	$ 260,000	
Engineering	$ 220,000	$ 255,000	
Contingency	$ 220,000	–	
TOTAL CAPITAL COST	$ 6,550,000	$ 7,300,000	
Design Capacity	11 MGD	14.5 MGD	
Cost per Gallon of Capacity	$ 0.56	$ 0.50	

nine months later (May 1999). The major factors contributing to the accelerated construction schedule are presented and discussed in the following list.

Major Factors Contributing to a Compressed Implementation Schedule for a New Microfiltration System at Manitowoc Public Utilities

- Prepurchase of Equipment

  Given that the US Filter–MEMCOR system had significantly better performance during the pilot study, MPU negotiated a sole source prepurchase of their equipment during the design of the new microfiltration facilities. This allowed the overall design to be accelerated and tailored to the specific equipment. This also took the microfiltration equipment fabrication off the critical path for the job.

- Reuse of Existing Facilities

  Reusing certain existing facilities with heavy structural work (i.e., pump stations, clearwells, backwash tanks) saved time associated with constructing these facilities from ground up.

- Minimizing Interference with the Existing Plant Operation

  Building a new microfiltration plant provided the opportunity to centralize most of the major construction work in a location remote from the existing operations. This minimized the amount of time spent coordinating construction activities with plant operations. For this project one new building was constructed that houses almost all of the components of the microfiltration treatment process.

- Allowing Substitutes

  There were several situations where a particular specified product was not readily available. For instance, stainless steel piping was used in several cases when fittings were not readily available for some other material. Working with the contractor to find a suitable alternative helped to keep the project moving on schedule.

FIRST YEAR OF OPERATING EXPERIENCE

The startup and testing of the new microfiltration plant took approximately one month. The plant was quickly producing high-quality effluent with typical particle counts of one to five particles/ml in the nominal size range of *Cryptosporidium* cysts (<4 μm). A summary of the operating data during the first year of operations is provided in Table 4.

Table 4 Summary of the first-year operating data

Operating Parameter	First-Year Operating Data
Effluent Turbidity	0.01–0.04 NTU
Flux Rate	62 gpm/sf
Backwash Interval	30–45 min
Chemical Cleaning Interval	5–10 days
Rejected	7–9%
Pressure Hold Integrity Test Results	Less than 0.2 psi/min
Typical pinning frequency	1 pin per CMF unit per month (12 CMF units installed at Manitowoc)

CONCLUSION

Manitowoc Public Utilities endeavored to provide upgraded water treatment to protect their consumers from the risks associated with *Cryptosporidium.* After comparing the cost and performance characteristics of ozone and membrane filtration, they selected membrane filtration to replace their conventional treatment process. A 14.5 MGD microfiltration system was designed and constructed in a very short time frame of only 18 months. The capital cost of the new facilities was $7.3 million, which resulted in a very low cost per gallon of capacity of only $0.50. The plant has been operating since May 1999 and has been producing high-quality finished water that meets all the treatment objectives.

CHAPTER · 17

The Use of Microfiltration for Backwash Water Treatment

Orren D. Schneider, Ph.D., P.E.
Hazen and Sawyer, P.C., NY, NY

Eric Acs
Metcalf and Eddy, Inc., Lebanon, NJ

Silvana Leggiero
NYC Department of Environmental Protection, Corona, NY

David Nickols, P.E.
Hazen and Sawyer, P.C., NY, NY

BACKGROUND

The use of microfiltration as a means for treating waste filter backwash water was investigated. During a pilot study, three membrane units from different manufacturers were evaluated. The results from this testing indicated that these units produced water low in turbidity (<0.1 ntu), color (<5 scu), and that could remove at least 4-log of pathogens. Based on these results, a treatment scheme is envisioned whereby the backwash water is treated by microfiltration and UV disinfection (as a secondary barrier) prior to being discharged to the plant clearwell, thereby eliminating a recycle stream. A conceptual cost analysis indicated that by removing the recycle stream from the main process flow, some of the cost of the membrane/UV system could be recouped through lower concrete costs for the main treatment process.

INTRODUCTION

As part of an extensive pilot program to evaluate treatment options for New York City's Croton System, the Joint Venture of Metcalf and Eddy/Hazen and Sawyer has been operating a pilot plant at the New Croton Reservoir since 1996. The goal of this pilot operation was to confirm possible treatment alternatives, select the best treatment alternative, and optimize design parameters so that a 290-mgd treatment plant can be designed. Based on the results of piloting, a treatment scheme including coagulation/flocculation, dissolved air flotation (DAF), intermediate ozonation, and high-rate biological filtration has been selected. A process flow diagram of this treatment train, including a conventional residuals treatment portion, is shown in Figure 1.

Because of the potential risk of recycling pathogens through treatment plants, the USEPA will regulate the handling of filter backwash water in the near future. This new regulation will likely prevent filtration plants from recycling waste backwash water to the head of the plant without some level of treatment. An alternative residuals treatment process was also considered to reduce the volume of water to be returned to the head of the plant, thus allowing for the design of smaller treatment process units while still providing an effective pathogen removal mechanism. The alternative treatment scheme included microfiltration units and ultraviolet light (UV) disinfection to treat filter backwash water. A process flow diagram is shown in Figure 2. As part of the pilot work at the New Croton Reservoir, studies on the treatment of backwash water have been performed using three different backwashable microfiltration units.

Pilot testing of treatment options was carried out at the Millwood Water Treatment Plant. The Millwood WTP operates using a treatment system similar to the scheme selected for New York City—DAF, intermediate ozonation, and filtration. In addition, one of the water sources available for treatment at the Millwood plant is chlorinated New Croton Aqueduct water, which the plant agreed to

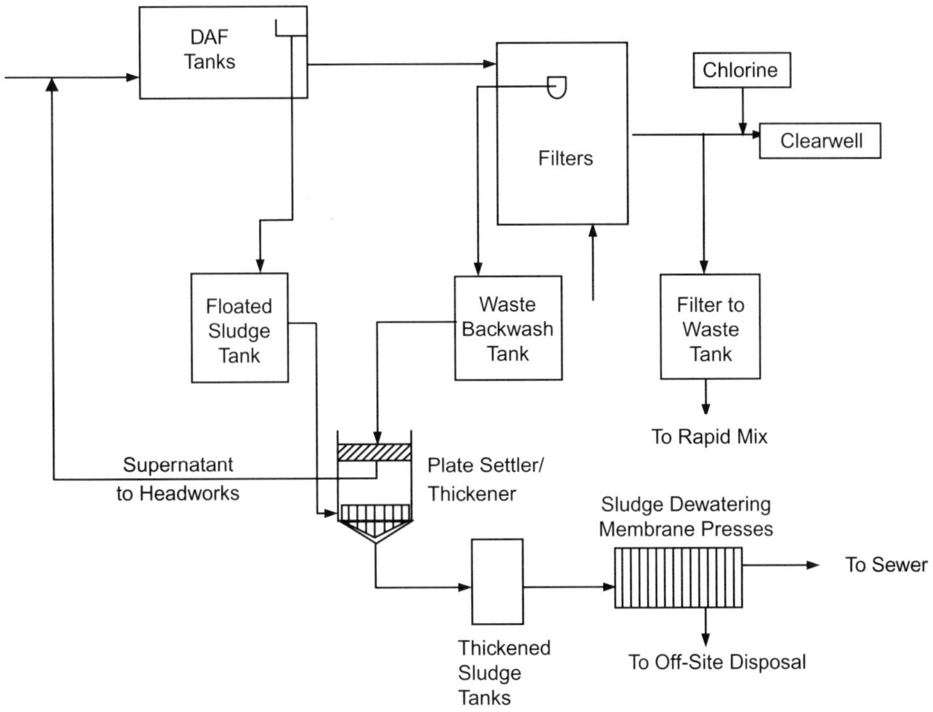

Figure 1 Residuals treatment process flow schematic

use during testing of the treatment alternatives. Thus, it was felt that the use of Croton water at the Millwood plant was a good simulation of backwash treatment for the New Croton Reservoir.

During the study the Millwood plant operated using potassium permanganate as a preoxidant, polyaluminum chloride as primary coagulant, and a nonionic filter aid polymer, along with intermediate ozonation. For the first half of the study (June through the middle of July), the plant operated using either chlorinated Croton Aqueduct water or a blend of chlorinated Croton and unchlorinated Catskill Aqueduct water. For the second half of the study (middle of July through the end of August), the plant operated exclusively with unchlorinated Catskill Aqueduct water.

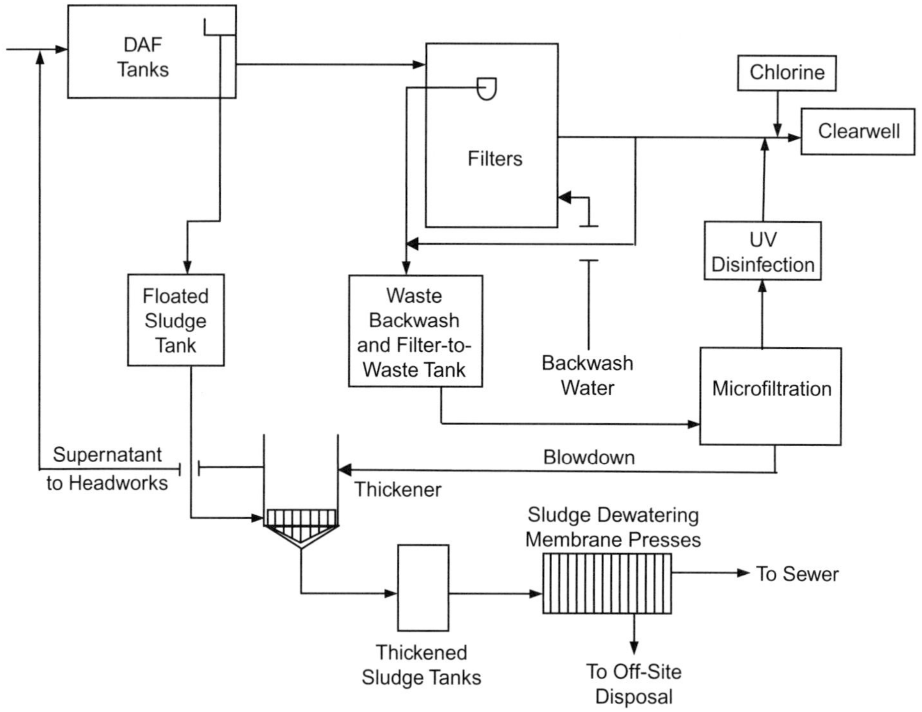

Figure 2 Alternative residuals treatment process flow schematic

The Millwood plant was designed with filters consisting of 24" of anthracite filter media over 24" of sand. Forty-eight to sixty thousand gallons of filtered water are used to backwash each filter. Backwashing typically occurs every other day and is initiated based on time rather than turbidity breakthrough or terminal headloss. The waste backwash water is directed to a concrete basin where it is equalized. In addition, 15,000 to 20,000 gallons of filter-to-waste water are directed to this equalization tank. The backwash water waste is pumped from the equalization basin at 128 gpm to plate settlers to remove solids from the backwash water. The effluent from the plate settlers is then pumped back to the head of the plant for treatment.

BACKWASH WATER CHARACTERIZATION

Because the treatment plant generally operates at a flow rate much less than the design flow, solids from the backwash water tend to settle to the bottom of the equalization basin. No mixing exists in the equalization basin to keep the solids in suspension other than that provided by the pumps that discharge into the tank. The plant personnel clean the basin every two years and remove an estimated one foot of accumulated solids, approximately 125 cubic yards.

The three pilot units were piped to receive the filter wastewater from the equalization tank. The feedwater was inconsistent in turbidity because of the solids settling to the bottom of the equalization basin and intermittent backwashing. The turbidity of the feedwater ranged from 0.8 to 2.0 ntu between filter backwashes. During a plant filter backwash, the turbidity spiked to 10 to 20 ntu. The spikes usually lasted two to three hours before returning to the lower, steady-state turbidity.

The equalization tank was pumped down to the bottom several times during the study. This resulted in very high-turbidity water, over 100 ntu, along with high levels of dissolved manganese (>0.77 mg/l) and iron (1.22 mg/l). The membranes were turned off once during the study due to lack of water in the equalization basin.

Several water quality parameters for the backwash feedwater are presented in Table 1. As expected with backwash water, the feedwater to the pilot units was high in parameters associated with particulate matter, including turbidity, particle counts, apparent color, manganese, total suspended solids, and total organic carbon. The median iron and soluble UV254 values were relatively low, 0.02 mg/l and 0.014 cm^{-1}, respectively.

BACKWASH WATER TREATMENT STUDY

Three hollow-fiber microfiltration pilot units (US Filter–Memcor, Zenon Environmental Systems, and Pall Corporation) were operated to treat backwash water produced by the Millwood water treatment

Table 1 Millwood equalized backwash water quality

	Turbidity (*ntu*)	UV254 (cm^{-1})	Apparent Color (*scu*)	Iron (*mg/l*)	Manganese (*mg/l*)
Minimum	0.87	0.002	0	0.01	0.03
10th Percentile	1.3	0.005	14	0.01	0.06
25th Percentile	1.5	0.009	20	0.01	0.09
50th Percentile (median)	2.7	0.014	35	0.02	0.23
75th Percentile	4.5	0.044	49	0.06	0.56
90th Percentile	36.1	0.082	363	0.43	0.68
Maximum	100	0.139	800	1.22	0.77
N	16	16	15	15	15

plant. The goal of this study was to determine the level of water quality produced by these pilot units and to establish preliminary design criteria for their possible use.

A comparison of physical and operational parameters for the three membrane units tested on the Millwood backwash water is shown in Table 2.

US FILTER–MEMCOR

Memcor Operations

The US Filter–Memcor unit was operated from May 18 to August 20 at a filtrate flow of 1.5 gpm. This corresponds to a flux rate of 50 gallons per square foot per day (gfd). The backwash timer was set at 20 minutes, which gives the machine a 90% recovery (90% of the water is treated, 10% is wasted during backwash).

Figure 3 presents the transmembrane pressure (TMP) values, flux rate, and permeate turbidity from May 18 to August 20. TMP varied with the inconsistent feed. During normal plant operations, with feedwater spikes during filter backwashes, the US Filter–Memcor

Table 2 Millwood WTP backwash water treatment study

	Memcor	Zenon	Pall
Filter Material	Polypropylene	Proprietary	PVDF
Type	Hollow-fiber microfilter	Hollow-fiber microfilter	Hollow-fiber microfilter
Pore Size	0.2 μm	0.1 μm	0.1 μm
Flow Configuration	Outside → In	Outside → In	Inside → Out
Flow rate (gpm)	1–3	7–15	5–10
Membrane Area (m^2)	4	56	26
Flux Range Tested (gfd)	33–100	25–46	25–50
Maximum TMP	22 psi	19 in. Hg	22 psi

unit TMP increased slightly, rising 4 psi in 30 days from 10 psi to 14 psi. At this flux rate and feedwater quality, the chemical cleaning interval is estimated at between 8 and 12 weeks. When the plant pumped down the equalization tank, the unit clogged with solids and it was not possible to restore the membrane to the original TMP values when normal feedwater was restored. TMP values ranging from 18 psi to 26 psi were recorded after an equalization basin pumpdown. The unit was chemically cleaned after it was realized that TMP values would not return with lower turbidity feed. Chemical cleanings occurred on July 8 and August 20. Chemically cleaning the unit using citric acid and Memclean™ solution (caustic soda with a surfactant) restored the TMP to the original values, indicating reversible fouling occurred.

It should be noted that the Memcor unit used for testing at Millwood has a looser fiber packing density and shorter fibers than a full-scale membrane module. Experience with side-by-side testing of small and full-size units has shown that when full-size Memcor membranes are used, the required cleaning frequency is higher than when the smaller fibers are used. This has been attributed to a more efficient backwash in the smaller pilot membrane.

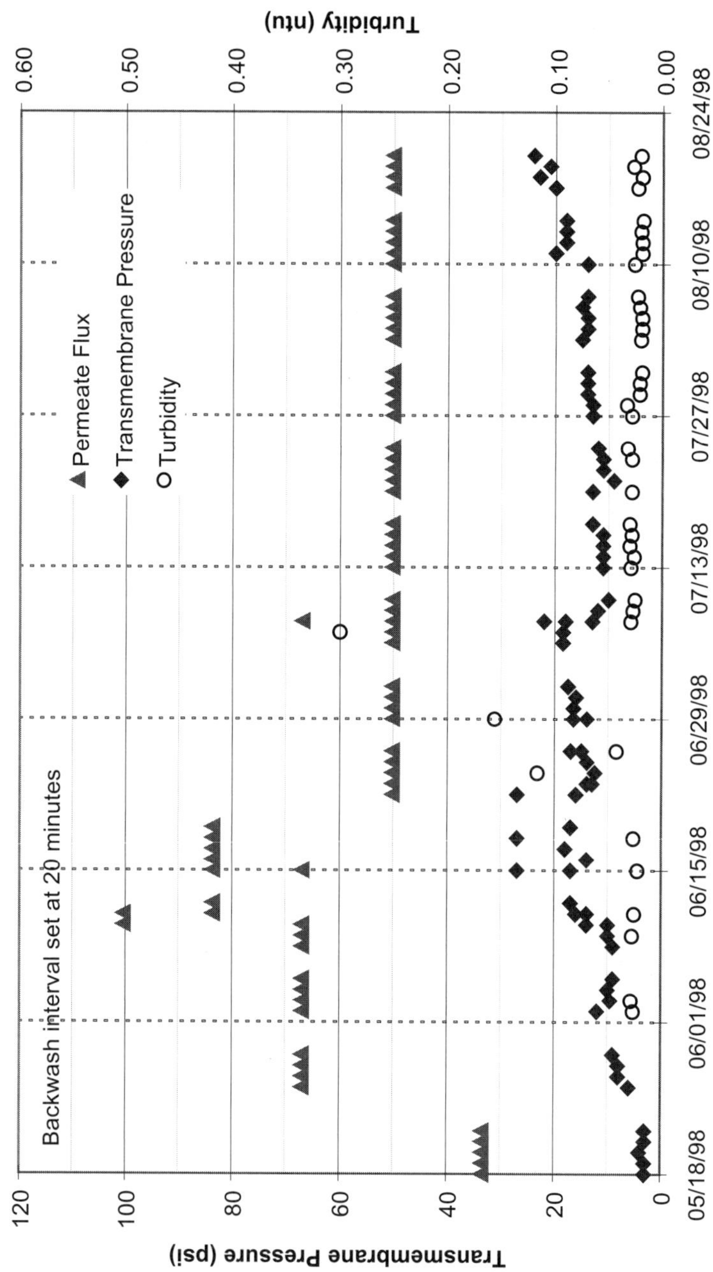

Figure 3 Memcor microfilter operating parameters

The US Filter–Memcor unit also had a 500-micron basket strainer as a pre-filter to the membrane. This was cleaned out twice, usually when the plant was pumping close to the bottom of the equalization tank.

Memcor Water Quality

Permeate

The filtered water quality is summarized in Table 3. The Memcor pilot unit produced water quality suitable for drinking during all feedwater conditions. All parameters associated with particulates were removed to levels below those set for the main treatment plant. These include turbidity with a median value of 0.028 ntu, apparent color with a median value of 0 scu, iron with a median value of 0.01 mg/l, and manganese values with a median value of 0.030 mg/l. The UV254, a measure of dissolved organic matter, had a median value of 0.014 cm^{-1}, the same as the raw water value. This indicates that dissolved material sorbed to particles captured by the filters did not desorb appreciably while the waste backwash water was in the equalization basin.

Concentrate

The water quality of the waste (concentrate) stream from the membranes was also analyzed. The average results of the analyses are presented in Table 3. As expected, the waste stream concentrates particulate matter. Thus, concentrate water from this unit is high in turbidity, particle counts, TSS, apparent color, DOC, aluminum, and manganese. As expected, the aluminum value, primarily from PACl floc, was very high (63 mg/l).

Table 3 US Filter–Memcor water quality

	Turbidity (*ntu*)		UV254 (cm^{-1})		Apparent Color (*scu*)		Iron (*mg/l*)		Manganese (*mg/l*)	
	Permeate	Con.*	Permeate	Con.*	Permeate	Con.*	Permeate	Con.*	Permeate	Con.*
Minimum	0.022	1.0	0.005	0.001	0	2	0.00	0.01	0.01	0.05
10th Percentile	0.024	7.5	0.007	0.005	0	6	0.00	0.02	0.02	0.16
25th Percentile	0.026	9.4	0.009	0.012	0	19	0.01	0.05	0.02	0.42
50th Percentile (median)	0.028	18.4	0.014	0.019	0	155	0.01	0.13	0.03	0.77
75th Percentile	0.05	35.6	0.018	0.026	0	361	0.01	0.47	0.05	0.77
90th Percentile	0.128	219.8	0.060	0.138	6	988	0.02	3.65	0.33	0.97
Maximum	0.156	407.0	0.143	0.326	10	2,080	0.02	7.38	0.77	1.34
N	15	14	15	14	15	15	14	14	14	14

*Concentrate

ZENON ENVIRONMENTAL SYSTEMS

Zenon Operations

Figure 4 presents the operating conditions of the Zenon membrane from May 27 to August 24. From June 1 to June 18 the Zenon pilot unit operated at a flux of 35 gfd. The maximum feed flow to the unit as supplied by the Millwood pumps was 8 gpm and was insufficient to operate at this desired flux. Therefore, in order to achieve this flux, only 4 gpm of the filtered water was sent to drain. The 11.75 gpm remaining was recycled back into the feed tank. The unit bleeds solids as waste from its feed tank using a peristaltic pump. Adjustment of this pump in relation to the water filtered controls the unit's recovery. This pump is set based on the solids content of the feed tank. For this time period, the recovery was set at 90%. Since the Zenon unit operates on a vacuum pressure, an increase in the vacuum pressure is registered when the membrane is clogging. During this period, the vacuum pressure lost only 2" Hg from –4.0" Hg to –6.0" Hg. This corresponds to a cleaning frequency of 150 days.

The Zenon membrane recovery was increased from 90% to 95% from July 10 to August 10. Vacuum pressure lost 3" Hg from –10.0 to –13.0" Hg. The unit was set up to backwash every 10 minutes for 15 seconds. The backwash was operated in this manner for the duration of the pilot test program. Operating at these conditions the projected run length is estimated at 4 months.

A booster pump was installed on August 10 to increase the flow to the Zenon membrane's feed tank. The increased feed from 4 gpm to 15 gpm would allow the machine to send all of its filtered water to drain with no recycling. The recovery of the machine was also increased to 98%. The unit ran at these conditions until August 28 when the pilot was shut down. The Zenon unit lost 2" Hg from –13.0 to –15.0" Hg during this time period, which is similar to its performance when it recycled water.

The Zenon membrane handled all feedwaters without clogging, including when the plant pumped down the backwash

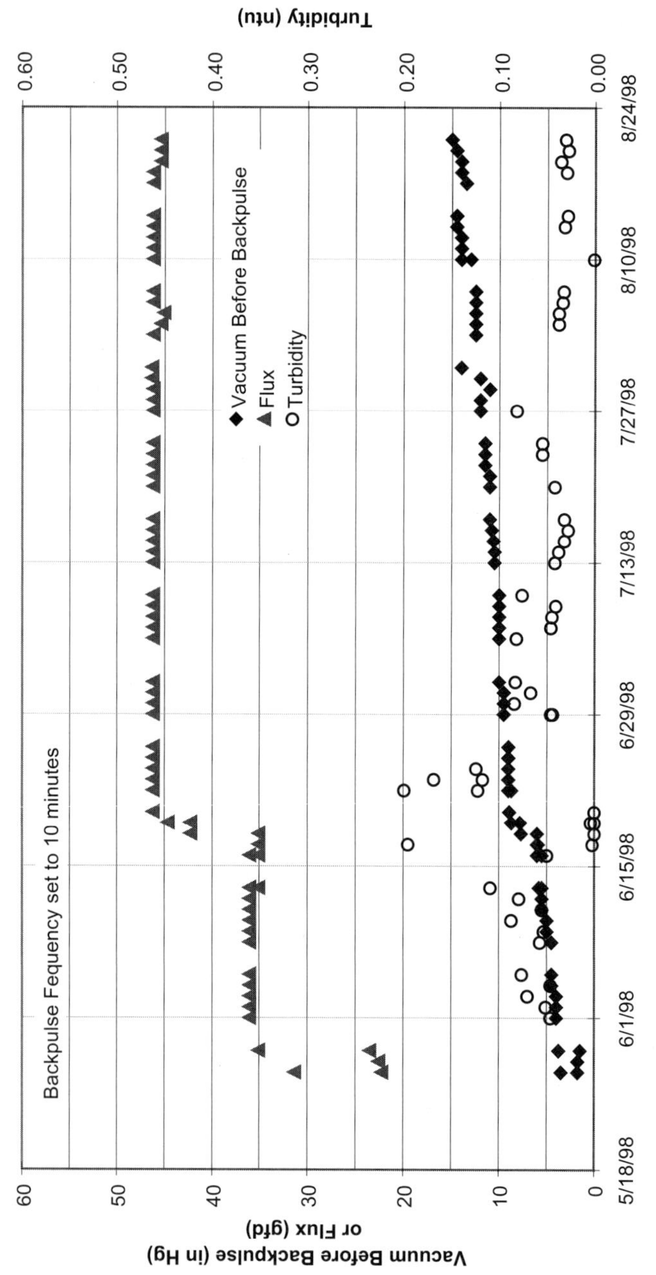

Figure 4 Zenon microfilter operating parameters

equalization basin, sending it water with a turbidity of over 100 ntu. This occurred when the water was not being recycled. The Zenon unit was not chemically cleaned while operating at the Millwood plant. Thus, it is not known if the fouling was reversible.

Zenon Water Quality

Permeate

The permeate water quality is summarized in Table 4. The Zenon pilot unit produced water quality suitable for drinking during all feedwater conditions. All parameters associated with particulates were removed to levels below the goals set for the main treatment plant. These include turbidity with a median value of 0.056 ntu, apparent color with a median value of 0 scu, and manganese values with a median value of 0.020 mg/l. The UV absorbance at 254 nm had a median value of 0.011 cm^{-1}, slightly less than the raw water value.

Concentrate

The wastewater stream (concentrate) from the Zenon pilot skid was also analyzed. The average results of the analysis are presented in Table 4. The concentrate stream from the Zenon membrane had the highest values of the three membrane systems for all of the measured parameters, indicating better removal from the feedwater. This is not unexpected as the Zenon has the smallest pore size (0.1 μm) and tightest pore size distribution of the microfiltration membranes studied.

PALL CORPORATION

Pall Operations

Figure 5 presents the operating conditions of the Pall membrane from May 11 to August 28. This unit uses an air scour as well as a filtered water backflush to clean the membranes along the outside of the membrane. This air scour, which is independent of the

Table 4 Zenon water quality

	Turbidity (*ntu*)		UV254 (cm^{-1})		Apparent Color (*scu*)		Iron (*mg/l*)		Manganese (*mg/l*)	
	Permeate	Con.*	Permeate	Con.*	Permeate	Con.*	Permeate	Con.*	Permeate	Con.*
Minimum	0.025	8.7	0.001	0.002	0	0	0.00	0.04	0.01	0.12
10th Percentile	0.040	12.3	0.002	0.003	0	2	0.00	0.06	0.01	0.21
25th Percentile	0.051	13.6	0.006	0.014	0	7	0.00	0.14	0.02	0.39
50th Percentile (median)	0.056	33.3	0.011	0.027	0	232	0.01	0.28	0.02	0.77
75th Percentile	0.063	53.8	0.021	0.046	0	500	0.01	0.41	0.04	0.77
90th Percentile	0.196	85.7	0.051	0.316	5	695	0.02	2.53	0.30	0.91
Maximum	0.382	88	0.099	0.984	6	1,104	0.04	2.71	0.77	1.24
N	14	15	15	16	15	15	15	15	15	15

*Concentrate

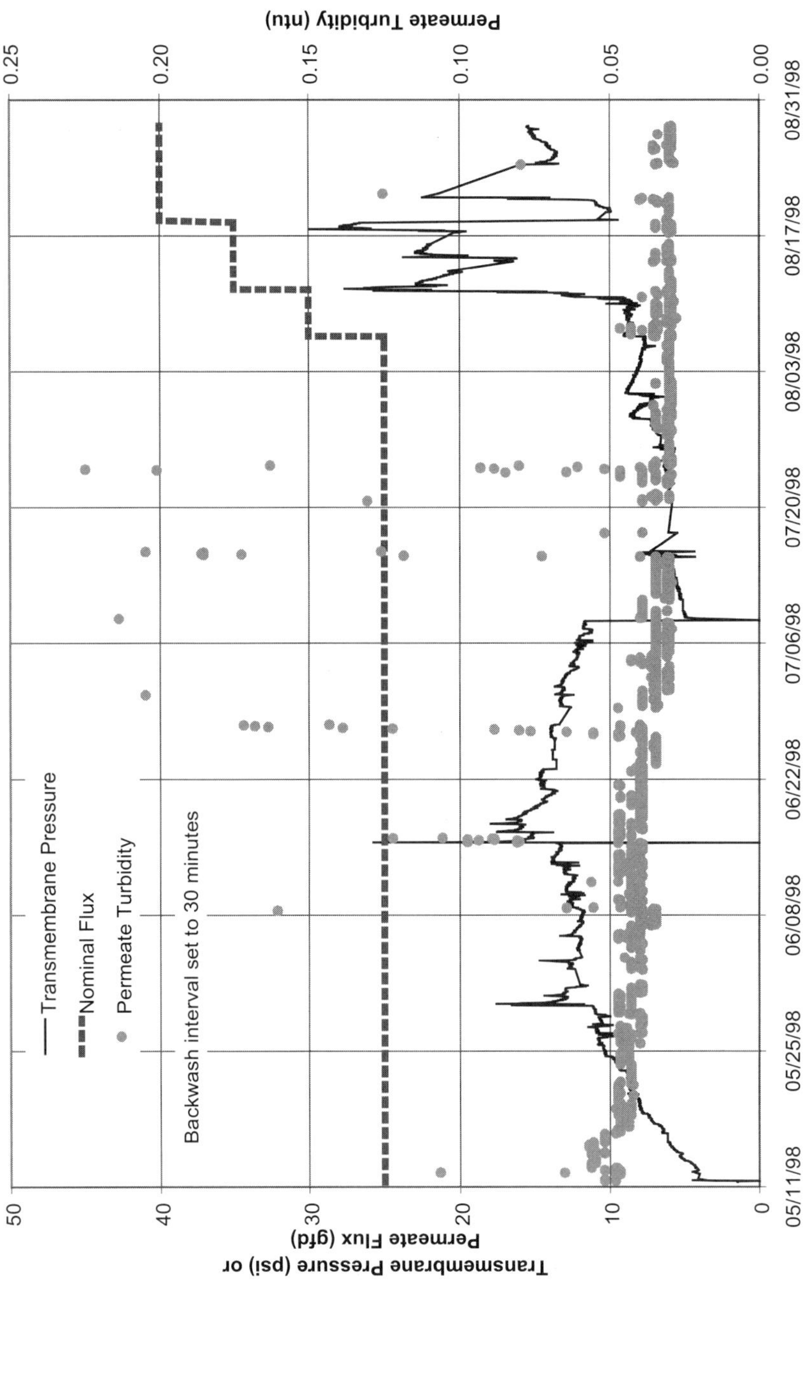

Figure 5 Pall microfilter operating parameters

backwash cycle, was set at an interval of 60 minutes, while the backwash interval was set at 30 minutes. The unit operates with a continuous flow (crossflow) of feedwater across the outside of the membrane. This crossflow is set at 10% of the filtered flow. The crossflow is directed back into the feed tank and is eventually filtered. The Pall unit operated at 5 gpm from July 1 to August 5, corresponding to a flux of 25 gfd. When the backwash water wastes are incorporated, the overall recovery is 97% for the 25-gfd flux. The Pall unit doses chlorine during backwash to prevent biogrowth. The dosage rate is adjusted until 10 to 15 mg/l of free chlorine is achieved in the permeate.

Additionally, the Pall unit had an 800-micron bag filter prior to the membrane. This filter was cleaned three times, usually when the equalization basin was pumped down, and high levels of solids were applied to the membrane.

Pall attempted to increase the flux through the membrane, but there was not enough feedwater supplied by the Millwood pumps to do this. Therefore, a booster pump was installed on August 10 to increase the flow to the membrane feed tank. The flow through the membrane was then increased gradually to 6 gpm, then 7 gpm and then finally to 8 gpm (41 gfd flux) from August 18 to August 28. At this flow rate, the overall recovery is 98%.

The TMP varied both with increasing fouling of the membrane and with the inconsistent feedwater quality. During normal plant operations, with feedwater spikes during filter backwashes, the Pall unit TMP increased at a steady rate of approximately 0.013 to 0.021 psi/hr. At a flux rate of 25 gfd, the anticipated cleaning interval is 8 to 9 weeks. During the times when the Millwood plant pumped down their equalization basin and water with turbidity greater than 100 ntu was fed to the machine, the TMP values increased significantly to 26 psi as the membrane clogged. The TMP values slowly improved upon normal backwashing when lower-turbidity water was fed afterwards to the unit, and prior values were restored.

The unit was chemically cleaned on August 5 before increasing the flux. The unit was cleaned again on August 18 because of high

TMP values. Chemical cleaning consisted of a citric acid solution followed by a caustic solution. After the chemical cleaning, the flux rate was increased to 30 gfd and then to 40 gfd. However, due to a high solids load in the feedwater, initially, the TMP was very high and actually decreased during operation. Because of this, it is not possible to calculate a cleaning frequency at this flux rate. Based on the results of the cleaning operations, it appears as if the membrane fouling was reversible and likely caused by iron and manganese.

Pall Water Quality

Permeate

The filtered water quality is summarized in Table 5. The Pall pilot unit produced water quality suitable for drinking during all feedwater conditions. All parameters associated with particulates were removed to levels below the goals set for the main treatment plant. These include turbidity with a median value of 0.024 ntu, apparent color with a median value of 0 scu, and manganese values with a median value of 0.03 mg/l. The UV254 had a median value of 0.013 cm^{-1}, slightly less than the raw water value.

Concentrate

The wastewater from this machine (concentrate) was analyzed and results are summarized in Table 5. The concentrate from the Pall unit is high in parameters associated with particulate matter. This indicates that the membrane is effectively removing these contaminants from the backwash water stream.

MILLWOOD PLATE SETTLERS

Operations

Backwash water from the plant's backwash equalization basin is pumped to the plate settlers at 128 gpm. A polymer is normally used to enhance solids settling; however, for the duration of the

Table 5 Pall water quality

	Turbidity (*ntu*)		UV254 (cm^{-1})		Apparent Color (*scu*)		Iron (*mg/l*)		Manganese (*mg/l*)	
	Permeate	Con.*	Permeate	Con.*	Permeate	Con.*	Permeate	Con.*	Permeate	Con.*
Minimum	0.020	0.8	0.001	0.001	0	0	0.00	0.01	0.00	0.02
10th Percentile	0.020	2.7	0.003	0.003	0	2	0.00	0.01	0.01	0.08
25th Percentile	0.021	3.3	0.009	0.011	0	9	0.00	0.02	0.02	0.14
50th Percentile (median)	0.024	5.7	0.013	0.022	0	55	0.01	0.03	0.03	0.33
75th Percentile	0.028	18.6	0.016	0.027	0	133	0.01	0.24	0.05	0.77
90th Percentile	0.079	68.2	0.064	0.195	1	628	0.04	1.52	0.34	0.98
Maximum	0.170	104.0	0.065	0.532	3	1,112	0.10	2.84	0.77	1.49
N	16	15	15	16	15	15	15	15	14	15

* Concentrate

pilot test, the polymer feed was turned off to prevent fouling of the membranes. The supernatant from the plate settlers is pumped to the head of the plant. The solids collected at the bottom are pumped to lagoons.

Solids appeared to be settling out in the equalization tank. Only when a backwash occurred or when the equalization basin was pumped down did the plate settlers see relatively turbid water. There was an estimated 6 inches of solids at the bottom of the plate settlers during the study. The solids pump to the lagoon is operated only once a year.

Plate Settler Water Quality

The results of the analysis for the plate settlers supernatant are presented in Table 6. Median iron and DOC values were approximately the same as the feed to the plates. Approximately 1 ntu of turbidity was removed by the plates (2.6 to 1.6 ntu). Apparent color was removed by approximately 67%, and manganese by approximately 33%. The median UV254 value was higher than the raw water feed. However, the UV254 from the plate settler was not

Table 6 Millwood plate settler effluent water quality

	Turbidity (*ntu*)	UV254 (cm^{-1})	Apparent Color (*scu*)	Iron (*mg/l*)	Manganese (*mg/l*)
Minimum	0.55	0.013	0	0.00	0.01
10th Percentile	0.64	0.014	4	0.00	0.03
25th Percentile	0.73	0.033	8	0.01	0.07
50th Percentile (median)	1.62	0.060	13	0.02	0.15
75th Percentile	4.01	0.082	59	0.06	0.17
90th Percentile	7.87	0.110	109	0.12	0.22
Maximum	9.24	0.115	125	0.14	0.24
N	7	7	7	6	6

filtered through glass-fiber filter paper while the raw water feed was filtered. The presence of particulate matter in the plate settler supernatant is likely the cause of the higher UV254 value.

MICROBIAL CHALLENGE STUDY

A microbial challenge study on the three backwash treatment processes was performed on June 8, 1998, by Dr. James Malley of the University of New Hampshire. Inactivated *Giardia lamblia* cysts, *Cryptosporidium parvum* oocysts, and viable MS-2 bacteriophage viruses were spiked into the feedwater for the four treatment units. Permeate samples from the units were collected over a period of several hours before being processed by a cyst recovery system. Results from these tests are presented in Table 7. The results indicate that the membranes removed almost all of the cysts fed into the backwash water. In addition, the Zenon microfilter removed a significant number of viruses. Based on discussions with the manufacturer, this has been attributed to the membranes' tight pore-size distribution.

During the challenge test, the viruses and cysts were injected directly into the backwash stream. In an actual plant, any viruses that were present in the backwash water would have gotten there as part of coagulant flocs. The likelihood that free organisms would be present in backwash water is minimal. Therefore, this challenge test shows rather conservative results and actual removals will likely be greater.

Table 7 Millwood microbial challenge study log removals

Treatment System	Operating Flux (*gfd*)	*Giardia* Cysts	*Cryptosporidium* Oocysts	MS-2 Bacteriophage
Memcor	67	>4.8	4.2	0.5
Pall	27	>4.8	>4.4	1.1
Zenon	36	>4.8	>4.4	2.3

ECONOMIC ANALYSIS

A conceptual cost analysis was performed on the alternative treatment scheme for the backwash and filter-to-waste streams, as shown in Figure 2. In addition to the costs of the membrane and UV systems, cost savings due to smaller tanks in the main treatment process and reduced chemical usage were also taken. Two membrane manufacturers were contacted for conceptual capital and operational costs for a 7-mgd treatment system. Using these capital and O&M costs, a net life-cycle cost difference for the membrane-UV system was developed using the same factors as those used for the development of the life-cycle costs for the conceptual design of the main treatment plant. The costs and savings for each line item are shown in Table 9. Life-cycle costs were developed using factors of 15% for engineering and construction management, 7% for legal and administrative costs, 3% inflation, 7% interest, a discount rate of 3.9% and a 30-year period.

The cost analysis indicates that a net total present worth of $2–4 million can be expected for a 7-mgd membrane/UV system. This cost analysis is very sensitive to the membrane system cost. The two membrane manufacturers contacted varied by approximately 150% in capital cost. It should also be noted that the long-term trend in membrane costs shows marked reductions in unit costs for membrane systems. In addition, as more membrane plants are operated for longer periods of time, better estimates of membrane life can be made, thereby aiding estimates of operational costs.

CONCLUSIONS

Based on both water quality and cost, the use of microfiltration and UV light for treatment of backwash water is easily justified for the treatment of waste filter backwash water. The results of testing conducted at the Millwood Water Treatment Plant (while using Croton Aqueduct water) indicated that several microfiltration

Table 8 Recommended changes to main treatment process

Item	Primary Design Criterion	Conceptual Design	Alternative Design	Change
Design Flow (w/recycle)	–	298-mgd	291 mgd	7 mgd
Average Flow (w/recycle)	–	153.9 mgd	150.3 mgd	3.6 mgd
Rapid Mix	HDT = 46 seconds	1,568 ft^2	1,506 ft^2	62 ft^2
Flocculation	HDT = 8.7 minutes	19,680 ft^2	19,200 ft^2	480 ft^2
Dissolved Air Flotation	6 gpm/sf	34,400 ft^2	33,644 ft^2	756 ft^2
Ozonation	HDT = 30 minutes	33,120 ft^2	32,400 ft^2	720 ft^2
Filtration	12 gpm/sf	21,120 ft^2	20,592 ft^2	528 ft^2
Solids Thickening	–	6 Plate Settler/Thickeners	6 Thickeners	6 Plate Packs
Chemical Usage				
Ozone	1.6 mg/l	2,053 lb/d	2,005 lb/d	48 lb/d
Alum	15 mg/l	19,246 lb/d	18,796 lb/d	450 lb/d
Cationic Polymer	1.5 mg/l	1,925 lb/d	1,880 lb/d	45 lb/d
Flocculant Aid Polymer	0.1 mg/l	128 lb/d	125 lb/d	3 lb/d
Filter Aid Polymer	0.15 mg/l	192 lb/d	188 ld/d	4 lb/d
Caustic Soda	6.1 mg/l	7,827 lb/d	7,644 lb/d	183 lb/d
Sodium Bisulfite	1.1 mg/l	1,411 lb/d	1,376 lb/d	35 lb/d

Notes: No adjustments have been made to tank depths to prevent changes in hydraulic grade lines.

Table 9 Life-cycle cost itemization

Item	Net Capital Cost (2004)	Present Worth of Operations (1997)	Net Change in Total Present Worth	Equivalent Annual Cost
Membrane System	$8,701,000	$479,000	$7,143,000	$407,000
UV System	$303,000	$312,000	$544,000	$31,000
Concrete & Excavation	–$4,162,000	$0	–$3,188,000	–$182,000
Ozone	$0	–$192,000	–$192,000	–$11,000
Pretreatment Chemicals	$0	–$671,000	–$671,000	–$38,000
Plate Settlers	–$605,000	$0	–$463,000	–$26,000
Membrane CIP Chemicals	$0	$96,000	$96,000	$5,000
Total	$4,237,000	$24,000	$3,269,000	$186,000

Notes: The total present worth assumes 15% for engineering and construction management, 7% for legal and administrative costs, 3% inflation, 7% interest, a discount rate of 3.9% and a 30-year period.

systems were capable of producing water of very high quality. This water would then be blended with filtered water from the main treatment process prior to final chlorination. The use of such a system would prevent the recycling of pathogenic cysts in the treatment process. Any cysts that survived ozonation in the main treatment process would be removed by microfiltration. Any pathogen that passed through the membranes would be irradiated with UV light in a germicidal wavelength. Thus, three barriers to cysts exist (ozone, microfiltration, and ultraviolet light) and a high degree of removal or inactivation is achieved.

The conceptual cost analysis has shown that such a system would have a net cost of $2–4 million. This cost represents an additional $0.011 per gpd of total plant capacity. For new plants, a system of this type can be cost effective due to potential cost savings in plant footprint. For existing treatment plants, these cost savings do not exist and the retrofit costs for microfiltration and UV treatment are greatly increased.

ACKNOWLEDGEMENTS

The authors would like to thank the following people for their assistance during this project: Gerry Moerschel, David Rambo, Eduardo Amaba, Jerzy Sobocinski, David Rokjer, Leo Fontana, Keith Muller, Collette Mannion, as well as representatives from Pall Corporation, Zenon Environmental Systems, and US Filter–Memcor. Without the effort of these individuals, this work would not have been possible.

CHAPTER 18

Implementation of a 25 mgd Immersed Membrane Filtration Plant: The Olivenhain Municipal Water District Experience

Kimberly A. Thorner, Esq., Project Manager
Olivenhain Municipal Water District, Encinitas, California

Douglas P. Gillingham, PE, Design Project Manager
Boyle Engineering Corporation, San Diego, California

Steven J. Duranceau, PE, Ph.D., Director of Water Quality & Treatment
Boyle Engineering Corporation, Orlando, Florida

Ernie Kartinen, PE, Vice President
Boyle Engineering Corporation, Bakersfield, California

Customers of the Olivenhain Municipal Water District (District) receive all of their water from the Colorado River and from northern California. Both of these sources are imported over distances of hundreds of miles before reaching the Olivenhain service area. In the event of an emergency interruption of these imported supplies, the District currently has approximately seven average days of treated water storage available. However, an extended interruption of the imported water system, such as could occur in a large earthquake, would jeopardize the health and safety of the District's customers and the economy of the San Diego region.

To enhance the reliability of the District's water supply, and to further sustain the increasing demand for a rapidly growing population, the District is joining forces with the San Diego County Water Authority to produce what has come to be known as the Olivenhain Water Storage Project. Together, they will construct a 24,000-acre-foot reservoir and 308-foot roller compacted concrete (RCC) dam — the tallest dam of its kind in the United States — resulting in a two-month supply of emergency water storage for residents in the District's service area and neighboring districts. This duration of supply is designed to sustain the region's water supply following an earthquake-induced failure of the imported water system until the system can be repaired.

As part of this initiative, the District has designed and initiated the construction of a 25 mgd (million gallons per day) immersed membrane water treatment plant (WTP). This plant, the largest drinking water ultrafiltration treatment plant in the world, will utilize membrane treatment technology that provides more certain removal of waterborne health threats while also benefiting the environment through less chemical usage. The immersed (sometimes referred to as submerged) membrane water treatment process is to be supplied by Zenon Environmental, located in Ontario, Canada. The District decided upon an ultrafiltration membrane process for its water treatment plant in order to ensure high-quality water that meets and exceeds regulatory standards.

A detailed history of this project, its significance, and the steps the District has taken over the years to implement this initiative will be provided herein. Environmental and other factors used in selecting the membrane treatment process as the preferred method of treatment will be enumerated. Pilot testing programs and membrane procurement procedures used by the District and its consulting engineer, Boyle Engineering Corporation, will be highlighted, in addition to quantifying lessons learned relative to its implementation, actual construction cost information, and financial impacts resulting from the membrane procurement process. Water purveyors seeking to investigate the use of immersed membrane processes or improve their current drinking water treatment processes will find the Olivenhain experience to be enlightening and informative.

OVERVIEW

The Olivenhain Municipal Water District is located in north San Diego County and comprises an area of around 48 square miles. The District serves approximately 49,000 residents throughout five cities, two counties, and a number of small communities (see Figure 1).

All of the treated water provided by the District to its customers emanates from the Colorado River or from Northern California and is delivered through aqueduct systems operated by the State of California, the Metropolitan Water District of Southern California (MWD), and the San Diego County Water Authority. The water purchased by the District is treated by MWD at its Skinner Filtration Plant located in Riverside County and is imported into San Diego County and the District's service area. Any interruption in the delivery of this imported water from the

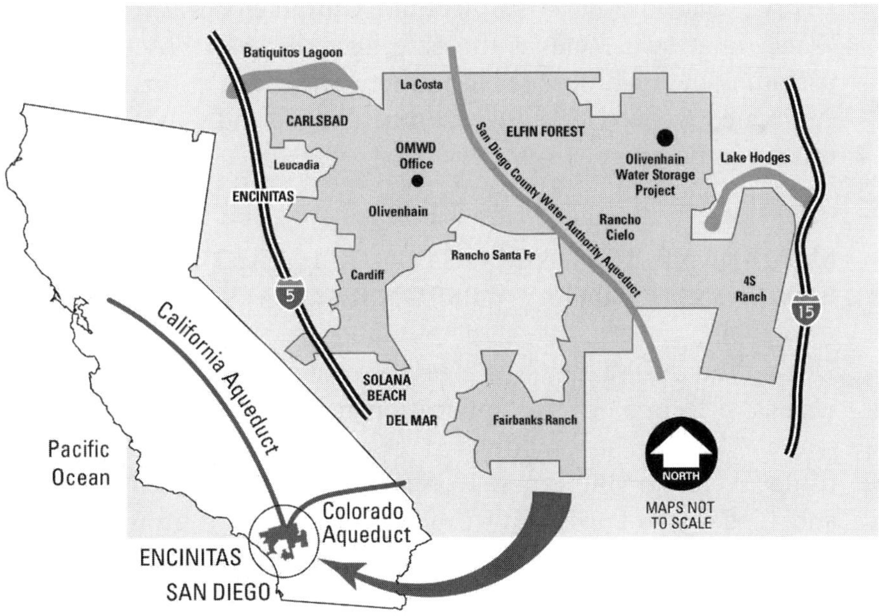

Figure 1

aqueduct would jeopardize the health and safety of the District's customers.

To minimize the possibility of water interruptions and to have water readily available in the event of an emergency, the District began to explore ways in which water could be stored and treated locally in a safe and cost-effective manner. It was anticipated that any water to be treated would be a varying blend of Colorado River water and California State Water Project water. Additionally, the possibility of treating local runoff from a nearby reservoir, Lake Hodges, was also considered. Each of these potential water sources are surface waters. Comparatively, Colorado River water tends to be quite hard and high in total dissolved solids (TDS), but relatively low in total organic carbon (TOC). California State Project water tends, on the other hand, to be relatively low in TDS, but has higher TOC levels. The local surface water from Lake Hodges is characterized by moderate TDS and high TOC levels.

A flexible water treatment technology that accommodates water of varying sources and quality was of great importance to the District. In addition to the need for compliance with the Surface Water Treatment Rule of the Safe Drinking Water Act (SDWA), providing the highest possible aesthetic water quality to its customers in a cost-effective and environmentally friendly manner was of utmost importance.

MEMBRANE TECHNOLOGY — A FLEXIBLE AND ENVIRONMENTALLY RESPONSIBLE ALTERNATIVE

During the 1970s, membrane processes emerged as a limited and expensive technology for application primarily to desalination of seawater and other saline and brackish sources using reverse osmosis (RO) membranes. However, the system-based approach and high cost of early RO processes limited the growth of this technology. The subsequent advent of spiral-wound membrane element configurations enabled competition and reduced costs.

Nanofiltration (NF) processes benefited from the spiral-wound configuration of the membranes, and the use of NF to control organic carbon and hardness has begun to grow tremendously. Membrane filtration (i.e., microfiltration and ultrafiltration) systems are now commonly used for microbial control and are becoming a standardized commodity, much like NF and RO membranes are today [1]. In fact, the membrane spectrum now encompasses a range of processes, as shown in Figure 2.

1996 WATER TREATMENT ALTERNATIVES STUDY

The District was aware that there were several types of plants that could be constructed to treat raw water in order to meet drinking water regulations. In 1996, the District requested that its consulting engineer undertake a Water Treatment Alternatives Study (1996 study) to determine which was the best type of water treatment technology for the District's needs.

The 1996 study evaluated both conventional (flocculation/sedimentation/filtration) and membrane technologies. The cost and the quality of both technologies were compared.

The findings of the 1996 study determined that a 20 mgd microfiltration membrane treatment plant could be built for costs at or below those for a conventional filtration plant. The operations and maintenance costs of a 20 mgd microfiltration membrane

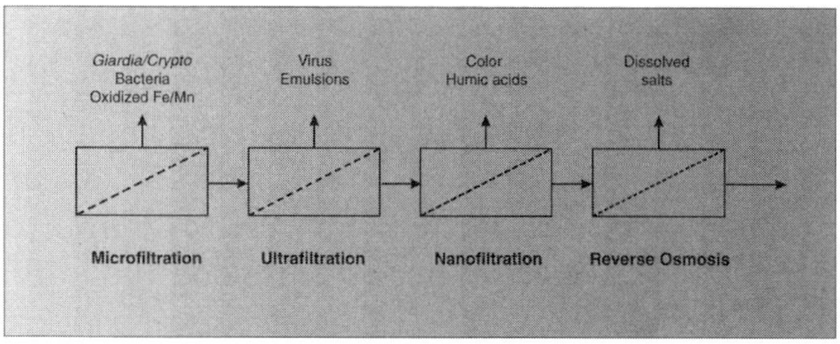

Figure 2

plant were also determined to be slightly lower than for a conventional plant of the same size.

In addition to cost issues, the 1996 study identified several other advantages of membrane systems for the Olivenhain plant. One of these was that, compared to conventional treatment, membrane-based systems provide flexibility for modular expansion in relatively small increments. This was important to Olivenhain because it allowed the District to plan to meet projected demand increases in smaller steps matched to actual needs, and with smaller capital commitments.

Environmental responsibility and compatibility with the plant's neighboring residents were also great concerns for the District. The 1996 study determined that these goals were best achieved through the use of membrane technology. A membrane filtration plant could be entirely enclosed using less land space than a conventional plant. In the affluent Rancho Santa Fe area where the District was considering building the treatment facility, the attributes of a facility enclosed within a building with no outside noises were appealing. Additionally, the ability to operate most of the time without coagulants meant that the handling of solids could be substantially reduced, another aesthetic benefit. Less sludge means fewer truck trips are required to haul waste to the landfill. This consequently results in less disruption to the environment, lower traffic impacts, and reduced labor costs.

Finally, whereas conventional filtration relies on "probability of capture" for filtration of particles, membranes act as absolute or near-absolute physical barriers. Micro- and ultrafiltration membranes are capable of achieving very high log rejections of *Giardia* and *Cryptosporidium*.

The water from which solids are removed pass through membranes during the treatment process. By varying the pore size of the membrane, microorganisms, suspended solids, and even dissolved minerals can be filtered and separated from drinking water. In operation, the feedwater passes along the outside of a hollow fiber that resembles a small-diameter straw. Particles larger than the pore size of the membrane cannot pass through the pores in the walls of the hollow fiber. Membrane filters are generally considered to be one of two types: microfilters or ultrafilters. The difference between the two types is the size of the pores in the membrane. Typically, microfilters are thought of as having a pore size of about 0.1 micron and ultrafilters a pore size of about 0.01 micron. However, these pore sizes are not "hard and fast." For example, the Zenon membrane filter to be used at Olivenhain's Water Treatment Plant has a pore size of about 0.035 micron (see Figure 3).

The 1996 study determined that a micro- or ultrafiltration membrane process treatment plant was best suited for the District's water treatment needs, with the possibility of a nanofiltration or low-pressure reverse osmosis component added at a later date downstream of the initial micro- or ultrafiltration plant to reduce hardness and total dissolved solids.

After careful consideration of the many appealing aspects of membrane technology and the findings of the 1996 Water Treatment Alternatives Study, the District decided to pursue this water

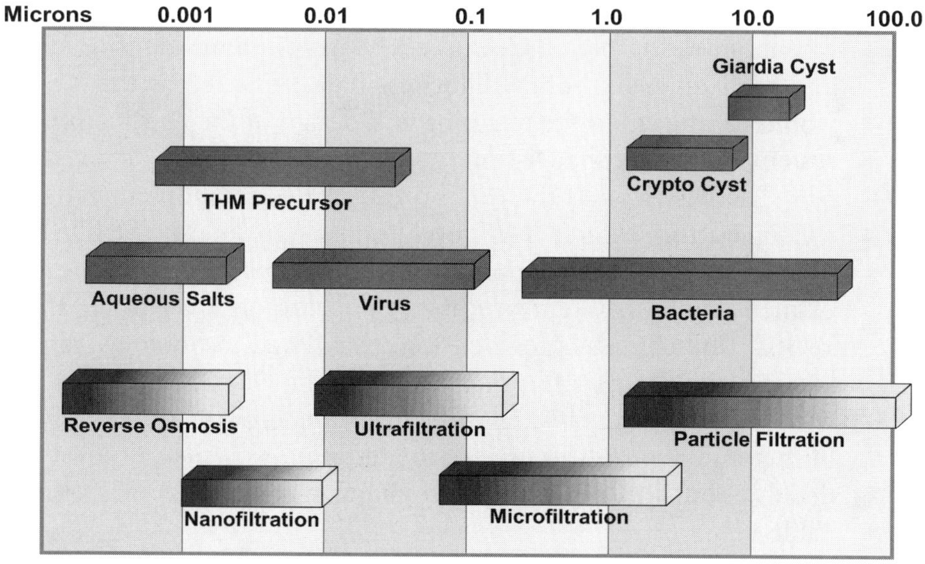

Figure 3 Filtration regimes

treatment alternative. It was determined, however, that pilot-scale test data would be required in order to evaluate treatment effectiveness of different manufacturers' membranes on Colorado River water.

PILOT TESTING AND INTEGRATED MEMBRANE SYSTEMS

The District chose to conduct the necessary testing and pilot-scale evaluations at the Water Quality Improvement Center (WQIC) in Yuma, Arizona. The WQIC is part of the U.S. Bureau of Reclamation and is a designated National Center for Water Treatment Technology. The Yuma facility proved to be a state-of-the-art testing arena with access to Colorado River water. The evaluations were conducted under the guidance and oversight of the District's consulting engineer.

The primary objective of the pilot-scale test was to examine the performance of microfiltration and ultrafiltration membrane processes for particle removal on Colorado River water. The secondary objective was to obtain performance and operation data of nanofiltration membrane processes for the removal of hardness and total dissolved solids. Pilot test data, collected at the WQIC, would be utilized in the planning and design of the new membrane system facility for the District.

The District and the U.S. Bureau of Reclamation entered into a Cooperative Research and Development Agreement (CRADA) for membrane system testing on July 29, 1997, to perform pilot testing of membrane treatment processes on Colorado River water at the WQIC. The CRADA provided for cost-shared technology transfer between the Bureau of Reclamation and the District.

The following table provides a summary of the different types of membrane processes used in the testing for particle removal and details about the specific units tested at the Yuma WQIC [2]. (see Table 1).

Table 1 Description of microfiltration and ultrafiltration membrane units

Manufacturer	Filter Type	Pore Size (*Micron*)	Membrane Area (*Square Meters*)	Membrane Type	Fiber Flow Path	Max. Flow per Mfg. (*gpm*)	Flow Tested (*gpm*)
Memcor	MF	0.2	45	poly-propylene	out-in	20	15, 18, & 20
Hydranautics	UF	0.05	25	polyether sulfone	in-out	12.5	7–9
Zenon	UF	0.03	42.7	(proprietary)	out-in	12	12
Pall	MF	0.1	52	PVDF	out-in	20	18
Aquasource	UF	0.01	14.4	cellulose acetate	in-out	8.5	7.2

PILOT TEST — SUMMARY OF FINDINGS

The five membrane pilot units were tested simultaneously, side by side, with the same source water over a several-month period. Colorado River water proved very similar in dissolved and suspended solids at the District and at the WQIC. Results of testing Colorado River water at the WQIC, therefore, represented conditions that could occur at the District. In addition, seasonal algae blooms in the canal system feeding the WQIC provided the opportunity to test membrane performance under challenging conditions. The data provided a solid basis for comparison of operation of microfiltration and ultrafiltration under variable conditions. This is significant since the data indicate that turbidities of less than 0.1 NTU can be achieved with membrane processes from a variable surface water supply under any feedwater scenario. Based on the data collected, it was also evident that microfiltration and ultrafiltration membrane processes can provide effective particulate removal.

Testing also showed that pH adjustment (to 7.0) of Colorado River water was necessary to prevent calcium carbonate scaling in piping and instruments, and a marked decrease in the performance of some of the membrane filtration units. In most cases, these units also performed best with feedwaters that were disinfected (0.5 mg/L chloramine residual) to avoid biological fouling. Last, the quality of the filtrate from the microfiltration and ultrafiltration processes was such that it could be used as feed to nanofiltration units. Water quality data confirmed that nanofiltration is very effective in removing hardness, total dissolved solids, and natural organic material (disinfection by-product precursor material) from Colorado River water that has been pretreated with microfiltration.

The ultimate disposition of the testing confirmed the District's decision to initially construct a micro- or ultrafiltration plant, followed at a later date by nanofiltration membranes downstream of the ultrafiltration plant. This setup of ultrafiltration followed by nanofiltration is referred to in the membrane industry as an "Integrated Membrane System" (IMS). The WQIC review of the various membrane technologies made it clear that the use of an IMS would be the best approach for the District, primarily due to its superior particle removal and hardness reduction capability. The ultrafiltration element of the treatment plant will allow the District to serve potable water to its customers. Subsequently, however, the nanofiltration element will allow the District to remove dissolved solids that cannot be removed by ultrafiltration, including salts, dissolved organics, and possible trace industrial organics such as MTBE.

MEMBRANE SYSTEMS SUPPLY PRE-PROCUREMENT AND DHS CERTIFICATION

The WQIC membrane pilot testing was completed in late 1998. At the same time, the District's consulting engineer completed the preliminary design of the membrane filtration plant. The District decided to undertake a membrane procurement process before proceeding with the final design of the treatment plant, thereby allowing the detailed design to conform to the specific membrane system selected and allowing the manufacturer of the membranes to participate in design review throughout the final design phase.

As membranes had become more affordable over the years, the District realized that it could purchase 25 mgd of membranes for the same price it had budgeted for 20 mgd only a few years earlier; hence, the plant was upsized another 5 mgd to 25 mgd.

The District and Boyle Engineering undertook a six-month membrane pre-procurement process. All the membrane manufacturers that had pilot tested at the WQIC were invited to submit initial statements of qualification for the procurement of 25 mgd of membranes and membrane processes. Financial stability, previous experience, and ability to provide the membranes and operate the plant, among other factors, were considered by the District during the initial qualification process. As a result, five manufacturers were invited to submit sealed proposals for the supply of membrane filtration equipment and operation of the water treatment plant. The District combined the contract for membrane supply and operation for the first one to five years into one contract, thereby achieving a single source for problem resolution during the initial operation of the plant and eliminating finger-pointing by the operator and the manufacturer in the event difficulties arose. The membrane manufacturer would now also be the operator of the plant.

Any membrane manufacturer that wanted to be considered by the District for use in its plant was required to be certified for water treatment use in California by the California Department of Health Services (CDHS). Only two of the membrane manufacturers were certified by CDHS at the commencement of the procurement process. The District sponsored all of the remaining manufacturers for CDHS certification. A certification protocol written by the District's consulting engineer and approved by CDHS was implemented at the City of San Diego's AQUA2000 Research Facility in San Pasqual, California. As a result, three new membrane manufacturers were certified in California and one manufacturer was upgraded in its certification. All shortlisted manufacturers were then eligible to participate in the proposal process.

The request for proposals (RFP) included the usual cost proposal forms, insurance, and bonding forms. In addition, the District included a copy of the comprehensive supply and operations contract that it would require the selected membrane manufacturer to enter. Any membrane manufacturer that took exception to the contract was requested to do so in writing during the proposal process. By including the future contract, all parties were made aware of the full extent of work that the District was requesting.

Manufacturers were required to submit guarantee information on chemical and electrical consumption of their processes as well as warranty information. Knowing how long a manufacturer would warranty its membranes was important to the District. Also, the District wanted guarantees that were bonded by the manufacturers as to how much electricity and chemicals would be consumed by their respective membrane systems. In addition, manufacturers were required to provide a guaranteed maximum annual membrane replacement cost ($/year) for 11 years.

Additionally, proposal instructions were set forth in the RFP that included details on items manufacturers needed to supply in their sealed bids. One of the most important elements that the District allowed for in its RFP process was *flexibility.* Several of the membrane manufacturers were capable of operating with high pressure, while others operated at medium to low pressure, and one operated under a vacuum scenario. Knowing that the high head was available from the 308-foot-tall dam upstream of the treatment plant, the District realized that the membrane manufacturer that could most effectively utilize the head would be given an edge in the life-cycle cost comparisons. The District set forth the available hydraulic gradient, the feedwater quality parameters, and the technical requirements expected of the plant. With those guidelines set, the membrane manufacturers had the flexibility to submit any scenario they could envision for the plant. In fact, membrane manufacturers were encouraged to be creative with their proposals during the District's pre-bid meeting.

The District formed a proposal evaluation panel consisting of key District staff, Boyle Design Engineers, Dr. Gregg Leslie of the Orange County Water Factory 21, and Paul McAleese of the Yuma WQIC. Of the five membrane manufacturers that submitted sealed proposals, four were shortlisted for interviews. The proposal evaluation panel traveled over a four-day period to the manufacturing facilities of the four shortlisted firms and conducted on-site interviews.

The panel unanimously concluded that Zenon Environmental of Canada was the preferred supplier for the District. Zenon's immersed membrane vacuum systems proposal eliminated the need for an upstream pump station as part of the treatment facility, thereby eliminating these costs. Energy recovery turbines were included in the Zenon proposal to utilize the available head and save energy costs for the District. Zenon's Zeeweed ultrafiltration system was certified by the CDHS with 4 log removal on *Cryptosporidium* and *Giardia* and 2 log removal on virus. (Testing showed *Giardia* and *Cryptosporidium* removal capability in the 6 to 9 log range.)

The contract for supply and operations of the 25 mgd primary process and 3.5 mgd secondary (backwash) process membrane treatment plant was executed with Zenon in March of 1999. Jointly, the District, the District's consulting engineer, and Zenon Environmental proceeded with final design of the plant based upon the Zeeweed ultrafiltration system.

To ensure the success of this treatment plant and further substantiate the capabilities of the Zeeweed ultrafiltration membranes, the District secured a 1 mgd membrane demonstration unit from Zenon Environmental (see Figure 4). A one-year trial of the membrane demonstration unit was undertaken during which time both Colorado River water and a blend of Colorado River (treated) water and Lake Hodges (raw) water were treated. The demonstration unit's performance exceeded design expectations and the cleaning frequencies were much lower than anticipated.

Figure 4 A Zenon membrane cassette

FINAL DESIGN AND OPERATIONS REVIEW

With the membrane manufacturer participating in the final design of the plant, the District benefited from a third-party review of the plans. Ensuring that all of the owner-furnished equipment (the membrane processes) integrated seamlessly with the rest of the plant became a focal point of design review. The District hired CH2M Hill, a third-party consultant, to review the plans for the plant at both 60% and 90% design from an operations perspective. As an added benefit to the District, CH2M Hill's sister company, OMI, was the operator proposed by Zenon to run the plant. Essentially, the same company that would eventually operate the plant for the District was undertaking an operations review of the plant during design. Final design was completed in December of 1999.

BIDDING AND CONTRACTOR PREQUALIFICATION

The District took great care during the design of the plant and pre-procurement of membranes to ensure that the best possible endresult was achieved. The same philosophy was followed during the bid process of the construction of the plant. Due to the highly technical nature of the start-up and commissioning of the plant, as well as the installation of the Zenon equipment and other owner-furnished material, prequalification of bidders on this project was essential to the District. Prequalifying contractors that were allowed to bid on the treatment plant construction would ensure that the low bidder on the project was also qualified to undertake the work.

All contractors interested in bidding the construction of the plant were required to participate in the prequalification process. Minimum requirements for qualification included the previous completion of a water or wastewater plant with a capacity of 15 mgd and with a construction cost of at least $7 million. Failure to meet this minimum requirement was mandatory grounds for rejection of a bidder. Contractors that met the minimum requirement were then scored in prepublished categories such as financial stability, treatment plant construction experience, safety records, previous claims, litigation, change order approval rate, equipment available, quality control, etc. Any overall score that fell below a predetermined minimum score was disqualified and not allowed to bid. Three of the 13 contractors that participated in the pre-qualification process were disqualified.

The end result of this contractor prequalification process was beneficial to the District. By allowing only qualified contractors to bid the construction, each contractor knew that it was bidding against other contractors with similar experience and resources. Hence, all contractors sharpened their pencils in the bidding process, leading to more competitive bids.

JW Contracting of California was the successful low bidder and construction was commenced in June of 2000.

COST ESTIMATE AND FEDERAL SUPPORT OF THE PROJECT

The total cost of the ultrafiltration treatment plant is budgeted at $25 million. The membrane process supply contract to Zenon Environmental is $11 million and the construction contract awarded to JW Contracting is $14 million. In recognition of the value of this project, and in order to aid the District in this monumental effort, the Environmental Protection Agency (EPA) approved a $1 million grant in 1998 and an additional $3 million grant in 1999 that the District utilized to procure ultrafiltration membranes. An additional $2.6 million was awarded in 2000. The District is optimistic about future EPA support that will allow the District to incorporate nanofiltration membrane treatment, thereby eliminating waterborne health threats that currently survive through more conventional treatment processes.

CONSTRUCTION COMMENCEMENT

Construction commenced on the Olivenhain Membrane Water Treatment Plant in June of 2000. The basins that will hold the immersed Zeeweed membranes were the first concrete portion of the treatment plant to be poured and can be seen in Figure 5.

As of the submittal date of this paper, the construction of the Olivenhain Water Treatment Plant is 20% complete. All Zenon process equipment was delivered to the site in December 2000. While the District was excited about watching the foundation literally being poured on the building, it had, in fact, been laying the foundation for many years. Through careful and deliberate consideration and effort over an extended period of time, the District was able to select the best treatment technology for its needs, procure the preferred membranes, design a superior plant, and construct the plant using a qualified and skilled contractor. Because of all of the work done at the outset, the construction of the treatment plant is proceeding on schedule, within budget, and without incident.

THE OLIVENHAIN EXPERIENCE AND MEMBRANE TECHNOLOGY

Today, pathogens are reduced in untreated water by either removal or inactivation via disinfection. Membrane technology points to a time when removal capabilities will significantly outweigh the need for inactivation, thereby completely changing the practice of disinfection. Membranes also promise to transform our use of chemicals. Coagulants and oxidants will become less important as ways are identified to produce water without them. The Olivenhain Municipal Water District is poised to lead this change in water treatment by utilizing membrane technology in the Olivenhain Water Treatment Plant.

The state-of-the-science treatment plant — currently under construction with a completion date of November 21, 2001 — will have the capability and flexibility of treating raw water from

Figure 5 The first concrete pour on the Olivenhain Water Treatment Plant

various sources. With the treatment plant built at the foot of a 308-foot-tall dam and by drawing water directly from the reservoir, the available hydraulic gradient from the dam will be converted to energy via the use of turbines in the treatment plant. This energy will meet up to approximately one-half of the plant's power needs, resulting in significant savings to the District in terms of electrical operating costs. Ancillary projects, including an electrical substation, pump station, and flow control facility, will work in tandem with the treatment plant to ensure that all elements of the Olivenhain Water Storage Project come together to form a first-class water facility, delivering a safe, continual water supply to all.

Figure 6 Rendering of the Olivenhain Water Treatment Plant

When the Olivenhain plant goes online late this year, it will be the world's largest ultrafiltration drinking water treatment plant. Although this status is likely to be short-lived, as even larger membrane filtration plants now in the planning and design stages come to fruition, the District will be proud to hold this title and to have played a role in the continuing development of this new treatment technology. The District is hopeful that its experiences will benefit others considering membrane filtration projects.

ACKNOWLEDGEMENTS

The authors wish to acknowledge the OMWD Board of Directors for their many years of support and vision that have made this project a reality, and Naomi Sabino, Project Coordinator, for her editing skill and assistance with this paper.

REFERENCES

1. Duranceau, S.J. (2000). The Future of Membranes. *Jour. AWWA* 92(2) 70–71.
2. Duranceau, S.J., K.A. Thorner, P.D. McAleese. (1998). The Olivenhain Water Storage Project and Integrated Membrane Systems. Microfiltration II Conference, November 12–13, 1998, San Diego, California.

CHAPTER · 19

Demonstrating the Integrity of a Full-Scale Microfiltration Plant Using a Bacillus Spore Challenge Test

Peter Trimboli, Applications and Process Manager
USF Memcor, Sydney, Australia

Jim Lozier, Director of Membrane Technologies
CH2M-Hill, Tempe, Arizona, USA

Warren Johnson, R&D Project Manager
USF Memcor Research, Sydney, Australia

BACKGROUND

In 1995 New Zealand (NZ) revised its drinking water standards and included, for the first time, compliance criteria for protection against *Giardia* and *Cryptosporidium*. The new standard notes:

> *"The incidence of protozoa is reduced substantially when the water treatment coagulation/filtration process results in drinking water with a turbidity below 0.1 NTU. However, a compromise value of 0.5 NTU is used in the present standards because the value of 0.1 NTU may not be presently attainable by many New Zealand drinking water supplies. The value of 0.1 NTU remains the target"* [1].

Consequently, to ensure compliance with current and future standards, Tauranga District Council (TDC) specified the new Joyce Rd Water Processing Plant (JRWPP) must be capable of meeting a 0.1 NTU turbidity standard. The NZ standards also state that the *Cryptosporidium* compliance criterion for a treatment plant without coagulation is that:

"All drinking-water leaving the treatment plant passes through a filter which removes all particles larger than 5 microns. The water supplier must be able to demonstrate that the plant is operating within specification."

Within the USA, *Cryptosporidium* oocysts are often considered to have a nominal size range of 2–5 microns. Again, to ensure compliance with such a removal capability, TDC specified the treatment plant must be able to demonstrate complete removal of all particles *1 micron* and larger. The contract further defined *complete removal* as log 5 rejection of such particles, and specified a bacillus spore challenge test to demonstrate this capability.

USF Memcor was the successful tenderer, and designed, installed, and commissioned a 36 ML/d (9.5 USMGD) microfiltration plant. The plant consists of 10 USF Memcor 90M10C Continuous Microfiltration (CMF) units. The plant is fed by gravity and includes the ability to dose an aluminium salt for colour removal and reprocess backwash water. These features are well detailed by others [2, 3], and at the time of the Challenge Test the plant was the largest operating USF Memcor CMF plant in the world. To the authors' knowledge, a bacillus spore challenge test had never been carried out on a full-scale operating water treatment facility. Carrying out such a test was in itself a significant challenge. The preparation, protocol, and the results of the test are the topic of this chapter.

THE CHALLENGE OF THE TEST

Bacillus spores have been used previously to investigate the integrity of a microfiltration unit [4] and are ideally suited to an integrity test of a water system, as:

- They maintain a stable population, as very specific conditions are required for the spores to germinate.
- They are robust, and will resist the rigours of dosing and a water treatment plant feed system.

- They are resistant to chlorine.
- They are resistant to heat, allowing the water samples to be pasteurised in the laboratory to minimise interference.
- They often occur naturally, and pose little hazard to public health.

Preliminary tests on the raw water showed very low spore counts (<20/L), hence the test will require spores to be generated. *Bacillus subtilis* has been used previously in integrity studies; spores of this species have a size range 0.5–0.8 microns [4]. This is below the specification requirement and close to the nominal pore size of the USF Memcor polypropylene membrane (0.2 micron). To provide a better match for the Joyce Rd plant, *Bacillus megaterium* (ATCC No. 14581) was selected as it was thought their spores were close to the 1 micron removal level stipulated in the plant specification.

Bacillus spore challenge tests have been carried out on a pilot scale previously, but not on a full-scale operating water treatment plant. Performing such a test on the JRWPP posed significant challenges, including:

- The treatment plant was to be providing finished water for potable water distribution during the test, hence approval from the health authorities was needed.
- A very large quantity of spores had to be generated to meet the log dose requirement at the plant flowrate.
- Operation of the secondary MF units reprocessing backwash water had to be inactivated.
- Backwash water produced by the MF during challenge testing required treatment to inactivate the retained spores prior to its discharge to a public waterway feeding a downstream water treatment plant. Again, regulatory approval was required and a protocol developed to inactivate rejected spores before discharge.

- The plant was at the time of the test the largest operating USF Memcor CMF plant in the world, and the results would be public information.
- Dosing the spores into the feed main, which is a gravity supply at high pressure (up to 1,040 kPa [150 psi]).
- Maintaining aseptic sampling techniques at multiple locations in a short time period.
- Modifying plant control logic to allow operation in "manual" mode.

PREPARING FOR THE CHALLENGE TEST

The Challenge Test was originally scheduled for March 1998. However, while preparing for the test it was found that:

- Insufficient spores were generated to enable a challenge at much above log 5.
- The *B. megaterium* spores were smaller than first understood.
- The integrity of one of the CMF units was unacceptable.

These findings increased substantially the chance of failing the test, and although the spores were generated and brought to site, it was decided to postpone the test. The test at the JRWPP was subsequently carried out in May 1998. The above findings and their impact will be discussed in the relevant section. USF Memcor was operating a pilot CMF unit, a 6M10C, near the JRWPP. It was located at the Oropi WTP, also owned by Tauranga District Council. The opportunity was taken to dose this plant with the spores generated for the March test. The results for this pilot challenge test are also presented here.

Spore Generation

The first step in preparing for the challenge test was to determine the number of spores required. The test needed to be of sufficient duration to be an effective challenge, while trying to minimise the number of spores required. Generating sufficient spores to challenge a 36 ML/d microfiltration plant was itself challenging. To minimise the number of spores required, key variables were set as follows:

The first attempt to generate spores yielded 2.3×10^{10} spores, which was only sufficient to challenge the plant at 1×10^5/L (based on a feedwater flow of 250 L/s, and a 1,000 mL sample size). At this marginal concentration, the detection of even one spore in the plant filtrate would result in a failed test. USF Memcor determined such a result to represent unacceptable risk of failure and subsequently established 10 spores/1,000 mL as the maximum passing concentration to satisfy the test requirement. This required a minimum of 3×10^{11} spores /L be dosed to the feedwater. With the second attempt, *600* Petri dishes were required to generate this magnitude of spores [5]! Spores were brought to the plant site for the test as a 1 litre solution containing 7.5×10^{11} spores.

Table 1 JRWPP key operating variables

Variable	Set Value	Comment
Plant Filtrate Flowrate	250 L/s (3,936 USgpm)	This is the lowest flowrate the plant will operate with all 10 CMF units filtering, to maximise the challenge dose.
Feedwater Spores Concentration	10^6/1,000mL	At a log 5 rejection, up to 10 spores can be detected in each filtrate sample and still pass the test. A large sample size of 1,000 mL limits the mass of spores required, and increases the sensitivity of the test to contaminants.
Dosing Duration	30 minutes	Spores were dosed for 30 minutes during the 60-minute test period.

Bacillus megaterium Spore Size

B. megaterium have a vegetative cell size of 2–6 microns [6]; however, the spore size was not well known. To confirm the spores were of the appropriate size for the test (i.e., less than 1 micron but greater than 0.5 micron) a size characterisation was made prior to the March attempt at the Challenge Test [5]. Fifty spores were randomly selected for examination (see Table 2).

This work showed the *B. megaterium* spore to be smaller than expected and in the size range of *B. subtilis*, with a width well below the 1 micron specification. In particular, the spores' smallest diameter was as small as 0.5 micron. This is close to the nominal pore size of the polypropylene CMF membrane (0.2 micron). When combined with the low spore number, this result increased the risk of failure, and the March test was postponed until a higher concentration of spores could be obtained.

Plant Operating System

The JRWPP is fed by gravity from a reservoir via a 4 km pipeline running at high velocity (up to 4 m/s), and made from various materials including uPVC. To limit the effect of water hammer on the raw water main, as CMF units are removed from service for backwashing, feedwater, at the same rate as that processed by an individual CMF unit, is diverted to a break tank for the duration of the backwash cycle. The break tank contents are then pumped to the CMF units for the sweep stage of the backwash. To prevent the introduction of spores into the break tank (and the subsequent need for their inactivation), no backwashing of CMF units was

Table 2 *B. megaterium* spore characteristics

Shape of spores: Generally oval with a number being completely spherical.		
Width (smallest diameter)	Average:	0.74 μm
	Range:	0.5 μm to 0.9 μm
Length (largest diameter)	Most in Range:	1.2 μm to 1.5 μm

performed during the 40-minute test. This also had the benefit of not causing changes in feedwater flow rate to the plant. Previous studies [4, 7] confirm spore and particle rejection is constant across the filtration interval (for particles larger than the pore size), hence running the plant at steady-state conditions for the Challenge Test represents normal operation.

The JRWPP is an automatic unattended facility. The plant flowrate setpoint responds to levels in the treated water reservoir. The plant can operate with between 4 and 10 CMF units filtering, at flowrates from 25 L/s (396 USgpm) to 43.5 L/s (690 USgpm) depending on reservoir level. The CMF units normally backwash after a set rise in filtration resistance. For typical water quality of less than 5 NTU, each unit may backwash every one to two hours. In order to ensure all units were filtering at a constant flowrate of 25 L/s, and not backwashing, the following changes were made to the control system:

- The plant was shut down on the morning of the test. This allowed the level in the treated water reservoir to drop so that the control system requested all units filtering. The CMF units were then converted from automatic operation to manual, to prevent them from stopping filtration as treated water reservoir levels increased.
- The plant flowrate setpoint for all treated water storage levels was changed to 250 L/s. This will set all CMF unit flowrates to 25 L/s.
- Finally, backwash of the CMF units was disabled by increasing the backwash resistance setpoint to a value that should allow the CMF unit to filter for well over an hour before requesting a backwash. In case a unit did request a backwash, the backwash pumps were manually shut down. In this case the CMF unit will continue filtering for a short while longer before shutting down in an alarm condition.
- To minimise the backwash production during periods of high feed turbidity, the JRWPP also has the facility to

reprocess backwash water using two CMF units (referred to as secondary units) that are capable of filtering either feedwater or backwash generated by the eight primary units. Again, to eliminate the need to inactivate spores in the backwash tank, backwash reprocessing was disabled, and the backwash tank was allowed to drain directly to the plant disposal tank.

Spore Dosing Arrangement

Bacillus spores were dosed into the feed using the aluminium salt dosing system. Aluminium salts are dosed to the feedwater for intermittent colour removal, during and directly following rainfall events. Concentrated aluminium chlorohydrate is dosed into a carrier water line, which pumps the solution to a jet mixer for dosing into the raw water main. The jet mixer is located approximately 200 metres upstream of the plant to provide a two-minute detention time between dosing and the microfiltration membranes at plant maximum flowrate.

The 1 litre solution of bacillus spores was made up into a 40 L dosing solution using plant filtrate. The plant design includes $2 \times 100\%$ duty/standby dosing pumps, each capable of 30 L/minute flow. The pumps' suction pipework was redirected to the 40 L container. At 60 L/minute, spores were dosed into the plant feedwater for 40 minutes. At a plant feedwater flowrate of 250 L/s and a spore concentration in the 40 L solution of 3.75×10^{10} spores/L, the theoretical challenge is 1.26×10^{6} spores per litre of plant feedwater.

Once into the raw water main, the bacillus spores must pass through a 1,000 micron strainer and a 300 mm (12 inch) pressure control valve, before entering the plant feed manifold. All 10 CMF units take feedwater directly from this manifold and produce filtrate into a common filtrate manifold.

Sample Points

Grab samples were taken at the following locations:

- **Dose tank.** This is a sample of the solution as it was dosed into the raw water main.
- **Feed.** Taken from the plant feed manifold (500 NS steel pipe). This is the key sample point to measure the concentration of spores in the feed to the CMF units.
- **Combined filtrate.** Taken from the plant filtrate manifold (500 NS steel pipe). This is the key sample point to measure the concentration of spores in the filtrate, and determine the log rejection of spores across the plant.
- **CMF unit.** Each individual CMF unit filtrate sampling point, immediately upstream of the CMF unit flow control valve. Used to detect a difference in rejection across units.

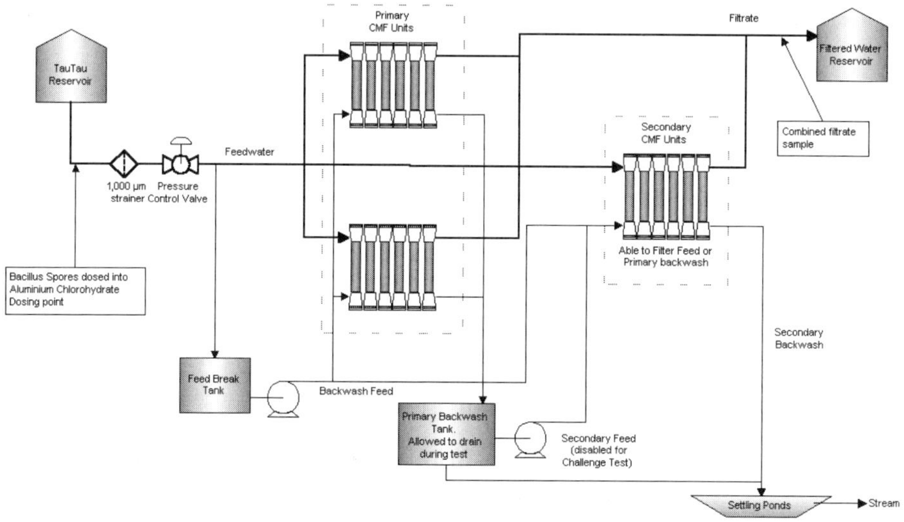

Figure 1 Joyce Rd Water Processing Plant flow schematic

All filtrate samples were 2,000 mL. Analysis was carried out using 1,000 mL of the sample. Calculations indicated it would take at least 12 minutes for the bacillus spores to reach the combined filtrate sampling point. The first set of samples was subsequently taken after 15 minutes of dosing. All sampling points were sanitised using a 70% (v/v) solution of methylated spirits in water 15 minutes prior to the commencement of the Challenge Test. The sampling point taps were then opened and adjusted to give flowrates suitable for rapid filling of sample containers. The taps were left open throughout the duration of the Challenge Test.

The JRWPP has an online particle counter on the filtrate; however, its supply is taken from the same line used for the Combined Filtrate sample point. During the test, the flow to the Combined Sample point upset the flow to the online particle counter. Consequently the particle counter results were not representative and hence not presented here. Batch particle counts of the feed, combined filtrate, and CMF unit filtrate were taken during the Challenge Test, with one batch particle sample for each spore sample, except the dose tank.

Control Samples

To minimise the potential for contamination of samples during testing, specific quality assurance measures were taken prior to and during sampling.

- Personnel access was limited during the test; only four individuals performed the sampling.
- Separate personnel were used to collect feed and filtrate samples to minimise the risk of filtrate sample contamination.
- All samplers wore surgical latex gloves, and these gloves were sanitised with 70% methylated spirits prior to taking each set of samples.
- Samplers used aseptic techniques in the handling of sample bottles and closures when taking the samples.

- All exit doors from the plant were closed, and plastic sheets were taped to the lower air ventilation screens to minimise air movement in the plant during sampling.

Notwithstanding these protections, a negative control check for aerial contamination during sampling was taken in conjunction with each combined filtrate sample. The control sample was taken by opening a 2 litre plastic container placed near the combined filtrate sampling point. Its closure was removed just prior to taking the combined filtrate sample, and replaced just after taking the sample.

System Integrity

During the week of the bacillus spore Challenge Test, USF Memcor personnel measured and confirmed the integrity of all CMF units using the Pressure Decay Test (PDT) and the Diffusive Air Flow (DAF) Tests. Both of these tests are discontinuous, that is, the CMF units must be removed from service to perform. Previous work shows these tests conservatively estimate the integrity of an operating CMF unit [8]. That is, both the PDT and DAF tests indicate a plant will achieve *at least* the level of integrity the tests estimate. Both tests rely on the principle of a membrane's bubble point.

If air is applied to one side of a completely wet membrane and the pressure slowly increased, surface tension forces initially resist the air pressure, and no air flows across the membrane. When the pressure reaches a particular level, air bubbles will be seen escaping from one or more of the pores. This pressure is known as the bubble point pressure and is related to the pore diameter [7, 8]. The bubble point of a membrane can be correlated to membrane pore size of a membrane. Table 3 compares and contrasts the PDT and DAF tests.

USF Memcor uses a PDT value of a 0.78 kPa/minute loss of filtrate side air pressure per minute (0.11 psi/minute) as the criterion for establishing the integrity the CMF units required to meet the specified log 5 spore rejection. This log reduction value approaches the limit of detection of the PDT [10]. Consequently, DAF tests were also performed on the CMF units of the JRWPP, to more accurately determine their integrity, and compare results between PDT and DAF tests.

The CMF units come equipped standard with the components (valving and controls) required for automated PDTs. No equipment is provided to conduct the DAF test. To perform DAF tests, each CMF unit was outfitted with a 15 NS ($^1/_2$ inch) manual valve connected to the lower feed side manifold. Plastic tubing, long enough to reach the top of the CMF unit membrane block, is attached to the outlet side of the valve. Once the test has commenced, the displaced liquid flow is allowed to stabilise and measured using a measuring cylinder.

Table 4 details the maintenance work carried out on each unit, and estimated integrity at the time of the test.

Table 3 Characteristics of USF Memcor integrity test methods

Test Method	PDT	DAF
Measurement Method	Filtrate side of membrane is pressurised to 80–120 kPa (12–17 psi) with the feed side of the membranes isolated. The air supply is shut off, the feed side is opened to atmosphere. As air passes across the membrane the air pressure on the filtrate side of the membrane is measured.	As per PDT, but air supply isn't shut off. The feed is vented via a small valve, and airflow across the membrane is measured by how much water the air displaces on the feed side of the membrane.
Key Characteristics	Accurate to approximately –5.2 LRV Sensitive to leaks on the filtrate side of the membrane Sensitive to temperature changes as air viscosity changes	Accurate to approx. 6–6.5 LRV, as actual membrane bypass flow is measured More expensive (extra equipment is needed) Too sensitive for LRVs below approximately 4.5

PDT results prior to the March Challenge Test attempt detected a significant loss in integrity in unit B4. The unit was tested further using a sonic test. This test was developed by USF Memcor to locate which module (or leaking valve) is the source of the integrity loss. Simply put, the sonic test hears the failure. A

Table 4 CMF unit integrity test results before/after maintenance

		Pressure Decay Test		Diffusive Air Flow Test	
CMF Unit #	CMF Unit Maintenance	Result* (*kPa drop/min*)	Estimated LRV	Result (*mL/s*)	Estimated LRV
A1	None	0.72	5.2	3.5	6.2
A2	None	0.60	5.1	NA†	–
A3	None	0.63	5.1	4.5	6.1
A4	None	0.55	5.1	4.4	6.1
A5	None	0.67	5.1	10	5.8
B1	None	0.63	5.2	5.9	6.0
B2	None	0.58	5.2	NA [†]	–
B3	None	0.6	5.1	9.5	5.8
B4	1 module repaired‡ and 2 modules isolated§	5.42/1.17	4.0/4.9	NA**/45	– /5.1
B5	6 modules repaired [‡]	1.15/1.00	4.9	NA/42	– /5.2

*PDT and DAF results are dependent on other factors apart from the kPa drop or DAF flow. These include the test pressure and CMF unit operating conditions. These factors are not as significant as the above factors, and hence have not been reported.

†CMF units A2 and B2 did not have the facility for manual DAF testing installed.

‡Modules are "repaired" by "pinning" the ends of damaged fibres, effectively isolating them.

§Unit B4 was found to have cracks in two module headpieces. The modules were isolated using the inbuilt filtrate isolation valve for the Challenge Test.

**A DAF test was not carried out on unit B4 prior to fixing the failed headpieces. The test result would have been approx. $^1/_2$ L/s flow — too high to measure.

sonic test is a manual test that follows a similar procedure to the DAF test; however, instead of measuring liquid flow (as in the DAF test) CMF modules are individually monitored by a sensitive accelerometer (microphone). The accelerometer is connected to an instrument that filters out background noise, amplifies the noise of air passing through the module, and can be heard by an operator wearing headphones. Thus, the sonic test can identify which of the modules have failed. The integrity loss in unit B4 was caused by cracks in the headpieces of two CMF module assemblies. This type of failure is extremely rare, and thus was not determined immediately. The modules were isolated using the integral filtrate isolation valves incorporated in the CMF module for this purpose.

Approvals

The JRWPP is an operating potable water treatment plant, hence approval was required for the Challenge Test. The New Zealand Ministry of Health is the regulatory authority, and they had requirements. They wanted an assurance that the bacillus spores were not harmful and, notwithstanding that, an inactivation protocol for the rejected spores. The assurance the spores were not harmful was provided by the client's consulting engineers. The New Zealand Ministry of Health also consulted their own advisors. The rejected spores inactivation protocol was to maintain the rejected spores in the plant settling ponds at a free chlorine residual of 2 mg/L for 24 hours. After 24 hours, the discharge of supernatant from the settling ponds could recommence. The performance of the inactivation protocol was not tested.

TEST PROTOCOL

Table 5 details the Challenge Test schedule.

SAMPLE ANALYSES

All samples were delivered to the laboratory within 24 hours of the Challenge Test, and were tested for mesophilic aerobic spore counts following the method as described in APHA, Compendium of Methods for the Microbiological Examination of Foods (Chapter 18, Mesophilic Aerobic Sporeformers) [5]. All untested portions of samples were held by the laboratory at 0° to 4°C, and all test plates from filtrate samples showing bacterial growth were retained for further identification. A 1 litre aliquot of each filtrate sample was heat treated at 80°C for 30 minutes. The large sample volume meant that they could not reach 80°C within the seven minutes

Table 5 JRWPP bacillus spore Challenge Test schedule

Time	Actions
Minus 1 hour	Open the backwash tank drain valve, and backwash all CMF units.
Minus 30 minutes	Confirm all CMF units have backwashed, and the backwash tank is empty.
Minus 15 minutes	Disable CMF backwash. Clean and alcohol swab all sample ports. Allow feed and filtrate sample points to run continuously.
0 minutes	Take initial samples Combined Filtrate Sample, Feed Sample and Dose Tank only.
5 minutes	Start dosing pumps and begin to dose spores into the feed stream.
20 minutes	Take first set of dosed samples (Combined Filtrate Sample and Feed Sample only).
30 minutes	Take second set of dosed samples (Combined Filtrate, Feed, and CMF unit samples taken).
40 minutes	Take third set of dosed samples (Combined Filtrate, Feed, and CMF unit samples taken).
45 minutes	Stop dosing pumps.
60 minutes	Take final set of samples (Combined Filtrate and Feed Sample only).

required by the method. (Typically 200 mL samples are used.) Consequently, each aliquot was placed in a boiling water bath, and mixed to prevent localised overheating. Once the water in the pilot bottle reached 80°C, the aliquots were transferred to the 80°C water bath for 30 min. They were then cooled and tested per litre by membrane filtration technique using 0.22 micron membrane filter pads.

Control samples were tested by adding 50 mL of sterile rinse solution to each control container. The closures were replaced, and the containers shaken to thoroughly rinse the inner walls of the containers and their closures. The rinse solutions were heat treated and tested as above. Laboratory quality control included monitoring of all heat treatment procedures. All mesophilic aerobic spore counts included appropriate positive and negative culture controls and media controls.

JRWPP CHALLENGE TEST RESULTS

Spore Counts

Table 6 summarises the spore counts for each of the sample times.

Particle Counts

The particle counters used during the study had a lower detection limit of 2 microns. Unexpectedly, bacillus spore dosing into the feed did not affect the particle count (Met One batch particle counter, model WGS 267), as shown in Figure 2. Combined filtrate particle counts are also shown, and as expected, were not affected by spore dosing.

OROPI WTP PILOT UNIT CHALLENGE TEST

The spore sample produced for the March attempt was used for a challenge test on the 6M10C pilot unit located at the Oropi WTP.

No maintenance work was carried out on the unit. It was dosed with the bacillus spores in an "as is" state. It had been operating continuously on the feedwater to the Oropi WTP for approximately five months. The integrity of the unit was confirmed immediately after the test, again using the integral PDT, and a manual DAF test (Table 7).

Table 6 JRWPP bacillus spore Challenge Test results

Time (minutes)	0	15	25	35	60
Dose tank concentration	6.4×10^9	–	–	–	–
Plant feed spore count (cfu/1,000 mL)	<1 per mL	8.7×10^5	1.1×10^6	8.3×10^5	5.0×10^3
Combined filtrate (cfu/1,000mL) No. Samples = 5	2*	<1	5†	<1	<1
Combined filtrate *B. megaterium* spore count (cfu/1,000 mL)	<1	<1	<1	<1	<1
Control Samples *B. megaterium* spore count (cfu/container)	<1	<1	<1‡	<1	<1
B. megaterium log rejection		>5.9	>6.0	>5.9	>3.7§
Each CMF unit filtrate *B. megaterium* (cfu/1,000 mL) No. Samples = 20			All <1**	All <1 [**]	

*The 2 counts were gram-positive cocci.

†The 5 counts were gram-positive bacilli, identified as *Bacillus brevis*.

‡Two colonies were found. They were gram-positive cocci.

§At time = 60 minutes spore dosing into the feed had already ceased reducing feed concentration and hence calculated rejection.

**Various colonies were found, including gram-negative bacilli, gram-positive cocci, and gram-positive bacilli.

††All CMF units were operating at 25 L/s (0.4 USgpm/em) with a TMP between 39 and 54 kPa at a feedwater temperature of 12°C.

The Challenge Test on the Oropi CMF unit, being a self-contained pilot unit, was much simpler. The 1 litre spore suspension was simply added directly to the CMF unit break tank. Once mixed, the break tank contents were filtered. Table 8 summarises the Challenge Test results.

DISCUSSION

Both Challenge Tests showed a complete rejection of *B. megaterium* spores, with a subsequent log removal of >5.9 at JRWPP, and >5.5 at the Oropi pilot. Low levels of bacillus spores were found in some of the filtrate samples of the test at the JRWPP; however, they were not *B. megaterium* (see Table 6). The significant rejection of

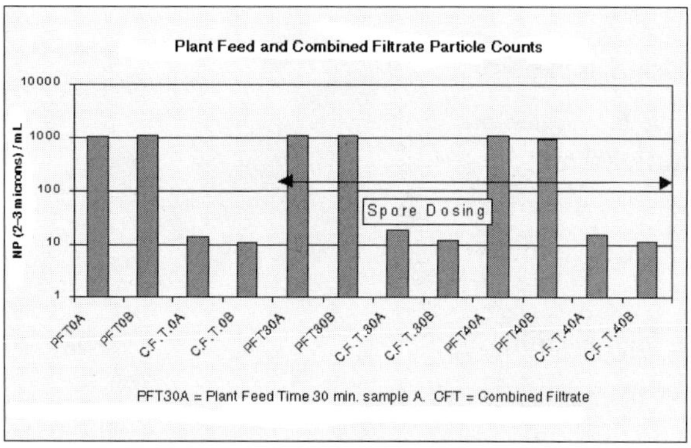

Figure 2 JRWPP bacillus spore challenge test

Table 7 Oropi 6M10C CMF pilot unit integrity test results

Pressure Decay Test		Diffusive Air Flow Test	
kPa drop per min	Estimated LRV	mL/s	Estimated LRV
1.7	4.8	5	4.9

B. megaterium makes it extremely improbable that they passed through the microfiltration membrane [5]. Possible sources of contamination were:

- The plant filtrate pipework. It may have been colonised by bacilli during plant construction or operation.
- The sample membrane filtration and plating-out process. It is difficult to maintain aseptic conditions while filtering a 1,000 mL sample.

The JRWPP plant was not expected to completely reject the *B. megaterium* spores, particularly as the two secondary CMF units showed a lower level of integrity than the primary CMF units. PDT results across the plant were reasonably consistent. This is expected as the PDT result of the CMF units was near its limit of sensitivity. However, the DAF test predicted an LRV approaching 6 for the primary CMF units, versus 5 for the secondary CMF units. The integrity test indicated an integrity level of LRV 5, but the plant(s) successfully rejected the spores at a log 6 challenge. The sample size was small; however, this result confirms other work showing the USF Memcor integrity testing methods conservatively estimate the

Table 8 Oropi 6M10C pilot unit bacillus Challenge Test results

Sample #	1	2	3	4	5
Spore suspension concentration	2.3×10^{10} / 1,000 mL				
Break tank spore concentration*	3.5×10^{5} / 100 mL				
Filtrate *B. megaterium* spore count (cfu/100† mL)	<1	<1	<1	<1	<1
B. megaterium log rejection	>5.5				

*The break tank volume is 380 L, which should give a theoretical spore count of 6.1×10^{6}/100 mL. The break tank contents were mixed by recirculating through the CMF unit. During this procedure the CMF unit filtrate valve was inadvertently left open, resulting in a number of the spores being caught on the membrane surface prior to taking the inoculated feed sample.

† 100 mL samples were used.

‡ CMF unit operating parameters were flowrate 2.9 L/s (46 USgpm), TMP 49 kPa, and feedwater temperature 13°C. Feedwater turbidity 1.2 NTU.

integrity provided by the CMF equipment. The logistics of the Challenge Test were substantial. They clearly show that while such a test may serve to demonstrate the integrity of a plant like the Joyce Rd Water Processing Plant, it is not a practical method for routine monitoring.

The results support the now substantial amount of work illustrating the excellent particle rejection properties of microfiltration membranes. In this case, they also show how those properties can be demonstrated on a pilot scale unit, and repeated on a full-scale operating water treatment plant. The work also shows how a failure of filtration components could be detected, and isolated, allowing the plant to maintain integrity. The results also illustrate the value of using aseptic sampling techniques and sterile containers. Without such precautions, it is highly unlikely that such low counts could be consistently obtained in a municipal environment.

CONCLUSION

The bacillus spore Challenge Test on the Joyce Rd Water Processing Plant illustrates:

- The effectiveness of such a technique in evaluating membrane integrity of a full scale treatment facility.
- The capability of a USF Memcor CMF plant to effectively reject particles smaller than pathogenic bacteria and protozoa of primary concern in drinking water applications.

Based on spore dosing at high concentrations the plant achieved a >log 6 rejection of *B. megaterium* spores. The test at the Oropi WTP illustrated the capability of a >log 5.5. rejection, and demonstrates the repeatability of the performance of USF Memcor CMF systems, and how pilot scale results can predict the performance of a full-scale facility.

The Challenge Test also demonstrated that where the bacillus spore levels in the feed are low, it is not practical as a monitoring technique, only serving to demonstrate integrity. Where natural bacillus spore levels exist, though, they may not be as useful as *B. megaterium* as an indicator because of their propensity to exist in the environment and represent an ambient source of contamination.

It is the goal of the Tauranga District Council to maintain the Joyce Rd Water Processing Plant in a condition that consistently provides complete removal of particles in the size range of *Cryptosporidium* and *Giardia*. The plant is currently maintaining a log 5 rejection *as determined* using the PDT and DAF tests. However, as both Challenge Tests completely rejected the bacillus spores, further studies are required to develop a quantitative correlation between indirect integrity test methods and degree of particle removal.

REFERENCES

1. New Zealand Drinking Water Standards (1995). NZ Ministry of Health.
2. Rundle C. et al., "Membrane Filtration for Control of Organic and particulate Contaminants," proceedings of the AWWA Annual Convention, Atlanta, June 1997.
3. Butler, R. and Warne, S. "Microfiltration Plant Design for Tauranga District Council's Joyce Road WTP," NZWWA, August 1997.
4. Freeman, S. et al., "Evaluating Microfiltration Performance with Bacillus Spore, Particle Count, and Particle Index Measurements," proceedings of the AWWA Annual Conference, Toronto, June 1996.
5. Richard Nowacki Consultancy, "Bacillus megaterium Spore Challenge Test on the Joyce Road Water Treatment Plant" Tauranga, NZ, 29 May 1998, Report No. 980529.
6. Stanier, R.Y., Doudoroff, M., Adelberg E., "General Microbiology." Prentice Hall 1975.
7. Trussell, R.R., et al., "Membranes as an Alternative to Disinfection," proceedings of Microfiltration II Conference, San Diego, November 1998.
8. Johnson, W.T. "Predicting Log Removal Performance of Membrane Systems Using In-Situ Integrity Testing," proceedings of AWWA Annual Conference, Atlanta Georgia, June 1997.

9. Hoffman F. "Integrity Testing of Microfiltration Membranes." *J. Parenteral Sci. Technology* 38(4): 148 (1984).
10. USF Memcor. "System Integrity and Log Reduction Values," Process Engineering Note 04001, USF Memcor Sourcebook, US Filter.

Part 5 RO and NF Applications and Operations

Membrane Replacement in Reverse Osmosis Facilities

Selection of a Nitrate Removal Process for the City of Seymour, Texas

Desalting a High-TDS Brackish Water for Hatteras Island, North Carolina

What Are the Expected Improvements of a Distribution System by Nanofiltrated Water?

Membrane Replacement: Realizing the Benefit of Low-Pressure RO in Existing Infrastructure

First-Year Operation of the Méry-sur-Oise Membrane Facility

CHAPTER · 20

Membrane Replacement in Reverse Osmosis Facilities

Steven J. Duranceau, Ph.D., PE
Boyle Engineering Corporation, Orlando, FL

INTRODUCTION

Membrane replacement requires careful planning, and should involve computer modeling, membrane probing, membrane profiling and hydraulic evaluations to document modifications in plant performance. Although many existing plants have undergone membrane replacement, little information has been published in the literature on this component of membrane plant operation. The purpose of this chapter is to present an overview of membrane replacement, and describe two case studies to document membrane replacement. One case study describes membrane replacement at a hollow-fiber membrane facility (Sarasota, Florida), and a second case study describes membrane replacement at a spiral-wound membrane facility (Marco Island, Florida).

BACKGROUND AND OVERVIEW OF CASE STUDY FACILITIES

This chapter presents the results and findings of two different water utilities that had recently undergone membrane replacement at their reverse osmosis water treatment plants (ROWTPs). The first case study describes membrane replacement activities at a hollow-fiber membrane facility owned and operated by the City of Sarasota, Florida. The second case study summarizes membrane replacement activities at a spiral-wound membrane facility owned and operated by Florida Water Services Corporation (Florida Water) at Marco Island, Florida. This section provides an overview of each of the two facilities.

The City of Sarasota ROWTP

The City of Sarasota Water Treatment Facility is composed of two major water treatment plants: an ROWTP component and an ion-exchange (IX) component. The 12 million gallon per day (mgd) capacity WTF results from a combination of 4.5 mgd from the ROWTP and 7.5 mgd from the ion-exchange plant, of which 2.3 mgd is a blended bypass water. The City's ROWTP system is supplied by a network of eight brackish wells, averaging approximately 2,250 mg/L total dissolved solids content. The historical operating pressure of the membrane process has been approximately 390 psi. The ROWTP was placed online in 1982 and was originally configured using DuPont® B-9 polyaramid hollow-fiber membrane assemblies. The City's reverse osmosis system is composed of three separate sections or trains (lettered A through C). Each train is designed to produce 1.5 mgd of permeate from 2.0 mgd of raw water at a recovery of 75 percent and a total plant capacity of 4.5 mgd. All trains have the capability of operating independently from the others, and each train is composed of two stages. Originally, the three process trains were composed of 10-inch diameter DuPont® B-9 membranes, where 54 membrane bundles comprised the first stage, and 21 membrane bundles comprised the second stage. The primary objective of the most recent membrane replacement program was to enhance process performance using DuPont's® newest double-cartridge hollow-fiber technology for one of the City's three process trains. A flow diagram of the City's facilities is shown in Figure 1.

The Marco Island ROWTP

Florida Water is the largest investor-owned private utility in Florida, and owns and operates a 6 mgd ROWTP that treats 10,500 mg/L brackish groundwater, and a 5 mgd lime-softening

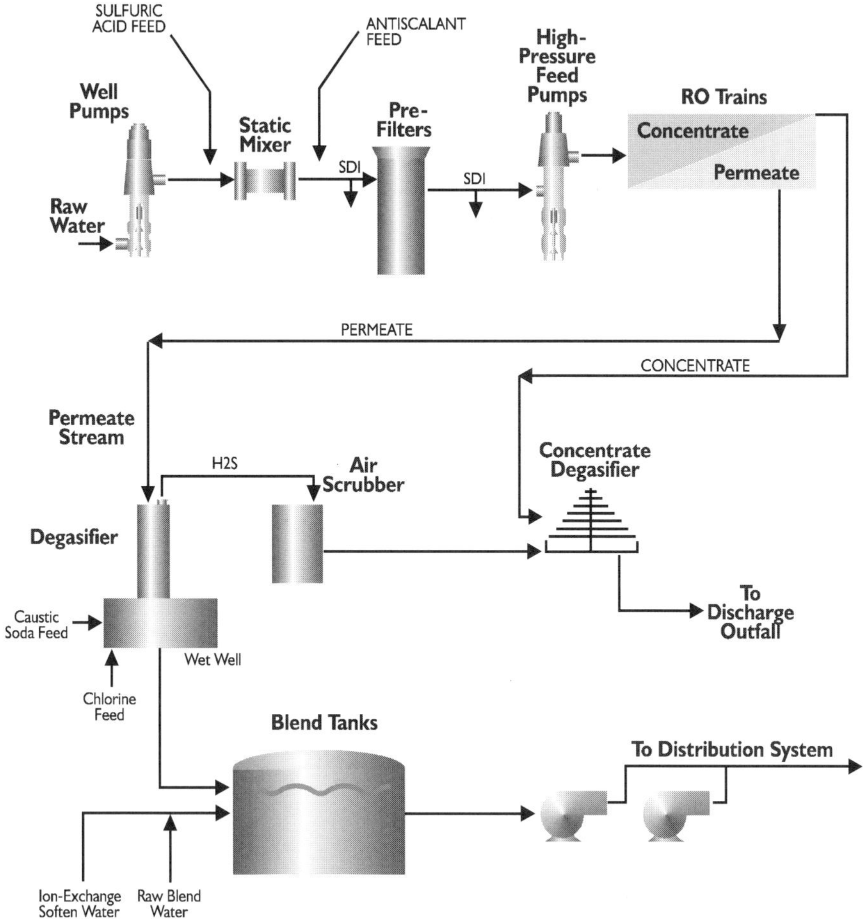

Figure 1 City of Sarasota ROWTP flow diagram

WTP that treats surface water on Marco Island. Marco Island is a popular tourist resort, having variable population changes, and is located off the southwest coast of Florida near Naples just west of the Everglades. The original membrane process arrays in Train Nos. 1–4 were designed as a 24:12 (first pass:second pass) system, with each pressure vessel (PV) containing six 8-inch-diameter, 40-inch-long membrane elements. Each of the process trains provides 1.0 mgd permeate capacity for a total of 4 mgd. The first two elements in each vessel are FilmTec® SW30-8040s and the remaining four elements are FilmTec® BW30-8040s. The second stage of each train contained 72 FilmTec® BW30-8040 brackish water membrane elements in 12 PVs. The original trains provided a total of 72,000 ft^2 of membrane surface area, and provided a water flux of 13.9 gallons per square foot per day (gsfd). Additional information about the original Marco Island ROWTP and its recent retrofit expansions can be found elsewhere in the literature [1, 2].

The ROWTP was recently expanded from 4 to 6 mgd, with the recently installed membrane skids No. 5 and 6 designed as a 22:12 (first pass:second pass) system. The 1.0 mgd permeate capacity membrane skid contained 132 Fluid Systems thin-film composite (TFC) 8822XR high-rejection brackish membrane elements in 22 pressure vessels in the first stage, and 72 Fluid Systems TFC 8822HR brackish membrane elements in 12 PVs in the second stage. Trains No. 5 and 6 were designed to provide 67,320 ft^2 of membrane surface area and to produce a water flux (F_W) of 14.8 gsfd. The primary objective of the most recent membrane replacement program at Marco Island was to enhance Train No. 1 process performance using Fluid Systems TFC membranes.

HOLLOW-FIBER MEMBRANE REPLACEMENT CASE STUDY: CITY OF SARASOTA

In August 1997, the City initiated studies into membrane fouling and reduced performance of its ROWTP operation. Investigations performed by the City and Boyle resulted in a change in scale inhibitor from polyacrylate-based scale inhibitor to sodium hexametaphosphate, which was found to have improved performance [3, 4]. In addition, based on work implemented by DuPont®, the City successfully implemented sodium hypochlorite cleaning (pH 12) to rectify the declining performance of the new permeators [5].

Train A continued to rely on standard DuPont® B-9 permeators (Bahrainian version) that had been installed in 1995.

However, in late 1997, DuPont® announced their new product line of low-pressure 8-inch-diameter cartridges [6]. The cartridges are available in double and triple internal staging configurations. As a result, the City initiated an investigation of DuPont's® newest, ultra-low-pressure double cartridge configuration, which included a side-by-side comparison to the latest spiral-wound and hollow-fiber technologies available in the marketplace [7].

In July 1999, based on the results of the double cartridge evaluation, train A was re-membraned from a 10-inch B-9 configuration to an 8-inch double cartridge configuration. Train A had to be retrofitted slightly to make this change possible. The City successfully implemented the required changes to the process train skid assembly to make Train A ready for replacement (rework stainless steel piping, recondition connection ports, other miscellaneous skid valve alterations). Table 1 provides data related to the double cartridge technology.

Water Quality Comparison

Table 2 provides a summary of water quality prior to and after membrane replacement for sulfate, total dissolved solids (TDS), chloride and conductivity. The information clearly indicates an improvement in water quality after retrofit membrane replacement had been implemented.

Table 1 DuPont® hollow-fiber double cartridge technology information

Membrane Type	HF Technology
Membrane Model	BW-L-8540, (double)
Nominal Membrane Area	4,475 ft^2 per cartridge
Nominal Permeate Flow	13 gpm per cartridge
Minimum Concentrate Flow	3 gpm
Design Feed Pressure	100–400 psi
Maximum Feed SDI	5 units
Design Recovery	75% per cartridge

Table 2 Summary of permeate water quality before and after replacement

Parameter	Before Retrofit Membrane Replacement	After Train A Was Replaced with Double Cartridges
Sulfate (mg/L)	44	3
TDS (mg/L)	450	77
Chloride (mg/L)	60	15
Conductivity (μmohs)	400–600	150

Hollow-Fiber Performance Results

Train A data represent the results of membrane replacement using the double cartridge hollow-fiber configuration. Presently, Train A operates with a feed pressure of approximately 150 psi, permeate flow of 940 gpm, and recovery of 66%. After membrane replacement, the pressure requirements had decreased from about 400 psi to 150 psi. The permeate flow increased from 850 gpm to about 940 gpm. Differential pressure, historically a problem with the original configuration, was reduced from 125 psi to less than 25 psi.

SPIRAL-WOUND MEMBRANE REPLACEMENT CASE STUDY: MARCO ISLAND

Based upon evaluation of current ROWTP operating performance and previous "probing" testing, Train No. 1 was scheduled for membrane replacement in August 1999. Overall plant performance of Train No. 1 indicated 34% loss of productivity over six years of operation. Computer software packages provided by membrane manufacturers were applied to various treatment scenarios for the ROWTP.

The original membrane Trains No. 1–4 contained a 24:12 array with six elements per pressure vessel. The first two elements of stage one were of seawater construction (FilmTec® Model SW30-80-40). The remaining positions in both stages were loaded with FilmTec®

Model BW30-80-40. These membranes in Train No. 1 were replaced with a 22:11 array using Fluid Systems® TFC 8822HR elements in the first two positions of stage one. The remaining positions in both stages were filled with Fluid Systems® TFC 8822XR membrane elements.

Membrane interconnectors, stub-tube adapters and end cap o-rings were also replaced during Train No. 1 membrane replacement. During Train No. 1 membrane replacement, lead elements of the original membranes were noted to have mechanical damage where the outer membrane layers of the elements had unraveled and extended downstream past the remaining layers. This effect is known as telescoping. It may have been caused by surges of high pressures, or higher-than-normal water velocities.

Spiral-Wound Performance Results

Table 3 shows that first-stage differential pressures averaged approximately 55 psi before the membrane replacement. After membrane replacement on Train No. 1, these pressures were lowered to 14 psi. Table 3 also presents additional information related to membrane replacement at Marco Island, providing comparison data before and after membrane replacement. Membrane productivity was estimated by temperature-corrected specific flux (Specific Flux = $Flux_{@25°C}$/ Net Driving Pressure), and the net driving pressure was corrected for osmotic pressure. Historically, Train No. 1's initial productivity level ranged from 0.125 to 0.100 gsfd/psi at 25°C in 1992 for the first stage of the system.

By June 1999, the production level was 0.076 gsfd/psi at 25°C, representing a significant decline in membrane productivity. The production level of Train No. 1 was increased from 0.078 to 0.086 gfd/psi after membrane replacement. Although the productivity was lower than the initial production, this was acceptable since an altered configuration and membrane were used. The altered configuration represented a balanced process train design that included interstage turbine control. Additional information on the interstage turbine impacts can be found elsewhere [8].

Table 3 Operating data before and after membrane replacement comparison

	Spiral-Wound Train No. 1	
Parameter	Before	After
Feed Pressure (psi)	344	330
1st/2nd Stage Pressure Drop (psi)	56/18	14/8
Total Permeate Conductivity (μmohs/cm)	621	448
Overall NaCl Percent Rejection	92.7	96.6
Water Flux$_{@25°C}$ (gfd)	12.68	13.9
Water MTC$_{@25°C}$ (gfd/psi)	0.078	0.086

SUMMARY

Although many existing plants have undergone membrane replacement, little information has been published in the literature on this component of membrane plant operation. As a result of rapid advances in membrane element film manufacturing and construction, the membrane elements initially supplied at plant start-up are often no longer available at the time of membrane replacement because either the model was discontinued or formulation improvements had been implemented during manufacturing.

Thus the newer or alternative membrane elements have to be used for membrane replacement. These newer-generation or modified membranes are thus placed into facilities that had been designed for an older model or version, which may require that newer operating conditions be employed to compensate for water transfer performance changes. Consequently, membrane replacement requires careful planning, and should involve computer modeling, membrane probing, membrane profiling and hydraulic evaluations to document modifications in plant performance.

ACKNOWLEDGEMENTS

This chapter could not have been completed without the combined support and contributions of several individuals. Special thanks go to Doug Taylor, Javier Vargas, John and Connie Morton, and the operations staff at the City of Sarasota's Public Works Utilities Division. Also, the efforts of John Losch and Ron Weis of Florida Water Services are appreciated. And last, the support of Jackie Foster, Jill Manning, Robbie Gonzalez, Will Lovins, Anke Backer and Orinthia Thomas at Boyle Engineering has been greatly appreciated.

REFERENCES

1. C.A. Kiefer, W.J. Conlon, L. Horvath and R. Terrero (1991). The Planning and Design of a RO Plant Using a Deteriorating Water Supply. *AWWA Proceedings Membrane Technologies in the Water Industry.* Orlando, FL: AWWA, March 10, pp. 149–160.
2. S.J. Duranceau et al. (1996). Coping with a variable TDS brackish water during the retrofit expansion of a desalting facility. *Proceedings of the ADA 1996 Biennial Conference Membrane and Desalting Technologies.* Monterey, CA: ADA, August 4–7, 1996.
3. S.J. Duranceau et al. (1997). Innovative Application of Off-the-Shelf Technology Pays Dividends for a Florida Utility. *Proceedings of the IDA World Congress on Desalination and Water Reuse.* International Desalination Association: Madrid, Spain, Oct. 6–9, pp. 455–468.
4. Boyle Engineering Corporation (1998). *City of Sarasota Hollow Fine-Fiber Reverse Osmosis Water Treatment Plant. Investigation, Pilot Study, and Assessment of Existing Conditions.* Public Works Utilities Division, 1750-12th Street, Sarasota, FL 34236.
5. R. Myers (1999). Biofouling Removal Techniques—Permasep Hollow-Fiber Membranes. *Proc. AWWA Membrane Technologies Conference.* Long Beach, CA, Feb. 38–Mar. 3, 1999.

6. Eckman, T.J., C.P. Shields, J.W. Strantz, and J.M. Wright (October 1997). The HF Cartridge TM—Development and Commercialization of an Entirely New RO Device/Technology. *Proceedings IDA 1997 Madrid Conference*. Madrid, Spain: International Desalting Association.
7. Duranceau, S.J., J. Foster, D. Taylor, J. Morton, J. Vargas (January 2000) Comparison of Hollow-Fiber and Spiral-Wound Ultra-Low Pressure Membrane Technologies on a Brackish Groundwater. *Florida Water Resources Journal*. Gainesville, FL: FS/AWWA; FWEA; FWPCOA.
8. Duranceau, S.J., et. al. (1999) Interstage Turbine: Innovative Use for Energy Recovery and Enhanced Water Production at a Membrane Desalting Facility. *International Desalination & Water Reuse Quarterly*, 8(4), 34–40.

CHAPTER · 21

Selection of a Nitrate Removal Process for the City of Seymour, Texas

Ernest O. Kartinen, Jr., PE
Boyle Engineering Corporation, Bakersfield, California

Christopher J. Martin, PE
Boyle Engineering Corporation, Bakersfield, California

INTRODUCTION

The City of Seymour, Texas, is located in central Texas in Baylor County 100 miles north of Abilene. The City of Seymour has a population of 3,200 people. The Baylor Water Supply Company (Baylor), which provides water service to the area around Seymour, services a population of 1,200 people. The water supply for both Seymour and Baylor is entirely from water wells.

The groundwater nitrate concentration exceeds the Maximum Contaminant Level (MCL) of 10 mg/L as nitrogen (45 mg/L as nitrate [NO_3]). In addition, the groundwater is somewhat hard and the total dissolved solids (TDS) concentration is higher than desirable. In 1994, the City authorized a study of means to reduce the nitrate concentration in the groundwater to less than the MCL. Processes that reduce hardness and TDS, as well as nitrate, were included in the study.

Providing sufficient capacity to serve both the City and Baylor was assumed in preparing the study. The present (1997) average water demand for the City and Baylor is 0.85 million gallons per day (MGD). The maximum day demand is 2.5 MGD.

Upon completion of the study in September 1994, the City began the search for a funding source. A combination grant/loan was secured from the Rural Development Administration (formerly the Farm Home Administration) in 1997. With funding in place, the City authorized design of a 3 MGD reverse osmosis treatment plant in late 1997. Design is complete and it is anticipated that the new treatment plant will be operational in late 1999 or early 2000.

TREATMENT PROCESSES CONSIDERED

Four treatment processes were considered in the 1994 study:

1. Ion exchange for nitrate removal only
2. Ion exchange for nitrate removal and softening
3. Reverse osmosis; and
4. Electrodialysis reversal

NITRATE REMOVAL BY ION EXCHANGE (IX)

Nitrate is an *anion* (negatively charged ion). It can be removed from water in an IX treatment plant by "exchanging" the nitrate anion for a more desirable (or less undesirable) anion such as chloride. While nitrate removal by anion exchange is frequently compared to softening by IX, the two processes are quite different and these differences must be accounted for if the least costly IX process for nitrate removal is to be designed. Softening by IX is accomplished by exchanging cations (positively charged ions—calcium and magnesium) for sodium typically.

An IX treatment vessel contains IX resins, which are synthetic compounds, usually in the form of small plastic beads. The "exchange capacity" of an IX resin is finite. That is, there is a finite number of exchangeable ions (anions or cations, depending on whether or not an anion or cation exchange is sought) on a ready-for-service resin. In the IX nitrate removal process, as the water being treated passes through the resin, the nitrate is exchanged for the chloride on the resin. The nitrate concentration in the water exiting the IX vessel is reduced and the chloride concentration is increased. As the treatment process proceeds, the chloride ions on the resin become fewer and fewer. Eventually, the resin must be regenerated. This is done by passing a salt (sodium chloride) solution through the "exhausted" resin. This regeneration process causes the nitrate (and other anions adsorbed during the treatment step) to be exchanged for chloride. The nitrate (and other anions) as well as the sodium are discharged to waste. The resin is then rinsed and the regenerated resin may be used to treat more water for nitrate removal.

The nitrate concentration in the treated water may be "controlled" by varying the quantity of salt used in the regeneration process. In Seymour's case, analyses indicated that a blend of about 70% treated water and 30% untreated water using about one ton of salt per MG of blended product (treated + untreated) water would be the most economical design. The quantity of wastewater produced by the regeneration process would be approximately 13,000 gallons per MG of blended product water.

Table 1 shows the estimated water qualities for an IX plant for nitrate removal only for Seymour. The figures illustrate the following points regarding IX treatment for nitrate removal:

1. The cation concentrations (calcium, magnesium, and sodium) are unchanged by the anion exchange process
2. The sulfate concentration in the water being treated is reduced to essentially zero
3. The bicarbonate concentration in the treated water is reduced slightly
4. The chloride concentration in the blended product water is almost twice that in the untreated water because the sulfate and nitrate removed are replaced by chloride
5. The TDS of the treated and blended waters is slightly lower than the TDS of the groundwater because chloride ions are lighter than the sulfate, bicarbonate, and nitrate ions that are "exchanged" from the water being treated

6. The Langelier Index of the blended water is more negative than that of the groundwater, indicating that the aggressiveness of the water is increased as a result of the changes in the concentrations of the ions in the water. The addition of 20 mg/L of caustic soda would raise the Langelier Index to about +0.02. (A negative Langelier Index indicates an aggressive water and a positive Langelier Index indicates a scaling [non-aggressive] water.)

(The treated water quality data in Table 1 are based on using a non-nitrate, selective IX resin. Nitrate-selective resins are available. However, they are generally more difficult to regenerate and more expensive to purchase than non-nitrate-selective resins. They should be considered for applications with high-sulfate-concentration waters.)

Table 1 Estimated water qualities for IX nitrate removal

Constituent	Untreated	Treated	Blend	Wastewater
Calcium	88 mg/L	88 mg/L	88 mg/L	88 mg/L
Magnesium	41	41	41	41
Sodium	110	110	110	7,800
Bicarbonate	415	351	369	4,000
Sulfate	99	0	29	5,600
Chloride	99	228	191	4,800
Nitrate (as nitrogen)	13	6	8	400
Hardness (as $CaCO_3$)	388	388	388	388
TDS	702	669	679	22,200
pH	7.1		7.0	
Langelier Index	–0.08		–0.2	

NITRATE REMOVAL AND SOFTENING BY ION EXCHANGE

For this option, IX softening (cation exchange of sodium for calcium and magnesium) would be included in the treatment plant ahead of IX nitrate removal. In this case, a hardness goal of 150 mg/L (as $CaCO_3$) and a nitrate concentration of 8 mg/L (as nitrogen) were selected. To meet these goals, analyses indicated that the blended product water should contain 63% softened water and 71% water treated for nitrate removal. It was estimated that 2.6 tons of salt per MG of blended product water would be needed for the softening process and one ton per MG would be needed for the nitrate removal process. A total of 64,000 gallons of wastewater per MG of blended product water would be produced by the regeneration processes.

Table 2 shows the estimated water qualities for an IX plant for softening and nitrate removal for Seymour.

Table 2 Estimated water qualities for IX softening and nitrate removal

Constituent	Untreated	Softened	NO_3 Removal	Blended Product	Wastewater
Calcium	88 mg/L	1 mg/L	11 mg/L	33 mg/L	1,800 mg/L
Magnesium	41	1	6	16	800
Sodium	110	286	265	220	7,400
Bicarbonate	15	415	351	369	1,800
Sulfate	99	99	0	29	2,300
Chloride	99	99	228	191	13,700
Nitrate (as nitrogen)	3	13	6	8	170
Hardness (as $CaCO_3$)	388	7	52	150	7,800
TDS	702	742	712	709	27,600
pH	7.1			7.6	
Langelier Index	–0.08			–0.7	

The figures in Table 2 illustrate the following points about IX treatment for softening and nitrate removal:

1. The hardness can be reduced to 150 mg/L (as CaCO3)
2. The sodium concentration in the blended product water would be twice that in the untreated water because the calcium and magnesium in the groundwater are replaced by sodium
3. The chloride concentration would be about twice that in the untreated water
4. As with IX treatment for nitrate removal only, a nitrate concentration in the blended product water of 8 mg/L (as nitrogen) can be obtained
5. There is little reduction in TDS because the nitrate, sulfate, and bicarbonate are exchanged for chloride and the calcium and magnesium are exchanged for sodium
6. The Langelier Index of the blended water is much more negative than that of the groundwater. Employing air stripping to remove excess carbon dioxide and the addition of about 10 mg/L of caustic soda would raise the Langelier Index to about +0.02

REVERSE OSMOSIS (RO)

Whereas IX exchanges undesirable ions for other more desirable (or less undesirable) ions, reverse osmosis (RO) separates contaminants from water rather than exchanging them for something else. RO is known as a "membrane process." That is, a membrane is used to separate contaminants from water.

The feedwater to the RO system is divided into two streams by the membrane: "permeate" and "concentrate." The water comprising the permeate stream is water that has passed through the membrane and is much lower in contaminants (such as dissolved minerals) than is the feedwater. The contaminants "rejected" by the membrane are carried out of the RO system in the concentrate stream. Since the concentrate stream flowrate is lower than the feedwater flowrate (feedwater – permeate = concentrate), the contaminants in the feedwater are concentrated in the concentrate stream.

RO is a pressure-driven membrane process. Pressure is applied to the feedwater to overcome the osmotic pressure and system hydraulic losses to produce a permeate stream. The amount of pressure needed depends on the feedwater quality, membrane characteristics, membrane flux rate, and other variables. An important consideration in RO is the "recovery." Recovery is the percentage of feedwater recovered as permeate.

While RO requires energy to create the driving pressure needed, RO does not require regeneration as does IX. Therefore, no salt is needed, although other chemicals are used to pretreat the feedwater, clean the membranes, and post-treat the permeate (or blend of RO treated and untreated water) if necessary.

In Seymour's case, and assuming a hardness goal of 150 mg/L as $CaCO_3$ (as was assumed for the IX alternative including softening), it is expected that the RO feed pumps will need to discharge at a pressure of 130 psi. The RO system recovery will be 81%—81 gallons of permeate and 19 gallons of concentrate for every 100 gallons of feedwater. The blended product water will consist of 38% untreated groundwater and 62% RO permeate. Overall system recovery (blended product water expressed as a percentage of the groundwater pumped) will be 87%. The concentrate stream is equivalent to about 144,000 gallons of concentrate (wastewater) per MG of blended product water. Table 3 shows the estimated water qualities for RO treatment.

Table 3 Estimated water qualities for RO treatment

Constituent	Untreated	Permeate	Blend	Concentrate
Calcium	88 mg/L	<1 mg/L	34 mg/L	440 mg/L
Magnesium	41	<1	16	200
Sodium	110	4	44	535
Bicarbonate	415	7	162	1,600
Sulfate	99	<1	38	840
Chloride	99	1	38	490
Nitrate (as nitrogen)	13	<1	6	62
Hardness (as $CaCO_3$)	388	3	149	1,900
TDS	702	14	275	3,700
pH	7.1	5.0	6.5	
Langelier Index	–0.08		–1.5	

The following observations about RO treatment of Seymour's groundwater may be made:

1. The concentrations of both cations and anions are substantially reduced by RO treatment.

2. The TDS of the wastewater (concentrate) from the RO process is no more than about 20% of the TDS in the wastewater streams from the IX processes.

3. However, the volume of wastewater from the RO process is substantially greater than the volume of wastewater from the IX processes.

4. The pH of the permeate will be considerably lower than the pH of the feedwater. This is because acid (and scale inhibitor) will be added to the feedwater to increase recovery. The acid lowers the pH and also produces carbon dioxide as a result of a reaction between the acid and the carbonate in the feedwater. Most of the carbon dioxide passes through the membrane and exits the RO equipment in the permeate.

5. The Langelier Index of the blended water is much more negative than that of the groundwater. Employing air stripping to remove excess carbon dioxide from the permeate and the addition of about 10 mg/L of caustic soda to the degassed permeate would raise the Langelier Index of the blended product water to about +0.02.

ELECTRODIALYSIS REVERSAL (EDR)

Electrodialysis (ED) is also a "membrane process." However, in RO the treated water (permeate) passes through the membrane, whereas in ED the dissolved solids removed from the water pass through the membrane. This difference stems from the fact that ED is an electrical charge-driven process.

An ED unit consists of a membrane "stack" with a cathode on one end of the stack and an anode on the other end. A typical stack consists of about 500 membranes with flow channels between the membranes. One-half of the membranes are made of IX anion resin and one-half are made of IX cation resin. Feedwater flows through one-half of the channels. "Concentrate" flows in the other one-half of the channels. As the feedwater flows through the stack, an electrical charge is imposed via the electrodes. The cations move toward the cathode, passing through the membranes made of cation resin. The anions move toward the anode, passing through the membranes made of anion resin. Two streams exit the stack:

- The "dilute" is the feedwater from which the cations and anions have been removed.
- The "concentrate" is the stream in which the cations and anions removed from the dilute stream have been concentrated.

A variation on the ED process is electrodialysis reversal (EDR). The difference between ED and EDR is that the electrical potential applied to the stack is reversed from time to time. The cathode becomes the anode and the anode becomes the cathode.

The ions then move in the opposite direction. This reversal reduces the tendency for precipitates to form on the membranes. Valving and piping are required to divert the water from what was the concentrate stream to the dilute stream and vice versa.

For Seymour, EDR was selected over ED for comparing potential treatment processes. At the time the report was written, ED did not have the operating history in the United States that EDR did.

In this instance, and assuming a hardness goal of 150 mg/L as $CaCO_3$, it is expected that EDR recovery will be 92%—92 gallons of dilute and 8 gallons of concentrate for every 100 gallons of feedwater. No blending of untreated and treated waters would be practiced with EDR. The concentrate stream would be about 87,000 gallons per MG of dilute (product water).

Table 4 shows the estimated water qualities for EDR treatment. The data were supplied by the manufacturer of EDR equipment.

Table 4 Estimated water qualities for EDR treatment

Constituent	Untreated	Dilute	Concentrate
Nitrate (as nitrogen)	13 mg/L	5 mg/L	114 mg/L
Hardness (as $CaCO_3$)	388	150	3,125
TDS	702	400	6,700
pH	7.1	6.8	7.0
Langelier Index	–0.08	–1.5	

The following observations about EDR treatment of Seymour's groundwater may be made:

1. The concentrations of both cations and anions are substantially reduced by EDR treatment, as evidenced by the much lower concentrations of nitrate, TDS, and hardness in the dilute as compared to the untreated water.
2. The TDS of the EDR concentrate would be nearly twice that of the RO concentrate. However, the volume would only be about 50% of that from RO.
3. The Langelier Index of the EDR dilute (product water) is much more negative than that of the groundwater. The addition of about 35 mg/L of caustic soda to the dilute would raise the Langelier Index to about +0.02.

TREATMENT PROCESS SELECTION

Table 5 summarizes the estimated costs, product water quality, and characteristics of the wastewater produced by each of the four processes considered. The capital costs in Table 5 are based on a treatment plant with a capacity of 2.5 MGD of blended product water capacity. The O&M costs are based on producing 310 million gallons per year (MGY) or an average of 0.85 MGD. The total annual costs were calculated assuming the capital cost would be repaid over 20 years at 8% interest and adding the annual O&M cost to the annual capital repayment cost. The $/1,000 gallon figures were calculated by dividing the total annual cost by 310 MGY.

As the figures in Table 5 show, the least expensive process is IX treatment for nitrate removal only. However, the hardness is not reduced and the TDS decreases only because sulfate, nitrate, and bicarbonate are replaced with chloride, which is lighter than the other anions. The chloride concentration in the product water was

Table 5 Comparison of processes

	Ion Exchange		Reverse	Electrodialysis
	Nitrate	Nitrate/Softening	Osmosis	Reversal
Estimated Costs				
Capital	$1,300,000	$1,900,000	$2,300,000	$4,100,000
Annual O&M	76,000	178,000	197,000	302,000
Total Annual	203,000	366,000	431,000	720,000
$/1,000 gallons	$0.65	$1.18	$1.39	$2.32
Product Water Quality (mg/L)				
Nitrate (as nitrogen)	8	8	6	5
Sodium	110	220	44	63
Chloride	191	191	38	56
TDS	679	709	275	400
Hardness (as $CaCO_3$)	389	150	150	150
Wastewater Characteristics				
TDS (mg/L)	22,200	27,600	3,700	6,700
% of Product Water	1.3%	6%	13%	8%

almost twice as high as in the groundwater. For these reasons, IX for nitrate removal only was not an attractive option.

IX for nitrate removal and softening was the second least expensive process considered. While the hardness goal of 150 mg/L (as $CaCO_3$) could be met, the TDS actually increases. In addition, not only would the chloride concentration in the product water be almost twice as high as in the groundwater, but the sodium concentration would also be twice as high. The concerns about high sodium concentrations adversely affecting public health also mitigated against using IX for nitrate removal and softening. Therefore, IX for nitrate removal and softening was not attractive either.

Considering that the RO product water quality is better than that anticipated from EDR (lower sodium, chloride, and TDS concentrations), as well as being less expensive than EDR, RO appeared to be the process of choice.

One final test remained. Seymour sent 25 gallons of water to Boyle's Bakersfield office to be treated. The water was passed through bench-scale RO and IX equipment to simulate the full-scale plant operation. The treated water was returned to Seymour in unmarked containers. A taste test by the City Council members and City staff decided the issue—the RO product water tasted much better than the IX product water. RO was selected as the process to be used to treat Seymour's groundwater.

DESCRIPTION OF PROPOSED REVERSE OSMOSIS TREATMENT PLANT

The plant has been designed to initially produce 3 MGD of blended product water (62% RO permeate and 38% untreated groundwater) using two RO trains. With the addition of a third RO train, the blended product water capacity will be 4.5 MGD. The major plant facilities are described below:

1. Groundwater from the wells will flow into a 100,000 gallon tank. The purpose of the tank is to provide storage ahead of the RO plant to dampen differences in flowrates between the wells and the RO plant. Provision for chlorinating the water flowing into the tank will be provided.
2. From the tank, water will flow by gravity to the suction side of the RO feedwater pumps—one for each RO train. Between the tank and the RO feedwater pumps, sulfuric acid and scale inhibitor will be added to the water. There will also be provisions for adding sodium bisulfite to the water to dechlorinate the RO feedwater if chlorine is detected in the water. It is anticipated that the RO feedwater pumps will discharge at about 130 psi. The feedwater pumps will be equipped with Variable Frequency Drives (VFD) so that the speed (and, consequently, the discharge pressure and flowrate) can be maintained to design conditions.

3. Downstream of each RO feedwater pump, there will be one 5 μm cartridge filter. The cartridge filters are intended as "guard filters" to protect the RO membranes from solids that may be in the feedwater.

4. Downstream of the cartridge filters will be the RO units. The RO equipment will consist of two stages. The concentrate from the first RO stage will be treated in the second RO stage. The permeate from both RO stages will be combined as they leave the RO equipment.

5. There will be an energy recovery device (ERD) between the first and second RO stages. Concentrate from the second RO stage will be under about 100 psi of pressure. The energy in the concentrate will be used in the ERD to increase the pressure of the first-stage concentrate as it enters the second-stage RO unit.

6. The RO permeate will flow into a degasser. In the degasser, carbon dioxide, produced by the reaction of the acid injected at the head end of the RO plant with the bicarbonate in the water, will be stripped from the permeate. The carbon dioxide will be vented to the atmosphere.

7. The degassed permeate will flow into a clearwell. Also flowing into the clearwell will be untreated well water. Sodium hypochlorite and caustic soda will be added to the untreated well water. The hypochlorite is for disinfection. The caustic soda is for pH adjustment to reduce the aggressiveness of the blended product water. The chemically treated well water and the degassed RO permeate will blend in the clearwell.

8. Low lift pumps will lift the water from the clearwell to an existing steel reservoir. An existing pump station will draw water from the reservoir and pump it into the distribution system.

9. A "clean-in place" system will be included in the RO plant to clean the RO membranes.
10. RO concentrate will be discharged through an eight-inch-diameter pipeline about 7,000 feet long into the Salt Fork of the Brazos River. The quality of the Brazos River is such that discharging the RO concentrate (about 3,700 mg/L TDS) is acceptable.
11. Sanitary wastewater and plant washdown water will be discharged into Seymour's sanitary wastewater system. The plant washdown water will pass through a "neutralization basin" loaded with limestone before it enters the sewer.
12. The RO plant will be housed in a prefabricated steel building, 75 feet by 100 feet by 14-foot eave height. Inside of the building will be:
    a. Electrical and control rooms, a laboratory, and a restroom.
    b. All of the chemical storage and feed equipment, including room for future installation of a fluoride system.
    c. Two 0.93 MGD (permeate) capacity RO trains with space for a third.
    d. The RO membrane clean-in-place system.
    e. The permeate degasser.
    f. The clearwell, which will be below grade with the degasser and the low lift pumps (to the existing reservoir) on top.

Design of Seymour's RO plant began in late 1997. It is expected that a construction contract will be awarded in early 1999 with construction being completed in late 1999 or early 2000.

CHAPTER 22

Desalting a High-TDS Brackish Water for Hatteras Island, North Carolina

Robert W. Oreskovich, Director
Dare County Water Department, Manteo, North Carolina

Ian C. Watson, PE, President
RosTek Associates, Inc., Tampa, Florida

Dare County, North Carolina, is located on the Mid-Atlantic seaboard of the United States. Given its easterly location, it was an early landfall for the 16th-century explorers, and is the site of the first English colony in the U.S. It is a county that consists primarily of water, and its land area is largely composed of barrier islands. Hatteras Island is the most southerly of the land area, and is well-known for its lighthouse, one of the largest on the East Coast.

Tourism is the basis of the economy, with some commercial fishing. The tourist industry has grown dramatically in the last 15 years, with the bulk of the growth located on the northern beaches. Hatteras Island is home to large tracts of federally owned land, so the growth has been limited to the three villages of Rodanthe-Waves-Salvo, Avon, and the southern end of the island.

Since about 1968, the residents of Avon, the southern end of Hatteras Island have had access to a potable water system. This system was originally owned and operated by the Cape Hatteras Water Association (CHWA), a user-owned cooperative utility. The source of water has been a shallow (40'–80') well field located in an area known as Buxton Woods.

The current water treatment system has been in operation since February of 2000, replacing the former 1.6 mgd system which dated from 1986. The CHWA tried in the early 1990s to expand the well field to meet future demand, and conducted several studies and pilot tests in order to bring the water quality into compliance with the then-current drinking water standards. However, the well field expansion was challenged on environmental grounds, since the well field is located in a sensitive maritime forest, and also the State determined that the well field was "groundwater under the influence." These two factors severely limited the ability of the CHWA treatment plant to meet or exceed the primary standards for drinking water quality, and to supply the quantity of water needed for the continued growth in the service area.

A 1992 study provided an overview of the treatment options available that would permit CHWA a reasonable opportunity of compliance with then-current standards, and those upcoming in the future, both known and speculative. The report covered both inadequacies of the current treatment process and process options for future compliance. It did not, however, address the question of raw water resources. And, in addition to treatment shortcomings, the transmission and distribution system was barely adequate to serve existing customers.

After imposing a moratorium on new connections, the CHWA Board of Directors attempted to find ways of meeting the challenges facing them. One of these efforts resulted in a study by Boyle Engineering Corporation* addressing the feasibility of desalting brackish groundwater. As part of this work three test wells were constructed to determine the occurrence, extent and quality of brackish water sources. The hydrogeologists, Missimer International, found high-yield limestone formations at 200–300 feet, but containing water of highly variable quality. On the basis of the third well, a treatment concept was developed. Although, by necessity, many assumptions had to be made, the treatment process presented was technically viable, with some confirmation of cost required through additional pilot testing. The key to the project was the

* *Future Water Supply Study*, Boyle Engineering Corporation, Santa Rosa, California, September 1995.

determination of both the extent and capacity of the limestone formation.

In July of 1997, the responsibility for and ownership of the CHWA potable water system passed to the Board of County Commissioners of Dare County. The system became part of the Dare Regional Water System. At this point, engineering contracts were signed with Missimer International for groundwater investigation and well field design; RosTek Associates, Inc. (formerly AEPI/RosTek, Inc.) for design and construction services for a Reverse Osmosis water treatment system; and Hobbs, Upchurch and Associates for distribution system upgrades and modernization, and for a new water treatment plant facility and structures, including ion-exchange treatment and filtration of the existing shallow raw water source.

Essentially, the project would provide a completely new water treatment plant for the residents, business and tourist facilities of South Hatteras Island, starting at the Village of Avon, and reaching south and west to the Ocracoke Ferry Terminal.

Finished water quality, before pH adjustment

Ion	IX Product	Permeate	Blend
Ca	92.00	0.90	28.23
Mg	8.80	2.00	4.04
Na	17.00	72.20	55.64
K	3.00	3.20	3.14
HCO_3	287.00	6.00	90.30
SO_4	10.00	2.70	4.89
Cl	70.00	115.20	101.64
TDS	344.00	202.20	244.74
Ca H	229.00	3.39	71.07
Total Alk	235.00	4.92	73.94
pH	7.60	5.74	6.97
CO_2	11.00	17.70	15.69
LSI	0.40	–4.10	–1.15

The Future Water Supply Study identified a treatment system that would combine the use of some of the existing shallow wells and treatment plant equipment with a totally new brackish water RO plant. The existing well water, highly colored and with variable iron content, would be treated for color/TOC reduction by anion exchange, and then oxidation and filtration for iron removal. Pilot testing confirmed both the process parameters and cost for both parts of the treatment process.*

By significantly reducing the THMFP and virtually eliminating the iron, the natural calcium hardness and the alkalinity of the shallow well water would be used advantageously by blending with RO permeate, which contains virtually no hardness or alkalinity.

As the hydro-geological studies progressed, two things became clear. First, the limestone aquifer was proving to be extremely productive, and second, the quality was very variable from site to site. Solute modeling predicted a steady increase in TDS with time, reaching a limiting value after about 15 years of about 15,000mg/l TDS! This required that the RO system be designed to operate in

Constituent	Design Water Quality (*mg/L*)	Projected Quality (*mg/L*)
Calcium	219	264
Magnesium	341	573
Potassium	78	153
Sodium	2270	4352
Strontium	8.95	10.1
Barium	0.41	0.31
Bicarbonate	272	240
Chloride	4749	8328
Fluoride	0.78	0.9
Sulfate (as SO_4)	166	787
Total Dissolved Solids (TDS)	8124	14,711
Silica as SiO_2	19.2	17.0
pH	7.20	7.40
Langelier Index	0.33	0.55
Silt Density Index	<2.0	<2.0

* *Planning a New Water Supply for Cape Hatteras, NC.* Robert W. Oreskovich et al., ADA Biennial Conference, Williamsburg VA, August 1998.

the early years at a higher recovery, with the recovery gradually declining as the feedwater TDS increased. In order to complicate the exercise even more, it was decided to incorporate energy recovery devices on each train, and to add membrane area as the TDS increased to maintain a relatively constant feedwater pressure.

As the pieces of the puzzle started to fall into place, the key decisions as to capacity, number of trains, membrane array, maximum salt passage, etc., were made. Earlier estimates and Planning Department studies showed that a build-out production capacity of 3.0 mgd would be required to meet the needs of the permanent residents, summer visitors, and fall fishermen. The shallow well field water production (to be used for blending with RO permeate after treatment) was limited to 1.0 mgd, as a result of several hydro-geologic studies prepared by CHWA during the Buxton Woods hearings. In fact, blending studies showed that a ratio of 2.33:1, permeate to shallow, would provide a well-buffered blend. This led to the selection of 2.1 mgd of RO permeate capacity, blended with 0.9 mgd of treated shallow well water. The RO was designed in three trains, with two installed initially.

As a result of these considerations, a decision was made by the County to limit shallow well production to 0.9 MGD maximum per day, and to provide for 2.1 MGD of RO capacity. It was further decided to install this capacity incrementally, with initial installed capacities of 0.6 MGD and 1.4 MGD respectively.

Selection of recovery, membrane type, and operating characteristics was very much dictated by the circumstances surrounding this project. Because the blend water from the shallow well field will have a TDS lower than 500 mg/l, with good calcium hardness and alkalinity, it was decided to use a standard-pressure high-rejection brackish water membrane for the RO portion of the treatment process. The projected permeate quality at start-up was relatively low, and as the feedwater salinity increased with time, the plant recovery would be lowered to maintain a relatively constant concentrate quality. This was fairly critical, since the concentrate discharge location is on the sound or north side of Hatteras Island at that point, and the sound water TDS averaged about 26,000 mg/l.

This became a limiting factor in the selection of the initial recovery, which turned out to be 70%. Based on the projections of future well water quality, at the limiting TDS of about 14,700 mg/l, recovery would have turned down to about 50%. Clearly this presented a compelling argument for the inclusion of energy recovery devices, which were installed in the plant.

In the design phase the decision was made to specify an RO feed pump that could accommodate both sets of conditions, initially 70% recovery, declining to 50% recovery. Further, the feed pressure to the membrane system was to remain fairly constant at around 400–425 psig. This was accomplished in the final plant configuration, with a relatively efficient vertical turbine pump, equipped with a 250 hp high-efficiency electric motor.

In selecting an Energy Recovery Turbine for this application, the primary criterion was the ability of the device to accommodate the future condition with major rework. The Pump Engineering Turbo™ was selected, with the obvious compromise that the initial operation would be less efficient than if the design point for the plant was fixed. However, using the projected rate of increase in feedwater TDS, and the corresponding change in operating conditions, the payback at the current power rate of $0.12/kwhr turned out to be about 4.5 years. If the feedwater TDS increased more rapidly (this in fact has happened!), the payback period would be shorter.

Because of the changing recovery, the decision was made to utilize a single-stage RO array, with seven elements per vessel. For the initial 70% recovery, the β-factor was less than optimal, but became acceptable at about 68% recovery. Elements of 400 square feet were used, with 16 vessels initially, increasing to 24 at final configuration. The average flux initially is 15.6 gfd, declining to a possible 10.4 gfd, if all 24 vessels are installed.

The membrane selection was left up to the system supplier, with Hydranautics CPA-3 elements being supplied.

On the shallow well-water side of the plant, the general contractor was given the choice of either supplying new vessels for the ANIX system and filters or using the vessels from the existing plant, removing them off-site for refurbishing, and reinstalling in the new building. The latter option was taken, which meant that for a period of time, a small amount of brackish wellwater had to be blended with the RO permeate to reintroduce some hardness and alkalinity. This blend could only be about 20 gpm/RO train, because of chloride limitations.

Because of the high TDS feedwater, a flushing system using permeate was included in the specification to evacuate the high-chloride water from the RO systems at shutdown. Included in this system is an ultraviolet sterilizer. After some discussion, this device was included to provide a measure of protection to the membranes in case the stored permeate became biologically active. Hopefully, this system will turn out to be an unused insurance policy, Fluid Systems, and Dow modeling software. The stored permeate, about 10,000 gallons, is also used as a chlorine-free source of water to backwash the ANIX beds and as makeup for the ANIX brining system.

Permeate flush tanks, pumps, and UV sterilizer

Anix vessels on right, sand filters on left

The RO permeate is blended with the treated shallow wellwater prior to the addition of post-treatment chemicals as the water leaves the plant building on its way to the 3.0 MG ground storage tank. Post-treatment systems include caustic soda, hydrofluosilicic acid addition, disinfection with sodium hypochlorite, and the addition of a corrosion inhibitor. Since no acid is used in the RO pretreatment step, scale control depending on a high-performance inhibitor, CO_2 stripping is not required, so there is no clear well or transfer pumping system. In fact none of Dare County's RO plants use CO_2 stripping.

The plant was installed during 1999, and started up with the usual euphoria and despair. A series of events tempered the rookie operating staff early on, including immediate leaks in pressure- and leak-tested piping systems, which always turned out to be acid, fluoride or some other unfriendly material; malfunctioning instruments; communications problems with well control systems; a blown 16" plastic pipe spool where the feedwater enters the building; and so on. When these normal but very tiresome problems were taken care of, and after several false starts, the RO units ran and made water. Two things were immediately noticeable: 1) the permeate quality was not as expected; and 2) the ERT was not providing the boost or the recovery promised. The first problem was solved by replacing the last two elements in each vessel with elements having higher chloride rejection than the CPA-3, but requiring a higher net driving pressure to produce the flow. This in turn resulted in an increase in feed pressure, which actually mitigated the ERT performance shortfall somewhat. The ERT is located on the RO feed pump discharge, and the increased boost resulted in only a small increase in pump discharge pressure.

Energy recovery turbine in foreground, RO feed pump in background

There are four wells supplying the RO plant, each having different quality and pumping rates. Therefore the combined feedwater entering the plant has variable TDS, depending on the wells in service at the time. The charts below show the various qualities available at start-up, pumping rates and available combinations for a two-train operation.

Well #	TDS, initial *mg/l*	TDS, current *mg/l*	Pumping Rate *gpm*	Well Combinations	Feed TDS, *mg/l*	Feed Conductivity, *μs*
RO-1	6,490	11,200	820	RO-2, 3	7,574	13,405
RO-2	8,687	12,760	900	RO-2, 4	6,605	11,690
RO-3	5,527	7,860	500	RO-1, 4	5,311	9,483
RO-4	3,894	6,840	700			

During well flushing and disinfection, the well water conductivity started to rise in wells RO-1 and RO-2. During this time SDI tests were conducted periodically on each of the wells, and on the combined water entering the plant. SDIs for RO-2 and RO-4 were consistently high, and soon after start-up both were inspected with a down-hole camera. It was clear from the video that the limestone in these wells was soft and interspersed with shell fragments and sand pockets. Some cavities had already developed. After consultation with the hydro-geologist, it was determined that the stress induced on the limestone surface when the wells started was causing sloughing of the material, and this was clearly seen in the video. RO-4 eventually cleaned up to the point that the SDI was fairly consistently below 4, but RO-1 was taken out of service shortly after plant start-up in February of 2000. Well control valves were installed at the wellheads, and the pumping rates were adjusted downward for wells 2 and 4. This has appeared to mitigate the problem. SDI values downstream of the 5 μ cartridge filters have been consistently lower than 2, but cartridge filter life has apparently not suffered.

The RO plant operation has been consistent since the start-up, with feedwater conductivity ranging from 13,000 μs to 21,000 μs. Permeate quality has consistently been below 500 μs. However, because of the rise in feedwater TDS, which now appears to stable, additional membrane area must be added to the plant this year, about three years ahead of schedule, and the recovery adjusted down to about 67%.

On the other side of the aisle, the shallow well system went into operation in May, and after the initial teething problems,

Control room with feed water panel in the background. Control system is the Delta V package, which is field bus based.

performed well. Color has been consistently reduced to below 5, from an average influent value of 30–80, depending on the wells in operation. Iron reduction has also been as expected, with iron in the effluent from the filters averaging 0.2 mg/l. Influent iron typically is around 1.0 mg/l. As the operators gain experience with regeneration cycles and start to optimize salt loading rates, the operation will become more consistent. Unfortunately, the ANIX system suffered from lateral failure after about four months of operation, and a significant amount of resin was carried into the filters. At writing, the repairs are being made, and the system will be put back into operation.

As yet it is not possible to accurately compile cost data for the operation. However, the estimated cost for operations, excluding debt service, was $1.25/kgal of blended water. The CHWA plant operating cost was between $5 and $6/kgal, a good part of which was the cost of hauling chemicals to the end of Hatteras Island, and hauling the water treatment plant sludge over 90 miles to a landfill. The total project was constructed for around $9 million, including wells, well housings, well site acquisition, outside piping, and the plant building and process equipment. A 3 MG ground storage tank was also included in the work, together with a separate high-service pumping system.

This was not an easy project, either to design or build. The location is remote, and the construction was interrupted in 1999 by one of the busiest hurricane seasons in many years. The hydrogeology of the limestone formation was unknown at the start, and the learning curve for predicting quality has been steep. However, the design goals of the RO portion of the plant have been met, and it is fully expected that the shallow well treatment system will be reliable and effective. The finished water quality is excellent, a vast improvement over that which was being delivered to the customers from the old treatment plant.

CHAPTER · 23

What Are the Expected Improvements of a Distribution System by Nanofiltrated Water?

Sandrine Peltier, Martine Benezet, Dominique Gatel
Vivendi-Generale-des-Eaux, Paris la Defense, France

Jacques Cavard
Sedif, Paris, France

SUMMARY

In the autumn of 1999, nanofiltration was introduced in the Méry-sur-Oise Water Treatment Plant (90 MGD), as an additional refining step downstream from flocculation, ozonation and dual-media filtration stages. NF 200 nanofiltration membranes were selected for their ability to remove NOM and atrazine. Eighty percent of the water produced by this treatment plant is nanofiltrated, with the remainder being biologically treated. Results of the full-scale data demonstrate that the distributed water quality was improved by the new treatment, with TOC at the plant outlet being lower than 1 mgC/L, total bacteria number decreased by 49% and THM by 50%, when compared with the results obtained over the same season prior to the installation of NF.

INTRODUCTION

Producing biologically stable water is difficult because finished water still contains Natural Organic Matter (NOM) and bacteria, both of which allow biofilm accumulation in distribution systems (LeChevallier *et al.* 1988, Mathieu *et al.* 1992). The Biodegradable

Dissolved Organic Carbon (BDOC) is the fraction of NOM that allows bacterial regrowth, and Servais *et al.* (1992) have shown that above a threshold of 0.1 mgC/L, bacterial regrowth can be observed. The reaction of chlorine with NOM also leads to the formation of trihalomethanes (THMs) at the end of the treatment and in the network (Sohn *et al.* 1997, Westerhoff *et al.* 2000). Consequently, scientific literature clearly recommends that the NOM concentration in treated waters be minimized, and that bacterial regrowth in distribution systems be further reduced with a resulting lower need to maintain a high chlorine residual in the system. Over and above biological treatments that have now become a standard method to reduce NOM (Bablon *et al.* 1988; Van der Kooij *et al.* 1989; Servais *et al.* 1991, Urfer *et al.* 1997), nanofiltration appears to be the most efficient way of minimizing the BDOC concentration at the plant outlet. The water produced by nanofiltration has very low NOM concentrations (DOC less than 0.2 mg C/L) and a BDOC content of under 0.1 mg C/L (Agbekodo *et al.*, 1996). The THM formation potential and the chlorine demand are also reduced (Legube *et al.*, 1995).

SEDIF chose to build a nanofiltration unit to treat the water from the Oise River, given that among the rivers in the Paris region, this is the one with the highest NOM load. A full-scale experiment was begun in Auvers-sur-Oise in 1994. The study carried out by Randon *et al.* (1995) showed that the TDC remained stable although the free chlorine residual decreased quite appreciably (from 1 $mgCl_2/L$ at the plant outlet down to 0.3 $mgCl_2/L$). Biofilm coupons were also used and revealed that the fixed bacterial biomass was reduced to near its detection limit (0.02 $\mu gC/Cm^{-2}$; Laurent *et al.*, 1999). Based on these conclusive results, SEDIF decided to commission a large-scale nanofiltration unit in Méry-sur-Oise in order to improve its total production capacity and the quality of the distributed water. This 56 MGD nanofiltration unit was started up in September 1999, providing a progressive increase of nanofiltrated water output that reached full production capacity in April 2000.

The study presented here covers the survey of bacteriological parameters such as Total Direct Counts (TDC), viable bacteria (VB), and trihalomethanes. Sampling and analysis began two years prior to the start-up of the new nanofiltration plant, in order to provide a sufficient database before the switch over to nanofiltration. This sampling will be continued for two years following the start-up of the plant to give a comprehensive assessment of the impact of nanofiltration on distributed water quality. At this point, the database for 2000 only covers the winter and spring periods, and, as a result, this chapter presents the results covering February to April for the years 1998, 1999 and 2000.

MATERIALS AND METHODS

The Méry-sur-Oise Water Treatment Plant

The Méry-sur-Oise Water Treatment Plant has two trains: the old train (nominal capacity: 40 MGD, average output: 10 MGD) produces finished water using coagulation, flocculation, settling, rapid sand filtration, ozonation, biological filtration on granular activated carbon, and chlorination (2 mg/L, 45 minutes minimum contact time) processes.

The new train (nominal and average capacity: 56 MGD) produces water that has undergone coagulation, flocculation with micro sand and polymer, lamellar settlement, intermediate ozonation, secondary coagulation, rapid dual-media filtration, micro filtration, nanofiltration using NF 200 membranes (molecular weight cutoff: 230 Dalton, TOC removal 90%, atrazine removal 90%, calcium rejection 40%), medium-pressure UV irradiation (40 mJ/Cm^2) and chlorination.

The water produced by the two treatment trains is blended, adjusted for pH and dechlorinated.

The Northern Paris Suburbs Distribution System

The distribution system fed by this water treatment plant covers 31 cities in the northern suburbs of Paris, and supplies water to over one million people. There are 9 one-line storage tanks and 12 booster chlorination facilities. The distribution system is divided into 3 main pressure zones.

Sampling points (Table 1) were chosen to represent various situations in terms of residence time and location (before and after residence in storage tanks, service lines and public network). For the needs of the study, residence times were calculated using the SWS hydraulic modeling software by STONER.

Table 1 Sampling points in the Méry-sur-Oise distribution system

Point n°	Average Residence Time (*range*)	Environment	Material and Size of the Pipe
1	0	plant	
3	36 (6–60)	tank	
4	35 (10–44)	public network	400 mm
5	33 (12–48)	public network	PE, 125 mm
15		public network	cast iron, 200 mm
2	40	public network	cast iron, 100 mm
6	5 (4–9)	public network	cast iron, 100 mm
7	21.5 (10–26)	public network	cast iron, 100 mm
8	28 (14–54)	public network	cast iron, 100 mm
9		private network	PEHD 52 m, copper 10 m
12	28 (14–54)	private network	PEHD 9 m, copper 12 m
11	54 (16–74)	private network	PEHD 4 m, copper 10 m
13	15 (8–28)	private network	lead 10 m, copper 8 m
14	6 (4–10)	private network	lead 5 m, steel 10 m, copper 8 m
10	18 (12–28)	private network	lead 20 m, steel 9 m, copper 10 m

Analytical Methods

Total Direct Counts (TDC) of the number of bacteria were determined by epifluorescence microscopy using DAPI staining. Viable bacteria (VB) were counted after the incorporation of CTC (chloride of 5-Cyano-2-3Dytolyl-Tetrazolium. Total Organic Carbon (TOC) and dissolved organic carbon (DOC) were determined using a Dorhmann DC-180 analyzer with a filtration using a 0.45 μm size membrane for DOC. Biodegradable dissolved organic carbon was determined using the Joret and Lévy method (1986).

The pH was measured using a mobile pH meter ref. 320 SET, and total and free residual concentrations were determined using the DPD (N,N diethyl-phenylenediamine) colorimetric technique with a portable HACH POCKET device. Turbidity was measured with a HACH 2100 P mobile analyzer.

The other chemical parameters were measured in accordance with the French standards: THM (NF T90-125), alkalinity (NF 90-036), conductivity (NF EN 27888).

RESULTS

Figure 1 shows the concentration of DOC in the Méry-sur-Oise finished water as a function of time. The variations prior to September 1999 are to be attributed to DOC variations in the raw water. After this date, the nanofiltration trains were progressively started up and reached the full capacity in April 2000. DOC concentration in the finished water fell over the same period from 1.5 mgC/L to 0.6 mgC/L, as a result of a higher proportion of nanofiltrated water in the water produced by the plant. The periods covered by the distribution system results, presented below, are characterized by a DOC concentration of between 2.0 and 2.5 mgC/L in 1998 and 1999 and less than 1.0 mgC/L in 2000.

The increased chlorine stability in the finished water also led to a decrease in free chlorine residual in the finished water, from 0.3 mg Cl_2/L prior to nanofiltration to 0.2 mg Cl_2/L after nanofiltration. As

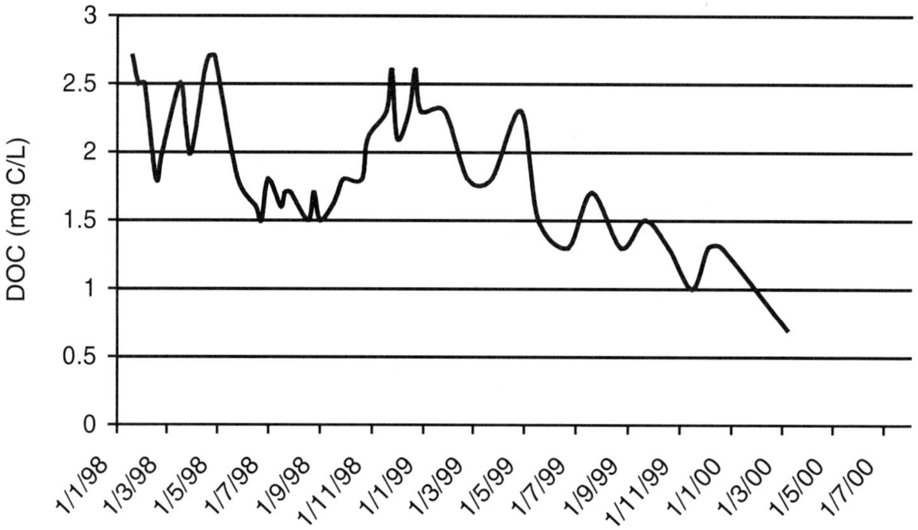

Figure 1 Concentration of DOC in the finished Méry-sur-Oise water

shown in Figure 2, this did not compromise the chlorine residual in the distribution system. The average free chlorine residual in the distribution system was 0.13 and 0.11 mg Cl_2/L in 1998 and 1999 respectively, and 0.15 mg Cl_2/L in 2000, with the operator's target being an average free chlorine residual of between 0.1 and 0.2 mg Cl_2/L. The injection of chlorine in the booster stations was also reduced over the same period, and the free chlorine residual remained stable throughout the system. The number of samples with low or no free chlorine residuals (i.e., < 0.1 mgCl_2/L) was also divided by a factor of 3, with less than 10% of the samples having a low chlorine residual. At the other end of the graph, it can be seen that the number of samples with an excessive chlorine residual (i.e., above 0.3 mgCl_2/L) was also minimized. This criterion is important for consumer satisfaction, given that free chlorine residuals are perceptible above this limit.

The evolution of the pH in the distribution system is presented in Figure 3. The pH increased because the saturation

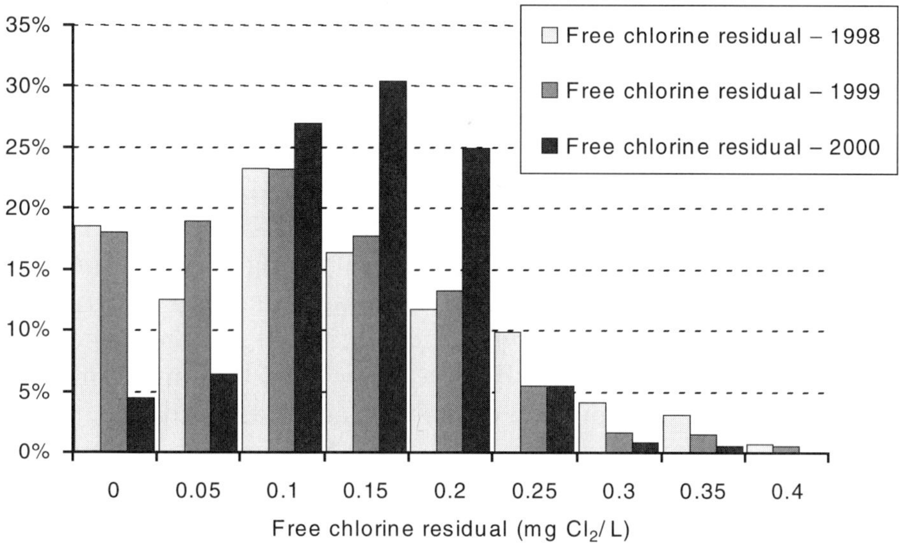

Figure 2 Free chlorine residual in distribution system samples

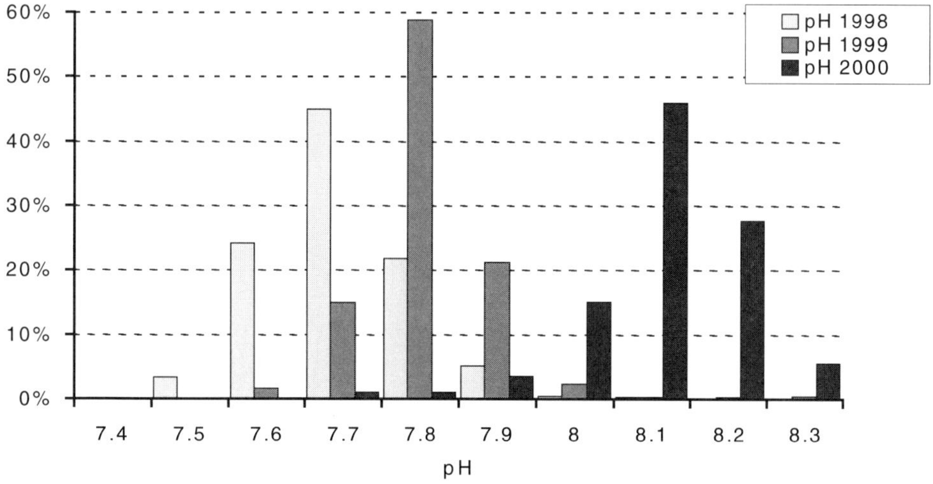

Figure 3 pH residual in distribution system samples

index of the water leaving the plant is maintained above 0.2 at all times for corrosion control. Since nanofiltration removes 40% of the calcium, the saturation pH is higher in nanofiltrated water, leading to a higher pH in distribution system samples.

The sum of the 4 THMs is presented in Figure 4. It can be seen that the average THM concentration was close to 20 µg/L in 1998 and 1999, with extreme values at 45 µg/L. These values comply with the current European standard of 100 µg/L. After nanofiltration, the average THM concentration was reduced to 10 µg/L, with maximum values at 30 µg/L. This result confirms that nanofiltration efficiently removes THM precursors, which is important given the potential introduction of more stringent standards.

The other main advantage of nanofiltration is the lower NOM content. This will result in a lower bacterial development in the distribution system. Figure 5 presents the TDC in distribution system samples. This figure shows a clear TDC reduction since the start-up of the nanofiltration system. In 1998 and 1999, the most

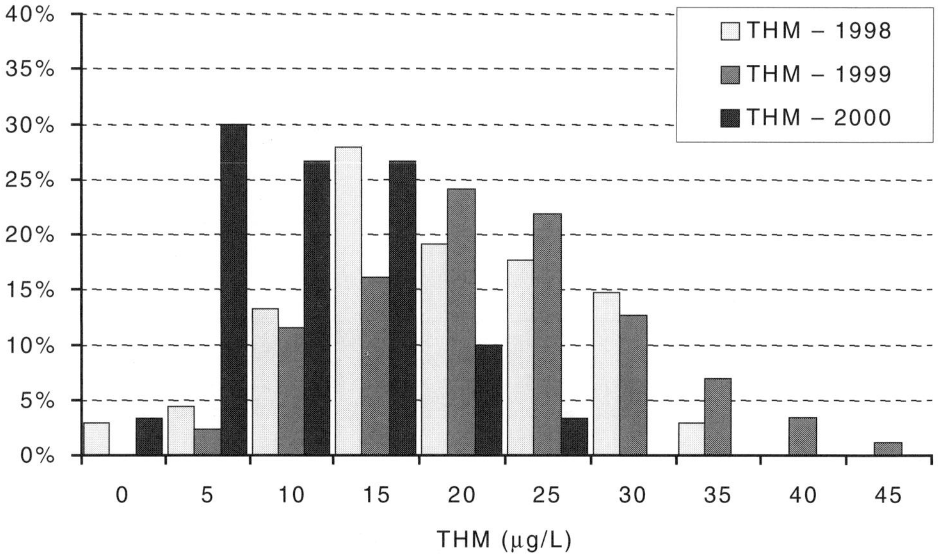

Figure 4 THM in distribution system samples

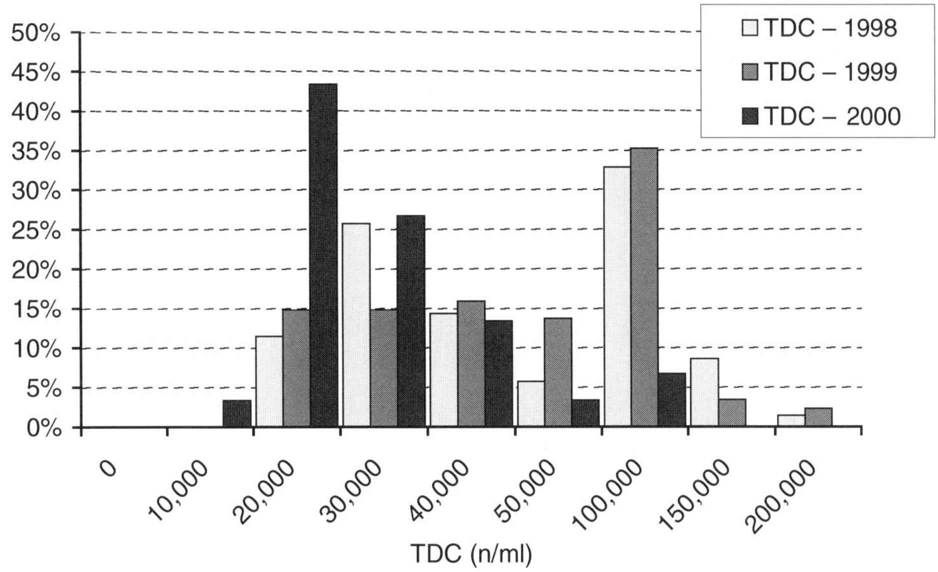

Figure 5 TDC in distribution system samples

abundant class represented between 100,000 and 150,000 cells/mL, in contrast with the situation in 2000 where the most abundant class was between 20,000 and 30,000 cells/mL. The TDC mean was 24,600 cells/mL in 2000, which should be compared with 54,800 cells/mL in 1998 and 48,700 cells/mL in 1999, representing a diminution of 49% TDC reduction, within comparable temperature and chlorine conditions. Since there are no TDC or free chlorine correlations covering the three years of the survey, the difference can be attributed to a lower NOM concentration. This result should be compared with earlier results of Sibille *et al.* at pilot scale (1997) and Laurent *et al.* at full scale (1999). In this later case where the conditions were very similar, the maximum amount of TDC was 40,000 cells/mL, which is in accordance with the current results.

Figure 6 presents the results for viable bacteria (stained with CTC), which confirm that the number of bacteria in distribution

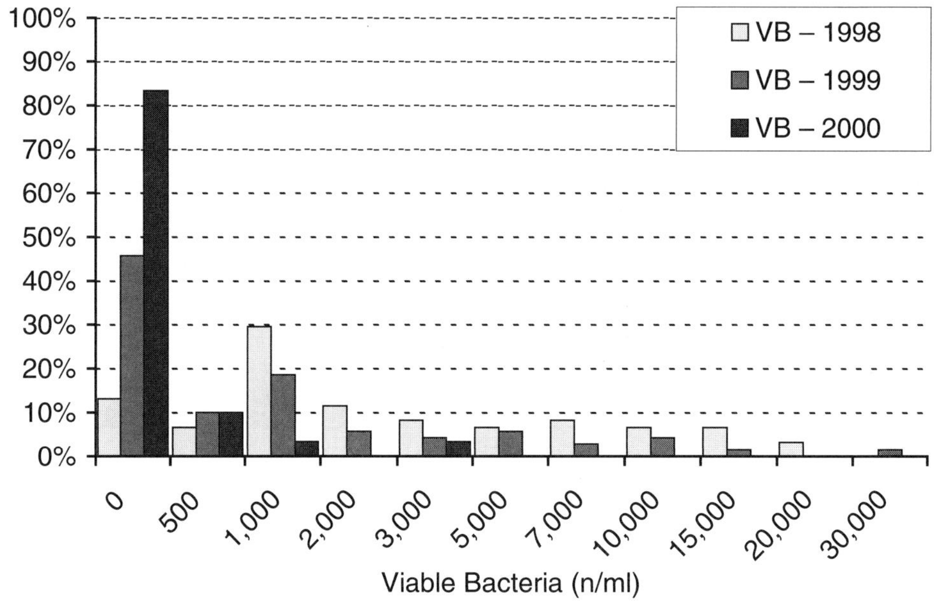

Figure 6 Viable bacteria in distribution system samples

system samples is reduced by nanofiltration. In 2000, most viable bacteria counts were lower than 500 cells/mL. In the two previous years, 97% of viable bacteria counts were greater than 500 cells/mL.

DISCUSSION

The reduction of NOM in water treatment plants is a key issue for water utilities, as this is the factor that controls the formation of disinfection by-products and regrowth in distribution systems. Nanofiltration is a new membrane process that dramatically reduces the NOM level. The results presented here confirm expectations for THM formation as well as regrowth. These results were obtained at a relatively low temperature, i.e., an average of 10°C. Sampling is under way to document the changes that take place at higher water temperatures.

Biofilm minimization was not directly measured in the results presented here, but indirect deduction leads to the conclusion that the biofilm was better controlled by nanofiltration, given that bacteria counts were lower in 2000. The other result that favours better regrowth control is the Biodegradable Organic Carbon consumption in the network (Figure 7 below). Prior to the introduction of nanofiltration, BDOC levels in finished water were between 0.3 and 0.8 mgC/L, but since the nanofiltration has been in service, BDOC levels in finished water are lower than 0.4 mgC/L. When all the results (before and after nanofiltration) are plotted together, a correlation can be established between the consumption of BDOC in the network and the BDOC level at the plant outlet. According to this correlation, if the threshold of 0.23 mgC/L in the finished water is not exceeded, there is no BDOC consumption in the network and some BDOC release can be observed. This complies with earlier results obtained by Servais *et al.* in 1995, which proposed a value of 0.15 mgC/L of BDOC as a threshold at which water could be considered biologically stable. In this particular reference, the BDOC method is that developed by

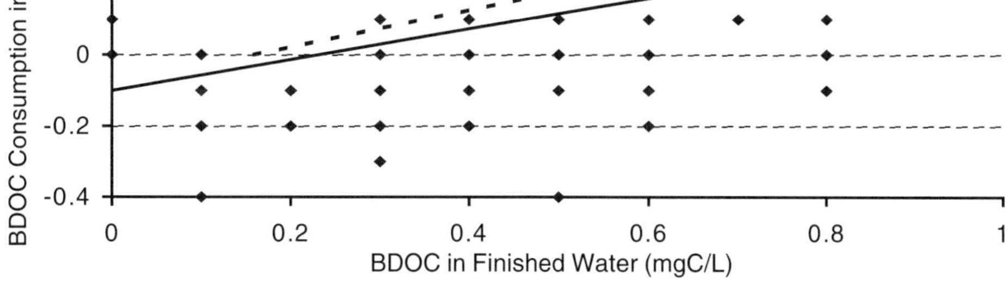

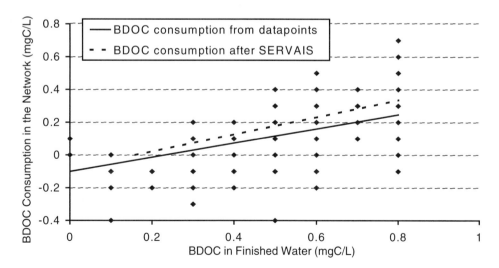

Figure 7 Consumption of BDOC in the network

Servais *et al.* in 1989 using a suspended biomass, which gives lower results than the Joret method used here.

CONCLUSION

Since September 1999, the northern suburbs of Paris have been supplied by a blend of biologically treated water and nanofiltrated water (20/80 ratio), being a combination that considerably reduces the NOM content in the finished water. In anticipation of the introduction of nanofiltration, the water quality in the distribution system was surveyed for two years prior to the changeover, and for one year thereafter. During the course of the study, the various microbiological and chemical water quality indicators were also closely monitored.

Results demonstrated that the use of NF treatment led to a reduced DOC content in the finished water, and, as a result, the chlorine residual in the network can be maintained with a lower chlorine dose at the plant with 33% less free chlorine residual at the plant outlet. Distribution system results indicate that the chlorine is more stable, with fewer points showing no or excessive chlorine residuals.

NF treatment produces water with a low THM formation potential, and the results presented in this chapter showed that the THMs fell by an average of 50% in distribution system samples, including samples taken at locations where there is a fairly long residence time.

The BDOC content of NF treated water is below the detection limit. The results indicate that this reduction permits a far better control of bacterial regrowth, with no observable BDOC consumption in the network. As a result, total direct counts are 49% lower than they used to be under comparable circumstances, with average counts under 30,000 cells/mL, which is very low. Viable bacteria counts are also drastically decreased.

These results were obtained in the winter and spring of 2000. This study will be continued to obtain a four-year database, covering all seasons before and after nanofiltration. This will allow the possibility for the observations to be confirmed over the longer term and cover the summer period.

ACKNOWLEDGEMENTS

The authors would like to acknowledge the works by Isabelle Romanillos, Olivier Chataigner, Claude Menge and Véronique Mary.

REFERENCES

1. LeChevallier M.W. *et al.*, Factors promoting survival of bacteria in chlorinated water supplies, *Appl. Environ. Microbiol.*, 54(3), 649–654, 1988.
2. Mathieu L. *et al.*, Parameters governing bacterial growth in water distribution systems, *Revue des sciences de l'eau*, 5, 91–112, 1992.
3. Servais P. *et al.*, Studies of BDOC and bacterial dynamics in the drinking water distribution system of the northern Parisian suburbs, *Revue des sciences de l'eau*, 5, 69–89, 1992.
4. Sohn J. *et al.*, In plant versus distribution system formation of chlorination by-products, *AWWA WQTC Proceedings*, Denver, 1997.
5. Westerhoff P. *et al.*, Applying DBP models to full scale plants, *JAWWA*, 92–3, 89–102, 2000.
6. Bablon *et al.*, Developing a sand-GAC filter to achieve high rate biological filtration, *JAWWA*, 80, 12–47, 1988.
7. Van der Kooij *et al.*, Effects of ozonation, biological filtration and distribution on the concentration of easily assimilable carbon (AOC) in drinking water, *Ozone Science and Engineering*, 11, 297–311, 1989.
8. Servais *et al.*, Microbiological activity in granular carbon filters at the Choisy-le-Roi treatment plant, *JAWWA*, 83–2, 62–68, 1991.
9. Urfer D. *et al.*, Biological filtration for BOM and particle removal: a critical review, *JAWWA*, 89–12, 83–98, 1997.
10. Agbekodo K., Legube B. et Coté P., Organics in NF permeate, *JAWWA*, 88(5), 67–74, 1996.

11. Legube B., *et al.*, Removal of organohalide precursors by nanofiltration. *Conférence AIDE "Disinfection of drinking water*, Kruger park (South Africa), *Water Supply*, 13(2), 171–181, 1995.
12. Randon *et al.*, Study of the behaviour of a distribution system supplied by a nanofiltration unit, *AWWA WQTC Proceedings*, New Orleans, 1019–1032, 1995.
13. Joret J.C., Levi Y., Methode rapide d'evaluation du carbone eliminable des eaux par voies biologiques, *Trib. Cebedeau*, 39, 3–9, 1986.
14. Laurent P. *et al.*, Microbiological quality before and after nanofiltration, *JAWWA*, 91(10), 62–72, 1999.
15. Sibille I. *et al.*, Microbial characteristics of a distribution system fed with nanofiltered drinking water, *Water Research*, 31–9, 2318–2326, 1997.
16. Servais P. *et al.*, Comparison of bacterial dynamics in various French distribution systems, *Jour Water SRT-Aqua*, 44, 10–17, 1995.
17. Servais P. *et al.*, A simple method for the determination of biodegradable dissolved organic carbon in waters, *Appl. Environ. Microbiol.*, 55, 2732–2734, 1989.

CHAPTER · 24

Membrane Replacement: Realizing the Benefit of Low-Pressure RO in Existing Infrastructure

Thomas Seacord, Kurt Frank, Gil Crozes
Carollo Engineers, Boise, ID

Greg Hill
Mount Pleasant Waterworks, Mount Pleasant, SC

OVERVIEW

The Mount Pleasant Waterworks (MPW) operates three reverse osmosis (RO) water treatment plants, with a total capacity of approximately 5 mgd, using the Middendorf Aquifer as a source. These three plants were constructed in the early 1990s and the membranes they use may be reaching the end of their useful lives. In an effort to reduce costs and maximize production capacity, the MPW evaluated new low-pressure RO membranes for use in their existing water treatment plants (WTPs) [1].

Recent advances in the production of composite membranes have led to the development of new low-pressure RO membranes. These membranes offer similar salt rejection and an increased permeability (i.e., 0.30 gfd/psi compared to 0.19 gfd/psi) compared to their predecessors. Additionally, new 400 ft^2, 8-inch by 40-inch elements are now available. Implementing these membranes offers the ability to increase production capacity if hydraulics are not limiting.

Increased permeability is a complicated issue associated with the use of these new low-pressure RO membranes. Increased permeability results in poor flow distributions and requires careful hydraulic consideration. Excess product flow from first-stage membrane elements (i.e., in a multi-stage array) results in lower flows to subsequent stages. Element recoveries and concentration polarization (β) factors must be less than 18% and 1.2, respectively, in order to prevent membrane fouling. Hydraulic remedies for distributing flow evenly between stages in a low-pressure RO membrane array include permeate throttling, interstage boosting, and hybrid membrane installations.

In the case of Mount Pleasant, infrastructure is existing, pumps are sized for traditional (high-pressure) brackish water membranes, and membrane equipment may be difficult to alter. This study consisted of pilot testing various low-pressure RO membranes from multiple suppliers. Modeling and cost analyses were conducted to determine the most economical way to implement low-pressure RO. This approach to low-pressure RO implementation and competitive procurement is anticipated to save Mount Pleasant approximately $1 million over the next ten years in operations and maintenance costs.

INTRODUCTION

The technology of RO membrane manufacture has advanced significantly since RO was first introduced in the 1960s. The introduction of composite polyamide membranes by DuPont in the 1970s inspired the desalination industry by lowering pressure requirements by approximately 25% to 40%. With increasing incentives coming from government agencies to improve the economics of membrane desalting, the RO industry responded by producing low-pressure *softening* membranes (i.e., nanofiltration (NF) membranes) in the late 1980s and low-pressure RO membranes in the mid-1990s [2, 3].

The MPW began operating their three RO WTPs in the early 1990s, before low-pressure RO membranes were available. Almost 10 years later, the original membranes have been paid for and they are still performing as they did initially. Now the question has become: will low-pressure RO offer any additional savings? Additionally, there exists the question as to whether these membranes may be installed in the existing infrastructure without the need for any significant upgrades or rehabilitation.

BACKGROUND

Reverse Osmosis Fundamentals

RO is a membrane filtration process that is capable of removing dissolved ions, salts, and organic compounds. The RO membrane treatment system is composed of three main components: membrane elements, pressure vessels, and the membrane array. A detailed description of how this configuration is assembled is described in detail elsewhere [3]. However, for the purposes of this discussion, it is important to understand the fundamental design criteria for configuring an RO membrane array. These design criteria consist of the membrane flux, recovery, and β factor.

Water quality resulting from filtration through RO membranes is governed by diffusion. When diffusion controls finished water quality, increasing the flux of water will decrease the permeate concentration of dissolved contaminants. *Flux* is a term used to quantify the flow of water through the surface area of an RO membrane and is calculated using Equation 1. Typical fluxes for groundwater RO range from 12 to 21 gallons per foot square of membrane per day (gfd). Membrane arrays must provide adequate membrane surface area to maintain flux to within this range.

$$F_w = \frac{Q_P}{A} \tag{1}$$

Where:

F_W: Flux of water through the membrane, gfd
Q_P: Flow of water through the membrane, gpd
A: Surface area of membrane, ft^2

Recovery is a percentage that represents the amount of water pumped into an RO system that is actually recovered as finished water. The remainder of water not recovered in the finished water is a waste stream that is referred to as *concentrate* and/or *by-product* for RO processes. Recovery is calculated as follows:

$$R = \frac{Q_P}{Q_F} \times 100\% \tag{2}$$

Where:

R: Recovery, %
Q_F: Feed flow, gpd

Recovery is an important factor for designing membrane systems since it relates to the economy of operation, membrane fouling, salt precipitation and concentration polarization [3, 4]. Concentration polarization is a process by which dissolved salts and organic compounds accumulate at the membrane surface at a concentration that is greater than the concentration in the bulk feedwater that passes across a membrane's surface. This process results in decreased product water flow and a degradation in the membrane's rejection characteristics. Membrane fouling may also result. The concentration polarization layer increases with increasing permeate flux, but may be reduced by increasing the flow of water across the membrane's surface. To limit the adverse affects of concentration polarization, membrane suppliers recommend a minimum element recovery (i.e., approximately 18%), which is used as a guideline for configuring an RO system array. This recovery corresponds to a β factor of 1.2, which is the ratio of solute concentration at the membrane surface to the concentration of solute in the bulk feed stream [5].

$$\beta = \frac{C_M}{C_B} = K_p e^{\left(\frac{Q_{Pi}}{Q_{Fi}}\right)} = K_p e^{\left(\frac{2R_i}{2-R_i}\right)} \tag{3}$$

Where:

β: Concentration polarization factor
C_M: Concentration of solute at the membrane surface, mg/L
C_B: Concentration of solute in the *bulk* feed stream, mg/L
K_P: Proportionality constant (dependent upon solute)
Q_{Pi}: Permeate flow from a single membrane element, gpd
Q_{Fi}: Feed flow to a single membrane element, gpd
R_i: Recovery from a single membrane element, %

RO is characterized as pressure-driven membrane filtration processes. The transmembrane pressure (TMP) for brackish water RO typically ranges from 60 to 350 psi. TMP, as presented in Equation 4, constitutes only the pressure required to overcome the headloss through the membrane barrier at a given flux.

$$TMP = \left(\frac{P_F + P_C}{2}\right) - P_P - \Delta\Pi \tag{4}$$

Where:

TMP: Transmembrane pressure, psi
P_F: Feed pressure, psi
P_C: Concentrate pressure, psi
P_P: Permeate pressure, psi
$\Delta\Pi$: Osmotic pressure, psi (see Equation 5)

$$\Delta\Pi = 0.01 \times \left(\frac{TDS_F + TDS_C}{2} - TDS_P\right) \tag{5}$$

Where:

$\Delta\Pi$: Osmotic pressure, psi
TDS_F: Total dissolved solids concentration in the feedwater, mg/L

TDS_C: Total dissolved solids concentration in the concentrate (by-product) water, mg/L

TDS_P: Total dissolved solids concentration in the RO permeate (product) water, mg/L

It is important to note that the TMP is not the membrane feed pressure. The membrane feed pressure is used to calculate energy costs at a given set of operating conditions (i.e., flux and recovery) and may be estimated by summing the terms in Equation 4 and adding any additional headloss or pressure drop through the membrane system [5, 6].

Osmotic pressure is the result of the naturally occurring phenomenon known as osmosis, where water containing few dissolved solids (i.e., TDS) will pass through a semi-permeable membrane to dilute a water containing high concentrations of dissolved solids. This phenomenon will continue to occur until an equilibrium concentration is reached on both sides of the membrane. Reverse osmosis is, in turn, pushing water that is highly concentrated in dissolved solids through a semi-permeable membrane to produce water of high purity via an applied pressure.

Membrane permeability is assessed by calculating the specific flux. Specific flux is presented in the units gfd/psi and represents the water produced by a membrane per unit of applied pressure. As the membrane fouls, the specific flux declines and is, therefore, a useful value used for tracking membrane fouling. Specific flux is calculated as follows:

$$SF = \frac{F_w}{TMP} \tag{6}$$

Where:

SF: Specific flux, gfd/psi

An essential component to the design, operation, and economics of a RO system is the membrane feed pump(s) and the power required to operate these pumps. Since RO is a pressure-driven membrane process, high-pressure membrane feed pumps precede the RO membranes, supplying water to the RO process at the pressure required to produce water at a given flux and recovery. Membrane feed pump horsepower is calculated using Equation 7. The membrane feed pressure, as discussed previously, is estimated by summing the terms in Equation 1 and adding any additional headloss through the RO system.

$$hp = \frac{\Delta PQ}{1714\eta_m\eta_p} \tag{7}$$

Where:

- hp: Required horse power, hp
- ΔP: Differential pressure (i.e., pump outlet pressure minus the pump inlet pressure), psi
- Q: Pump flow, gpm
- η_m: Motor efficiency, %
- η_p: Pump efficiency, %

Once determined, membrane feed pump horsepower can be used to estimate power costs as follows:

$$\begin{aligned} Power\ Cost\left[\frac{\$}{yr}\right] &= hp \times \frac{0.7457\ kW}{hp} \times \frac{24\ hr}{day} \times \\ &= \frac{365\ day}{yr} \times \frac{\$0.065}{kW\text{-}hr} \end{aligned} \tag{8}$$

Where the unit cost for power is assumed to be $0.065/kW-hr.

Hydraulics of Low-Pressure Reverse Osmosis

Low-pressure RO membranes have a specific flux that is 20 to 60% greater than the specific flux of traditional brackish water RO membranes. This increase in membrane permeability has changed the way that water will flow through a membrane array. Due to the high specific flux of low-pressure RO membranes, more water will flow through the first-stage membrane elements, leaving less water for production in stage 2. As presented previously, membrane arrays are designed to minimize membrane fouling and concentration polarization by maintaining high velocities of water across a membrane surface. The β factor for any given element in the membrane array should always be ≤ 1.2, which corresponds to a single element recovery of $\leq 18\%$. When low-pressure RO membranes are installed in a traditional manner, since more water will permeate through the stage 1 membrane elements, the β factor will often exceed the maximum design limit of 1.2.

To realize the benefit of low-pressure RO while preventing membrane fouling, the hydraulics of water flowing through a membrane array must be optimized to limit the water permeating stage 1 membrane elements. This limits first-stage recovery, producing a $\beta \leq 1.2$, and enhances the flow of water to stage 2 of the array. Hydraulic enhancements for membrane array design used to facilitate these conditions include:

- hybrid membrane configurations
- permeate throttling
- interstage pressure boosting [2, 7]

These hydraulic enhancements are illustrated in Figure 1. Compared to the conventional, high-pressure RO membrane example, each low-pressure membrane alternative requires less feed pressure. Therefore, money is saved.

The hybrid configuration, presented in Figure 1, utilizes two different types of low-pressure RO membranes. The membranes installed in stage 1 of the hybrid configuration have a lower specific flux than those installed in stage 2. This limits the water flux through stage 1 membrane elements and promotes the movement of water to stage 2. Alternatively, one type of low-pressure RO membrane may be used if there is a method for limiting first-stage recovery. Permeate throttling, as presented in Figure 1, involves applying a back pressure to the permeate stream from stage 1 in the membrane array, thereby reducing stage 1 recovery and promoting more water to move through to stage 2. Interstage pressure boosting, also presented in Figure 1, uses an additional pump placed between stage 1 and stage 2. This increases stage 2 recovery by increasing the second-stage feed pressure, which increases the flux and allows more water to be filtered by stage 2 membrane elements.

Facility Review

The MPW operates three RO WTPs that treat Middendorf Aquifer water. The primary objective of this treatment is to remove TDS and fluoride to acceptable levels. Figure 2 illustrates the process flow diagram for a typical RO WTP in Mount Pleasant. As indicated, scale inhibitor is added prior to cartridge filtration. Sulfuric acid was eliminated from the treatment process in 1998 and a new inhibitor, with an LSI threshold of 2.5, was implemented [8]. No fouling has been observed since the elimination of acid. Pretreated water is boosted in pressure by a membrane feed pump prior to filtration by RO. Each RO WTP has two membrane feed pumps that are controlled by variable frequency drives (VFDs). One pump is operated while the other is provided as a redundant measure. In each RO WTP, the operating RO feed pump supplies water to two or more trains. RO production capacity is held constant for each train by monitoring product water flow (i.e., from each train) and varying the speed of the membrane feed pump and the position of a flow control valve, as illustrated in Figure 2.

RO treated water is blended with raw Middendorf Aquifer water prior to disinfection, as indicated in Figure 2. Blending is performed to increase the mineral content of RO treated water and to promote chemical stability. Prior to blending, calcium chloride ($CaCl_2$) is added to the permeate water. In addition to blending, this helps to adjust the finished water Langelier Saturation Index (LSI) to approximately + 0.15, indicating that the water is depositing in nature with respect to calcium carbonate. Cascade tray aeration was performed previously to remove gaseous carbon dioxide produced as a by-product of acid pretreatment. Currently the MPW is evaluating the continued need for aeration since carbon dioxide is no longer formed due to the elimination of acid [1].

Table 1 presents the existing infrastructure and design criteria for each of the MPW's three RO WTPs. Process flows, components of the membrane treatment system, and chemical requirements are included. RO capacities, total and capacity by train, vary by WTP location. Train capacities range from 350 gpm in RO WTPs No. 1 and No. 2 to 450 gpm in RO WTP No. 3. Array configurations also vary accordingly. Membrane feed pump performance curves and pump characteristics are presented in Figures 3, 4, and 5. Prior to implementation of a replacement membrane or increasing capacity, these figures should be consulted to make certain the existing pumps support the hydraulic requirements of operations alternatives.

While the source water and infrastructure of RO WTPs No. 1 and No. 2 are similar, the current operation of these facilities is different. As indicated in Table 2, the RO feed pressure in RO WTP No. 1 is greater than that required for RO WTP No. 2. Additionally, the TDS and fluoride produced by RO from RO WTP No. 1 is less than that produced by RO WTP No. 2. As a result, more water may be blended in RO WTP No. 1 than in RO WTP No. 2. The reason that RO WTP No. 1 has a higher feed pressure and produces higher quality water than RO WTP No. 2, which has the same membranes and operation design criteria, is that RO WTP No. 1 was temporarily exposed to chlorine when this plant was brought online and commissioned. The CPA2 membranes that are currently used by the MPW in all three of their RO facilities are sensitive to oxidants like chlorine. The temporary exposure of the CPA2 membranes to chlorine during the start-up of RO WTP No. 1 altered the specific flux, decreasing productivity by approximately 40%. As a result, the current operation of the RO process in WTP No. 1 requires higher pressures and produces higher-quality water at a greater cost than RO WTP No. 2. Membrane replacement may offer the opportunity to reduce feed pressure and energy costs for RO WTP No. 1.

As indicated in Table 2, RO WTP No. 3, Train 4, only achieves a 400 gpm permeate flow. This is less than the design flow of 450 gpm. Train 4 was added as an expansion to RO WTP No. 3 in 1994. The membrane feed pump originally selected for this installation does not support the hydraulic requirements for operating four trains of CPA2 membranes at a capacity of 450 gpm per train. This issue is evident upon examination of the membrane feed pump performance curve for RO WTP No. 3, presented previously as Figure 5. To achieve a 450 gpm capacity from each of the trains in RO WTP No. 3, the feed pump is required to produce 2,250 gpm. At this flow rate, as indicated in Figure 5, the feed pump's hydraulic capacity is 165 psi. Due to the lower specific flux of the CPA2 membrane supplied for the RO WTP No. 3 Train 4 expansion, a higher feed pressure (i.e., > 170 psi) is required. Chemical cleaning did not produce any increases in the membrane's productivity. The programmable logic control (PLC) for the membrane feed pump VFD and rate of flow control valves has produced the conditions observed in Table 2. To increase the flow rate from the feed pump in RO WTP No. 3, and to realize the full capacity of this facility, a lower feed pressure is required. Membrane replacement with a low-pressure RO membrane may offer the ability to realize the full capacity of this facility.

STUDY OBJECTIVES AND PROJECT GOALS

The objective of this study was to determine the feasibility and benefits of implementing low-pressure RO membranes in the MPW's three RO WTPs. Feasibility will be assessed through pilot testing to determine if low-pressure RO is more sensitive to membrane fouling. Benefits that may be realized through membrane replacement include:

- Operations and maintenance cost savings (i.e., accounting for the cost of the membrane purchase
- Increased water treatment plant capacity
- Increased operational control in RO feed pumps due to lower pressures

MATERIALS AND METHODS

Pilot Testing

Low-pressure RO membranes have been noted to be more sensitive to fouling than traditional brackish water membranes due to an increase in the surface roughness, which creates a medium for fine particles to become permanently trapped [9, 10]. With this in mind, single-element membrane pilot tests were conducted to confirm the viability of low-pressure RO for treating the Middendorf Aquifer supply at the MPW.

The procedures for conducting pilot tests are described in detail elsewhere [11]. The equipment used for these pilot tests consisted of 2 single-element RO pilot plants, manufactured by Harn R/O Systems (Venice, FL), that utilize a PLC system to maintain a constant permeate flow and flux. A single-element recovery of 80% was maintained by implementing a concentrate recycle flow that facilitated a β factor ≤ 1.2. In such a test, water quality data may not be representative of full-scale operations; however, the membrane's sensitivity to fouling can be adequately determined. Hydraulic and water quality data were monitored online and recorded electronically. Feedwater for the pilot tests was taken directly from the full-scale plant after all pretreatment had taken place. Pretreatment chemical doses were summarized previously in Table 1. Manual measurements were also conducted to provide quality assurance in the online data. All electronic data were accessible via a remote telemetry system.

Four membranes from three manufacturers were screened on the pilot scale. These membranes and their characteristics are presented in Table 3. Also included in Table 3 are the characteristics of the CPA2 membrane, currently used by the MPW. It should be noted that each of the low-pressure RO membranes listed has a specific flux that is 20 to 60% greater than the CPA2 membrane. Additionally, more square feet of membrane area are available per element, which may allow the MPW to increase their production capacity after replacement, should the feed pump hydraulic capacity allow.

Modeling

For membranes successfully demonstrated on the pilot scale, replacement alternatives presented previously in Figure 1 were modeled using RO manufacturer supplied design software. These software programs included *RODESIGN V.6.4, ROSA V.4.0,* and *ROPRO V.6.1.* Membrane feed pressure and power costs can be determined from these models. A reasonable prediction of water quality may be attained from modeling data as well.

RESULTS AND DISCUSSION

Pilot Testing

Due to the potential for irreversible fouling, pilot tests were conducted to assess the viability of low-pressure RO for treating the Middendorf Aquifer supply at the MPW. Table 4 presents a summary of the operating conditions and fouling rates observed during these pilot tests. Fouling rates were determined as a linear regression of the hydraulic data presented in Figures 6 through 9. As indicated in Figures 6 through 8, there was no appreciable fouling detected from the operation of the ESPA1, ESPA2, or BW30LE membranes. While the fouling rates for these membranes, reported in Table 4, were extremely low, although cleaning frequencies are reported, 30 days of pilot test data do not provide enough information to support the low cleaning frequencies reported. However, it may be assumed that these membranes (i.e., ESPA1,

ESPA2, and BW30LE) are not sensitive to particle fouling and that cleaning to remedy such fouling may not be required.

As indicated in Figure 9, the TFC-ULP membrane did experience fouling throughout the entire testing period. Based upon the manufacturer's recommendation that a membrane be cleaned when there is a 20% decline in productivity (i.e., specific flux) the TFC-ULP membrane would require cleaning every 24 days. This level of required cleaning is too frequent and exceeds the 30-day cleaning frequency that is typically used as a design criterion for municipal RO facilities. Therefore, based on pilot testing, only the ESPA1, ESPA2, and BW30LE membranes are recommended for replacement at the MPW.

It should be noted that there was a difference in the permeability, or specific flux, of the membranes used for pilot tests when compared to the manufacturer specification reported previously in Table 3. Care should be provided when implementing hybrid configurations to specify the specific flux of full scale elements to ensure that the flow is distributed appropriately and $\beta \leq 1.2$ for each stage of the RO array.

Modeling and Cost Analyses

Table 5 presents a modeled evaluation of the operation and costs associated with the current treatment and membrane replacement alternatives for each of the MPW's three RO WTPs. To provide a representative review, train capacity was held constant and is identical to that currently employed at each WTP. This analysis indicates that there are power cost savings that may be realized by replacing the membranes at each RO WTP. However, when the amortized replacement costs are considered, only replacement at RO WTP No. 1 will provide an annual cost savings. An estimated \$19,000 per year may be saved by replacing the membranes in RO WTP No. 1 with a hybrid membrane configuration. This alternative provides the lowest annual cost considering only the power costs and the cost for replacing membranes (i.e., power cost, membrane costs, and additional infrastructure costs). While the power costs for the interstage boost replacement alternative were lower than those for the hybrid membrane configuration, the additional infrastructure required to implement this alternative (i.e., upgrades to membrane equipment) requires an additional $20,000 of capital per train, which makes this alternative more expensive.

Replacement Recommendations

RO WTP No. 1. As presented in Table 5, there exists the opportunity to save money by replacing membranes in RO WTP No. 1 with low-pressure RO membranes in a hybrid configuration. However, it should be noted that these replacement membranes have more square feet of membrane area per element. Currently, the CPA2 membrane elements utilized by the MPW have 365 ft^2 of membrane and operate at an average system flux of 16.4 gfd. Replacement of membranes in RO WTP No. 1 with new elements that have 400 ft^2 of membrane area would reduce the average system flux to 15 gfd, if the capacity was held constant. Capacity may be increased by 8% by increasing the production rate to maintain approximately the original design flux. Estimated annual costs for the current operation, replacement with a hybrid membrane configuration that produces the same capacity, and replacement with a hybrid configuration at an increased capacity are as follows:

- Current: $83,000/yr
- Hybrid replacement, same capacity: $64,000
- Hybrid replacement, 8% increase in capacity: $79,000/yr

These estimated costs include only the annual power costs, amortized membrane replacement cost, and annual chemical costs. As indicated, there are cost savings that may be recognized by replacing membranes at the current capacity or at an increased capacity. Compared to other alternatives to providing extra capacity, this alternative will save the MPW $42,000 per year [1]. Due to the need for more water in the Mount Pleasant service area, it is recommended that the capacity of RO WTP No. 1 be increased by 66 gpm to 866 gpm. This capacity requires 400 gpm of production per train and a reduction in the blend ratio (i.e., to meet finished water quality goals) from 7:1 to 12:1. Blending must be reduced since replacement membranes will not remove as much TDS and fluoride as the current membranes.

RO WTP No. 2. As indicated previously in Table 5, there is no economic incentive for replacing the CPA2 membranes in RO WTP No. 2. Additionally, the benefits of increasing water treatment capacity through membrane replacement at RO WTP No. 2 may not be realized due to limitations posed by this facility's by-product disposal permit. RO WTP No. 2 discharges by-product to a surface water, and by increasing the rate of by-product discharge, an alteration to the Nation Pollution Discharge Elimination System (NPDES) permit would be required. It is anticipated that it would be difficult to receive acceptance for this permit alteration and therefore, no replacement is recommended.

RO WTP No. 3. As indicated previously in Table 2, the design capacity for each of the four trains in RO WTP No. 3 is 450 gpm. However, Train No. 4 only achieves a 400 gpm production rate. This low production rate was experienced upon installation and is not the result of fouling. Chemical cleaning had no effect on improving productivity. The data presented in Table 4 indicate that the membrane feed pump supplying water to all four trains in RO WTP No. 3 operates at a pressure of 170 psi while producing only 2,200 gpm of the 2,250 gpm required to produce 450 gpm of permeate from each train. This condition is a combined result of the membrane's specific flux in Train 4, the operating PLC system, and the hydraulic limits of the membrane feed pump. From a design perspective, this is a very poor operating condition because if any fouling were to occur in Train 4, a further decrease in production rate would be experienced.

There are three solutions that will remedy this operational limitation: (1) replacement of the membrane feed pump; (2) replacement of the feed pump impeller (i.e., with a larger impeller diameter); or (3) membrane replacement with low-pressure RO membranes. Pump impeller replacement is not possible since the 10.375" impeller currently installed is the largest size that will fit in the existing pump casing. Therefore, membrane replacement and feed pump replacement remain the only options. As opposed to feed pump replacement, membrane replacement offers the opportunity to realize not only the full design capacity, but also an increase in this capacity as well as an increase in operational control. Table 6 presents membrane replacement alternatives for RO WTP No. 3 that include capacity optimization alternatives as well as current operations. Replacement alternative Options A, B, and C all involve the replacement of existing membranes with a hybrid membrane configuration. Note that replacement of only the stage 2 membranes was considered, but discarded from consideration due to the fact that feed pressures would remain at 150 psi, which does not provide significant benefit with respect to an increase in operational control. Additionally, an increase to plant capacity could not be realized. Replacement alternative Option A involves replacing only 2 of the 4 trains in RO WTP No. 3. Options B and C involve replacing all four trains at the current design capacity or at an increased capacity, respectively. Modeling results indicated that the resulting water quality is similar for each alternative and is therefore not provided. The blending ratio remains constant at 12:1.

Operational control may be gained through membrane replacement. Each replacement scenario presented in Table 6 lowers the required membrane feed pump pressure while producing a production rate that meets or exceeds the design capacity for this facility. Option A achieves operational stability by reducing the production rated from trains 1 and 2 where the membranes have not been replaced. The membranes in trains 3 and 4 are replaced and a production rate of 500 gpm may be achieved from these trains at a feed pressure of 140 psi, which as indicated in Figure 5, provides about 5% variation in VFD speed control. Control may be gained further through Option B or C, where all 4 trains are replaced with a low-pressure RO membrane and feed pump requirements are 127 and 137 psi, respectively. This increases the control of operations by as much as 10% over the current operations scenario.

In terms of replacement costs, which includes the amortized capital costs plus the annual power costs, very little change in cost is recognized for any of the replacement alternatives. This indicates that there are benefits in terms of gaining operational control and increasing capacity of either Option A, B, or C. Finished water capacity may be increased by as much as 50 gpm through Option A or B. However, due to a required by-product flow (i.e., to maintain $\beta \leq 1.2$) from each RO train, decreasing the capacity of trains 1 and 2 under Option A produces more by-product than Option B. Option B is preferred over Option A since more control is gained by lowering the required membrane feed pump pressure and producing less by-product at the same finished water capacity for a cost increase that is estimated to be negligible. However, the capacity of RO WTP No. 3 may be increased by approximately 250 gpm as described in Option C, where all four trains are replaced and the capacity of each train is increased, still providing more operation control (i.e., as demonstrated by the membrane feed pressure of 137 psi) at a water cost increase of only $0.01 per thousand gallons. Therefore, Option C provides the most benefit in terms of increased capacity and control without sacrificing finished water quality. This additional capacity represents a savings of approximately $95,000 annually when compared to other alternatives for providing capacity [1].

CONCLUSIONS

Low-pressure RO has helped to reduce the energy requirement for membrane desalination. However, low-pressure RO may not be applicable for every situation. The work conducted during this study has demonstrated the importance of pilot testing new applications prior to implementation. The TFC-ULP membrane was eliminated from further consideration at the MPW due to fouling observed during pilot tests. However, three low-pressure RO membranes did prove applicable for treatment of the Middendorf Aquifer by the MPW.

A hybrid low-pressure RO configuration was ultimately chosen for implementation. This configuration provided the easiest and most economical option for low-pressure RO implementation in existing infrastructure. Competitive procurement, between the remaining suppliers, of the replacement membranes produced a low bid of $480 per replacement membrane element. It is estimated that with this price, the MPW will save approximately $19,000 per year by replacing membranes in their RO WTP No. 1. Additionally, membrane replacement will offer the MPW an additional 0.5 mgd of capacity, valued at approximately $140,000 per year. Therefore, membrane replacement is estimated to save the MPW approximately $1 million over the next 10 years.

ACKNOWLEDGEMENTS

The authors would like to acknowledge Perry Alexander, Ken Burlison, Doug Phipps, Cindy Tyner, Rusty Walker, and Tony Wilder for their contributions toward the pilot testing portion of this work.

REFERENCES

1. Carollo Engineers. 2001. *Mount Pleasant Waterworks—Membrane Replacement Study and Water Treatment Master Plan*. Carollo Engineers, Boise, ID.
2. Nemeth, J.E. 1998. Innovative System Designs to Optimize Performance of Ultra-Low Pressure Reverse Osmosis Membranes. *Proc. 1998 IWSA Conference on Membranes in Drinking and Industrial Water Production*. Amsterdam, Netherlands.
3. AWWA. 1999. *Manual of Water Supply Practices—M46: Reverse Osmosis and Nanofiltration*, 1st Edition. American Water Works Association, Denver, CO.
4. Seacord, T.F., K. Frank, G. Crozes, and G. Hill. 2001. Process Design Implications of No-Acid RO/NF Treatment of Groundwater Con-

taining Hydrogen Sulfide. *Proc. 2001 AWWA Membrane Technology Conference*, San Antonio, TX.

5. Hydranautics. 1998. "Section 2: Principals of RO and System Design" in *RODESIGN V.6.4*. Hydranautics, Oceanside, CA.
6. Wilf, M. 1997. Effect of New Generation of Low Pressure, High Salt Rejection Membranes on Power Consumption of RO Systems. *Proc. 1997 AWWA Membrane Technology Conference*, New Orleans, LA.
7. Oklejas, E. 1992. The Hydraulic Turbocharger for Interstage Feed Pressure Boosting. *Proc. 1992 National Water Supply Improvement Association Conference*, Newport Beach, CA.
8. Ning, R.Y. and P. Alexander. 2000. Elimination of Acid Injection in Reverse Osmosis Plants. *Proc. 2000 AWWA Water Quality and Technology Conference*, Salt Lake City, UT.
9. Devitt-Mackey, E. 1999. *Fouling of Ultrafiltration and Nanofiltration Membranes by Dissolved Organic Matter.* Ph.D. Dissertation. Rice University, Houston, TX.
10. Seacord, T.F., G. Crozes, and E. Mackey. 2000. Treating a High Turbidity High Hardness Surface Water Supply with an Integrated Membrane System. *Proc. 2000 AWWA Water Quality and Technology Conference*, Salt Lake City, UT.
11. USEPA. 1997. *ICR Treatment Studies Data Collection Spreadsheets and User's Guide*. EPA 815-B-97-002.

Table 1 Existing infrastructure and design criteria

	Unit	RO WTP No. 1	RO WTP No. 2	RO WTP No. 3	Total
Process Flows					
Well Water	gpm	975	1,400	2,346	4,721
RO	gpm	700	1,050	1,750	3,500
Blend	gpm	100	88	146	334
Finished Water	gpm	800	1,138	1,896	3,834
Concentrate	gpm	175	263	450	888
Membrane System					
Membrane Feed Pumps	No.	2	2	2	–
Train	No.	2	3	4	–
Capacity	gpm/train	350	350	450[a]	–
Array	–	8:4	8:4	11:5	–
Elements per Vessel	No.	7	7	7	–
Flux	gfd	16.4	16.4	16	–
Specific Flux	gfd/psi	0.22	0.22	0.22	–
Recovery	%	80	80	80	–
Membrane[b]	–	CPA2	CPA2	CPA2	
Membrane Area	ft^2/element	365	365	365	–
Chemical System					
Inhibitor Dose[c]	mg/L	2	2	2	–
Inhibitor Demand[c]	lb/day	23	34	58	115
$CaCl_2$ Dose	mg/L	32	32	32	–
$CaCl_2$ Demand[d]	lb/day	964	1,446	2,406	4,816
Total Cl_2 Dose	mg/L	1	1	1	–
Cl_2 Demand	lb/day	13	15	25	53
NH_4 Dose	mg/L	0.22	0.22	0.22	–
NH_4 Demand	lb/day	3.7	4.3	7.1	15

a RO WTP No. 3, Train 4 only achieves a 400 gpm production capacity
b Hydranautics
c King Lee Technologies Pretreat Plus™ 0100
d 32% $CaCl_2$ solution; 6,000 gallon $CaCl_2$ storage tank existing

Table 2 RO water treatment plants–current conditions

RO WTP	Feed Pressure (*psi*)	Capacity (*gpm*)	Specific Flux (*gfd/psi*)	Blending Ratio (Q_p:Q_b)	Recovery (*%*)	TDS (*mg/L*)	Fluoride (*mg/L*)	Chloride[a] (*mg/L*)
RO WTP No. 1	210	800[a]	–	7:1	82	226[a]	0.6[a]	38[a]
Train 1	175	350	0.16	–	80	41	0.1	–
Train 2	200	350	0.13	–	80	38	0.1	–
RO WTP No. 2	155	1,138[a]	–	12:1	81	244[a]	0.6[a]	37[a]
Train 1	140	350	0.22	–	80	75	0.2	–
Train 2	150	350	0.22	–	80	70	0.2	–
Train 3	140	350	0.22	–	80	70	0.1	–
RO WTP No. 3	170	1,896[a]	–	12:1	81	276[a]	0.7[a]	38[a]
Train 1	150	450	0.22	–	80	89	0.3	–
Train 2	150	450	0.22	–	80	64	0.2	–
Train 3	150	450	0.22	–	80	79	0.2	–
Train 4	170	400	0.16[b]	–	71	37	0.1	–

a After blending and post-treatment
b Low specific flux is not due to membrane fouling. Insufficient hydraulic conditions exist to provide adequate flow to Train 4.
Q_p RO permeate flow
Q_b Blend flow

Table 3 Membrane characteristics

Membrane	Manufacturer	Membrane Surface Area[a] (*ft²*)	Specific Flux (*gfd/psi@18gfd*)	Salt Rejection (%)
CPA2	Hydranautics	365	0.19	99.5
ESPA1	Hydranautics	400	0.30	99.0
ESPA2	Hydranautics	400	0.23	99.5
BW30LE	FilmTec	440	0.24	99.0
TFC-ULP	Koch Fluid Systems	400	0.30	98.5

a 8-inch (diameter) × 40-inch (long) membrane element

Table 4 Summary of pilot test data

Membrane	Initial Transmembrane Pressure (*psi*)	Flux (*gfd*)	Specific Flux (*gfd/psi*) Clean Membrane	Specific Flux (*gfd/psi*) Before Cleaning[a]	Fouling Rate (*gfd/psi/day*)	Cleaning Frequency (*days*)
ESPA1	55	18	0.30	0.24	–0.00002	3,000
ESPA2	110	18	0.15	0.12	–0.00008	375
BW30LE	55	18	0.30	0.24	–0.0002	300
TFC-ULP	140	18	0.12	0.10	–0.0010	24

a Not based on pilot testing. Based on manufacturer recommended cleaning when specific flux declines by 20%.

Table 5 Membrane replacement alternatives

Alternative	Membrane (*s*)	Flux (*gfd*)	Capacity by Train (*gpm*)	Feed Pressure[b] (*psi*)	TDS (*mg/L*)	Fluoride (*mg/L*)	Cost[a] Capital[c] (*$/train*)	Power (*$/yr*)	Annual (*$/yr*)
Currently									
RO WTP No. 1		16.4	350	210	40	0.09	–	69,000	69,000
RO WTP No. 2	CPA2	16.4	350	155	72	0.16	–	63,000	63,000
RO WTP No. 3		16	450	170	67	0.18	–	119,000	119,000
Baseline									
RO WTP No. 1		16.4	350	155	55	0.19	40,320	50,000	60,000
RO WTP No. 2	CPA2	16.4	350	155	55	0.19	40,320	64,000	79,000
RO WTP No. 3		16	450	150	62	0.22	53,760	119,000	144,000
Throttle[d]									
RO WTP No. 1		15	350	145	52	0.18	50,320	47,000	59,000
RO WTP No. 2	ESPA2	15	350	145	52	0.18	50,320	60,000	78,000
RO WTP No. 3		14.5	450	140	51	0.21	63,760	111,000	141,000

NA Data not available
a All costs presented in 2001 $
b Modeled using RODESIGN V.6.4
c Replacement cost = $480/element (December 2000) [1]
d 15 psi stage 1 permeate throttle; $10,000 per train is added to capital cost to account for required upgrades
e 40 psi interstage pressure boost; $20,000 per train is added to capital cost to account for required upgrades

Table 5 Membrane replacement alternatives *(continued)*

Alternative	Membrane (s)	Flux (gfd)	Capacity by Train (gpm)	Feed Pressure[b] (psi)	TDS (mg/L)	Fluoride (mg/L)	Cost[a] Capital[c] ($/train)	Cost[a] Power ($/yr)	Cost[a] Annual ($/yr)
Hybrid									
RO WTP No. 1		15	350	130	120	0.41	40,320	40,000	50,000
RO WTP No. 2	ESPA2:ESPA1	15	350	130	120	0.41	40,320	54,000	68,000
RO WTP No. 3		14.5	450	127	115	0.48	53,760	101,000	126,000
Booster[e]									
RO WTP No. 1		15	350	120	48	0.17	60,320	40,000	54,000
RO WTP No. 2	ESPA2	15	350	120	48	0.17	60,320	53,000	74,000
RO WTP No. 3		14.5	450	115	47	0.20	73,760	94,000	128,000

NA Data not available
a All costs presented in 2001 $
b Modeled using RODESIGN V.6.4
c Replacement cost = $480/element (December 2000) [1]
d 15 psi stage 1 permeate throttle; $10,000 per train is added to capital cost to account for required upgrades
e 40 psi interstage pressure boost; $20,000 per train is added to capital cost to account for required upgrades

Table 6 RO WTP No. 3 replacement alternatives

Alternative	Train (No.)	Flux (*gfd*)	Capacity (*gpm/ train*)	Train Feed Pressure (*psi*)	Membrane Feed Pump Pressure (*psi*)	RO Feed Flow (*gpm*)	RO By-Product Flow (*gpm*)	Blend Flow (*gpm*)	Finished Water (*gpm*)	Capital Cost (*2000 $*)	Power Cost (*$/yr*)	Unit Cost[a] (*$/1000 gallons*)
Current[b]	1	16	450	150	170	2,200	450	145	1,896	–	119,000	0.12
	2	16	450	150								
	3	16	450	150								
	4	14	400	170								
Replacement Option A[c,d]	1	14	400	140	150	2,275	475	150	1,950	107,520	113,000	0.12
	2	14	400	140								
(Replace 2 Trains)	3	16	500	127								
	4	16	500	127								
Replacement Option B[c,d]	1	14.5	450	117	127	2,250	450	150	1,950	215,040	101,100	0.12
	2	14.5	450	117								
(Replace 4 Trains–Same Capacity)	3	16	450	117								
	4	16	450	117								

a Includes only the membrane replacement cost (i.e., amortized over 10 years, 6% interest) and annual power cost; costs of chemicals are not included, but remain constant at $0.03/1000 gal

b 12:1 blending (blending (i.e., RO treated water:Middendorf well water); 32 mg/L $CaCl_2$ for post-treatment stabilization and corrosion control

c Modeled using RODESIGN V.6.4., the Rothberg, Tamburini & Winsor Model V.4.0

d 12:1 blending (i.e., RO treated water:Middendorf well water); 32 mg/L $CaCl_2$ for post-treatment stabilization and corrosion control

Table 6 RO WTP No. 3 replacement alternatives *(continued)*

Alternative	Train (*No.*)	Flux (*gfd*)	Capacity (*gpm/train*)	Train Feed Pressure (*psi*)	Membrane Feed Pump Pressure (*psi*)	RO Feed Flow (*gpm*)	RO By-Product Flow (*gpm*)	Blend Flow (*gpm*)	Finished Water (*gpm*)	Capital Cost (*2000 $*)	Power Cost (*$/yr*)	Unit Cost[a] (*$/1000 gallons*)
Replacement	1	16	500	127								
Option C[c,d]	2	16	500	127								
(Replace 4	3	16	500	127	137	2,500	500	167	2,167	215,040	126,000	0.13
Trains–	4	16	500	127								
Increase												
Capacity)												

a Includes only the membrane replacement cost (i.e., amortized over 10 years, 6% interest) and annual power cost; costs of chemicals are not included, but remain constant at $0.03/1000 gal

b 12:1 blending (blending (i.e., RO treated water:Middendorf well water); 32 mg/L $CaCl_2$ for post-treatment stabilization and corrosion control

c Modeled using RODESIGN V.6.4., the Rothberg, Tamburini & Winsor Model V.4.0

d 12:1 blending (i.e., RO treated water:Middendorf well water); 32 mg/L $CaCl_2$ for post-treatment stabilization and corrosion control

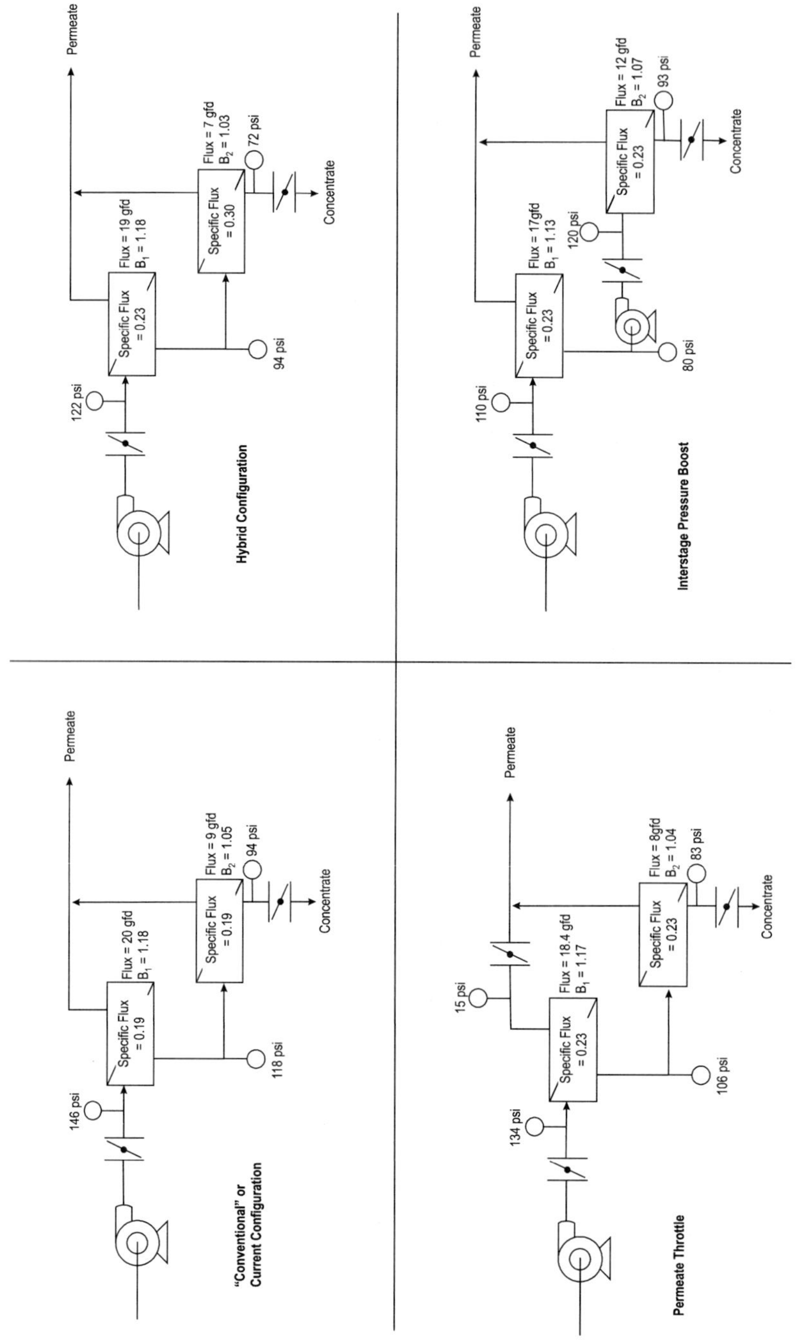

Figure 1 Membrane replacement alternatives

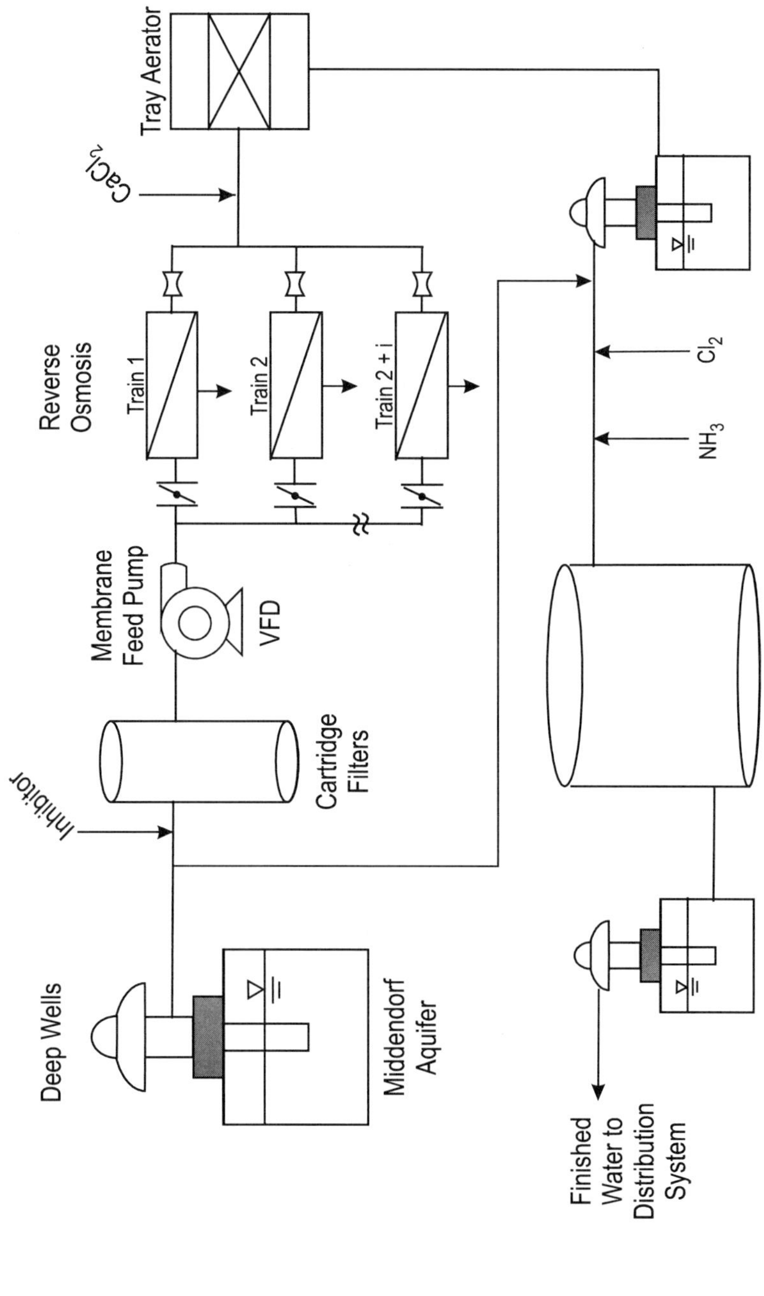

Figure 2 Water treatment process flow diagram

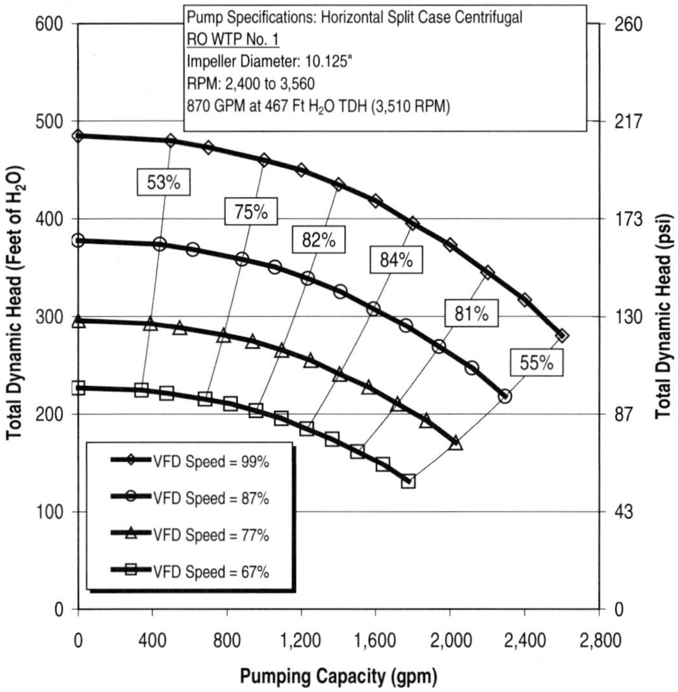

Figure 3 Membrane feed pump curve, RO WTP No. 1

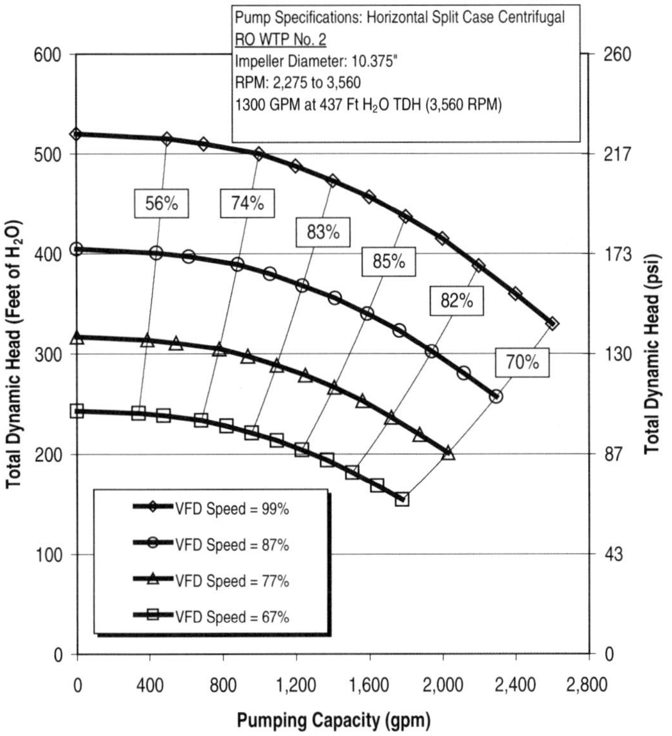

Figure 4 Membrane feed pump curve, RO WTP No. 2

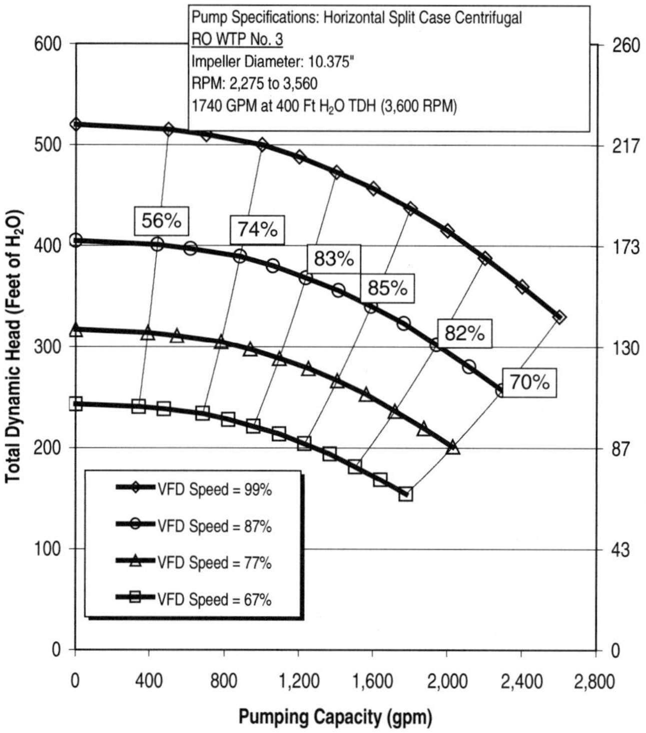

Figure 5 Membrane feed pump curve, RO WTP No. 3

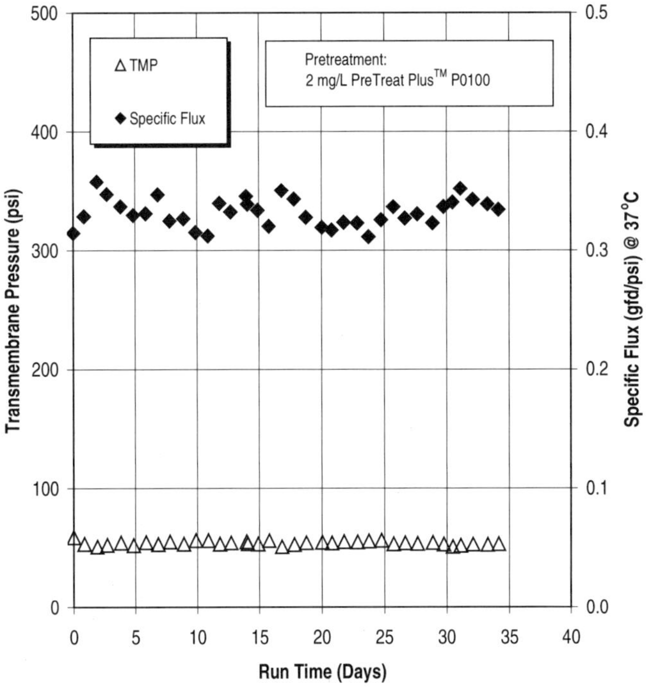

Figure 6 ESPA1 transmembrane pressure and specific flux at 18 GFD @ 37°C, 80% recovery

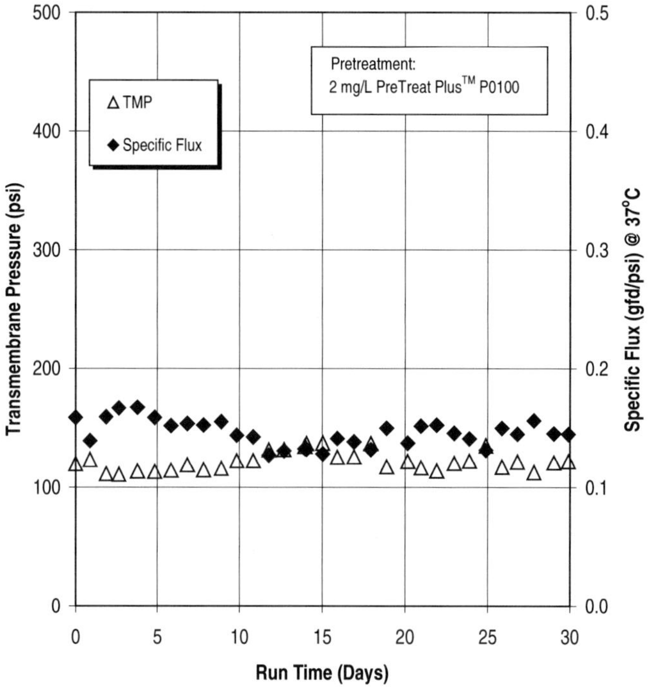

Figure 7 ESPA2 transmembrane pressure and specific flux at 18 GFD @ 37°C, 80% recovery

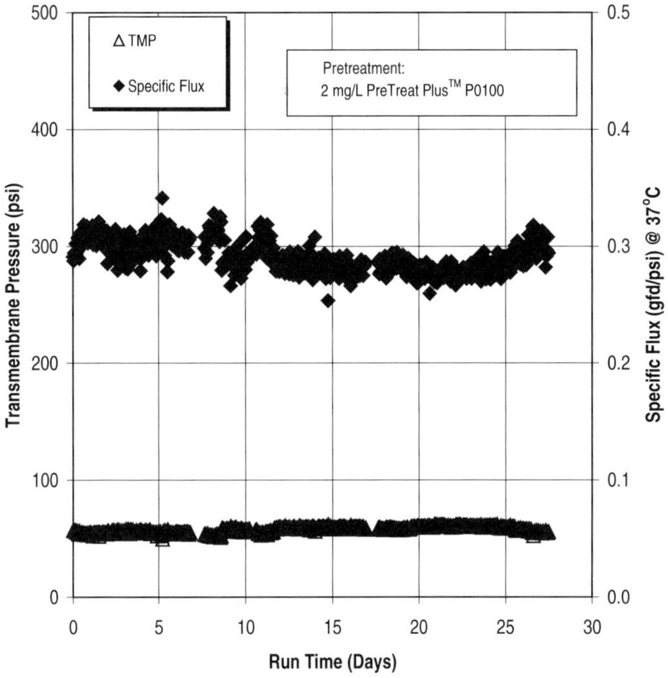

Figure 8 BW30 transmembrane pressure and specific flux at 18 GFD @ 37°C, 80% recovery

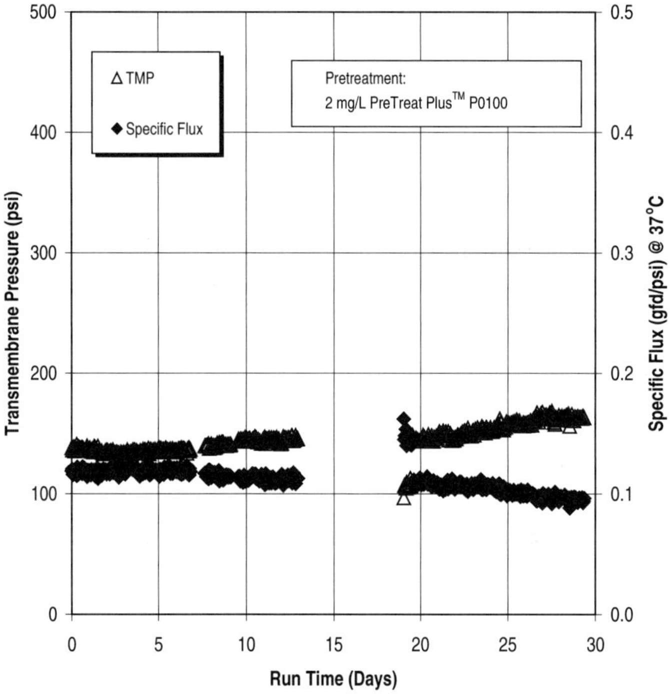

Figure 9 TFC-ULP transmembrane pressure and specific flux at 18 GFD @ 37°C, 80% recovery

CHAPTER · 25

First-Year Operation of the Méry-sur-Oise Membrane Facility

Claire Ventresque
Vivendi / Générale des Eeaux, Paris la Défense, France

Valérie Gisclon
Vivendi / Générale des Eeaux, Méry-sur-Oise, France

Guy Bablon
Vivendi / Générale des Eeaux, Paris la Défense, France

Gérard Chagneau
Syndical des Eaux d'Ile de France, Paris, France

For over a year the SEDIF (Syndicat des Eaux d'Ile de France) Méry-sur-Oise water treatment plant has supplied 800,000 inhabitants of the northern Paris region with water having undergone membrane treatment. This new facility was started up in 1999. A large number of publications [1, 2, 3, 4] have already described why membrane filtration was chosen to treat water from the River Oise. The water to be treated is a polluted river water whose temperature can vary between 1 and 25°C (33 to 77°F) and whose organic matter load can considerably increase according to the season. This facility (Figure 1), highly innovative in the water treatment sector, uses the nanofiltration membrane filtration process. It is innovative both in the process it uses and because of all the process controls that have been introduced.

The facility has over 1,000 computerised synoptic control panels, 450 online sensors, and the processes are entirely automated. This means that the facility operation systems are able to provide a real-time evaluation of the degree to which the nanofiltration membranes are clogged. Membrane cleaning is entirely automated

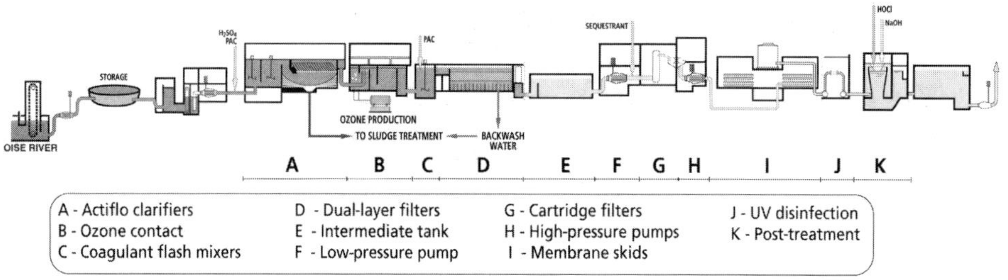

Figure 1 Process, Tranche 2

and the operator has the possibility of using his own chosen cleaning formulas.

The year during which the facility has been operational now provides us with a basis on which to establish a membrane efficiency assessment. As membrane clogging is closely linked to the quality of the supplied water, pre-treatment performance levels are also covered in this article.

The various improvements concerning the use of the membranes and their cleaning are also described.

DESCRIPTION OF THE TREATMENT STEPS

Pre-treatment

The pre-treatment [5] is engineered to supply the nanofiltration system with a maximum capacity of 180,000 m^3/d (45.5 MGD). It comprises a first flocculation settling stage using ACTIFLO weighted floc, followed by ozonation and filtration on dual-layer filters (sand and anthracite). Coagulation is carried out using WAC HB (polyaluminium chloride) and an anionic polyelectrolyte that, during periods of heavily polluted raw water, permits a considerable reduction of injected PAC levels.

Coagulation is carried out at a pH of 6.9 by adding sulphuric acid to reduce the quantity of dissolved aluminium in the water supplying the membranes.

Prefilters

Following pumping, prefiltration using 6 µm cartridges utilises 8 prefilters*, each with 410 cartridges†. These entirely automated prefilters backwash at frequencies calculated according to the quantity of particles (particles larger than 1.5 µm) contained in their supply water. Except when pre-treatment incidents occur, these frequencies range between 24 and 36 hours.

In addition, chemical cleaning is periodically carried out to extend the service life of the cartridges. This cleaning is automatically triggered as soon as a prefilter head loss attains approximately 650 mbars (10 psi). In these conditions, the expected life of the cartridges is 5 years.

The purpose of the prefiltration is to capture the particles that might clog or even damage the membranes. The prefilters act as a circuit-breaker should the water leaving the dual-layer filters contain an overly high level of suspended matter. The prefilters provide a reduction of at least 85% of particles larger than 1.5 µm.

Filtration on Nanofiltration Membranes

The 9,120 spiral-wound membranes of the facility are arranged in rows of 6 in 1,520 pressure vessels (PV). Each of the 8 membrane trains (one of which is presented in Figure 2) receives 860 m^3 of water per hour (3,700 gpm). The water is pumped by a variable-speed high-pressure pump able to deliver a pressure of between 5 and 15 bars (72 to 217 psi) at the membrane intake. The 85%

* Prefilters provided by Pall France, 3 rue des Gaudines, Boite Postale n°5253, 78175 Saint Germain en Laye, France

† Septra from Pall France

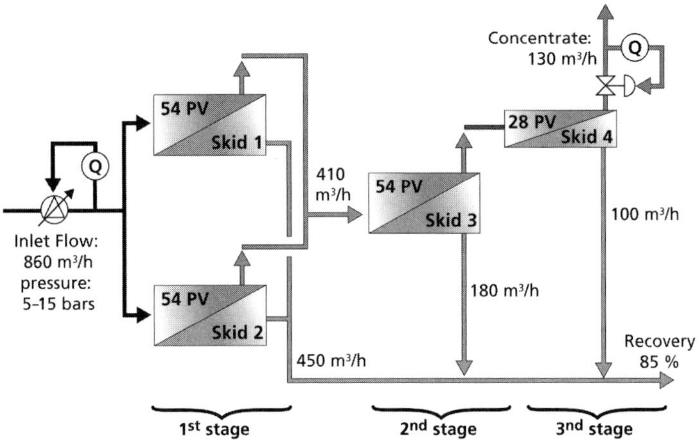

Figure 2 Diagram showing a nanofiltration membrane train

production yield is maintained at a constant level through the use of a control valve placed on the third-stage concentrate.

A membrane train comprises 4 skids: two 54 PV skids for the stage, one 54 PV skid for the second stage, and a 28 PV skid for the final stage. The water flow rates, the intake and outlet pressure levels, the longitudinal head losses and conductivities are individually measured for each of the skids. The longitudinal head loss* (noted as delta L) is the pressure loss between the inlet pressure and the concentrate pressure of a skid.

The spiral-wound Filmtec NF200 B-400 membranes†, specially engineered to treat water from the River Oise, retain almost all organic matter while permitting the passage of calcium [6].

* delta L = P at inlet of the skid – P concentrate

† DOW Filmtec, Minneapolis

Post-treatment

To begin with, post-treatment comprises degassing of the CO_2 contained in the permeate (using degassing towers). Five UV reactors are installed as a safety measure (radiation is 25 millijoules per cm^2). Finally, the permeate is rebalanced using sodium hydroxide. There is no need for adding minerals.

PRE-TREATMENT AND PREFILTER PERFORMANCE LEVELS

River water undergoes considerable quality variations over the year. Pre-treatment ensures the production of a constant quality of water throughout the year, no matter what the turbidity of the intake water (between 8 and 60 NTU). Figure 3 shows that the injected coagulant level (PAC) essentially depends on the turbidity of the supply water. The coagulant rate is controlled to maintain a turbidity of 1.1 NTU at the ACTIFLO outlets. As a result, the number of particles larger than 1.5 µm per millilitre is below approximately 5,000.

The ozonation at the ACTIFLO outlets acts as a pre-ozonation, meaning that it improves the recoagulation of particles that were not removed during settling. A second PAC injection is then carried out prior to the dual-layer filtration. When associated with ozonation, this results in a very high-quality water (Table 1). The rate of this second coagulation [7] is between 5 and 10 $g.m^{-3}$ for an ozone rate of approximately 1 $g.m^{-3}$. Online aluminium analysers*, and particle† and turbidity‡ counts allow the water filtered by the dual layer filters to be monitored in real time.

* AL9000 aluminium analyser from Environnement SA—111 Boulevard Robespierre 78300 POISSY France.

† HIAC Royco 2000

‡ Turbidimeter SIGRIST CT65IR

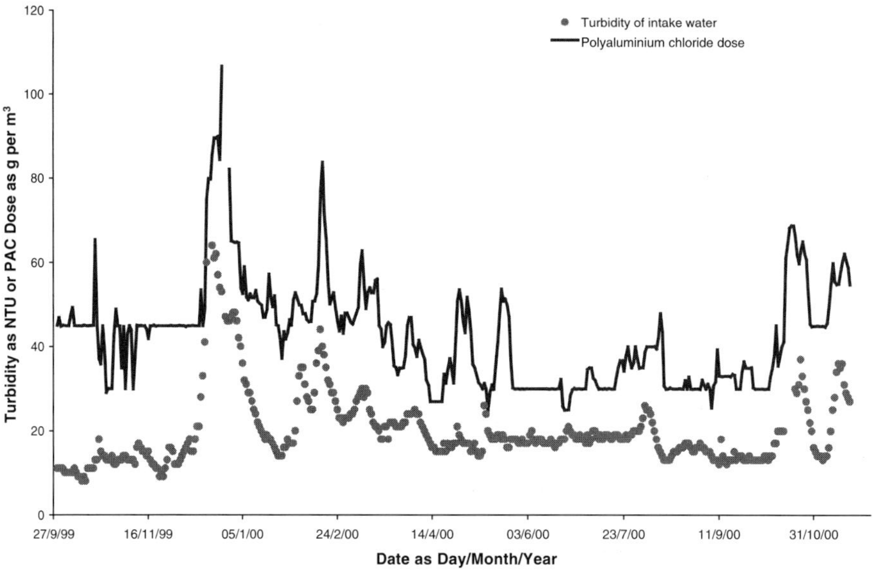

Figure 3 Rate of WAC HB (g.m^{-3}) injected into the ACTIFLO according to the turbidity of the supply water

Table 1 Water quality along the pre-treatment steps. Period Sept 99 to Nov 00—Mean values for the period.

Parameter	Intake Water	Outlet ACTIFLO	After Dual-Layer Filters	After Prefiltration
Particles-related parameters:				
SS (mg/l)	18.2	2.21	nm	nm
Turbidity (NTU)	19.8	1.1	0.05	nm
Particle count > 1.5 μm per ml	nm	around 5,000	24.8	3.8
Particle count > 0.5 μm per ml	nm	nm	nm	7,987*
Total aluminium (μg/l)	nm	nm	< 20	nm
Organics:				
TOC (mgC/l)	4.2	nm	2.2	2.2
UV adsorption (10^3 cm^{-1})	106.8	nm	nm	37.1

*Particle counts above 0.5 μm varied between 5,000 and 12,000
nm: No measurement

Dissolved aluminium is drastically reduced thanks to acidification during coagulation and this limits membrane clogging. Pre-treatment also reduces the TOC by between 30 and 60%. The pre-treatment bacteriological reduction presented in Figure 4 is approximately 4 log.

As can be seen, the prefilters are supplied with a water that has a low particle content (except during peak periods); turbidity and aluminium are reduced to a minimum. However, as shown in Figure 5, there is a high TOC variation in the river that is reflected on the water at the inlet of the nanofiltration.

The effect of prefiltration on the microbiological quality of the water was checked by a measurement campaign. No microorganism retention was revealed. Figure 6 shows that the microbiological

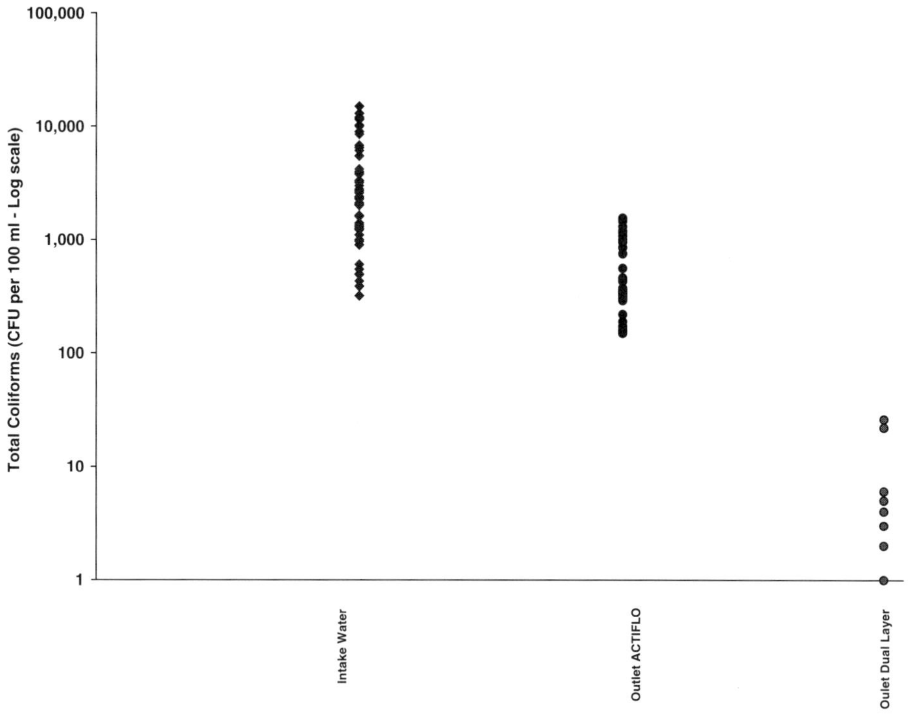

Figure 4 Effect of pre-treatment on the number of total coliforms (CFU per 100 mL)

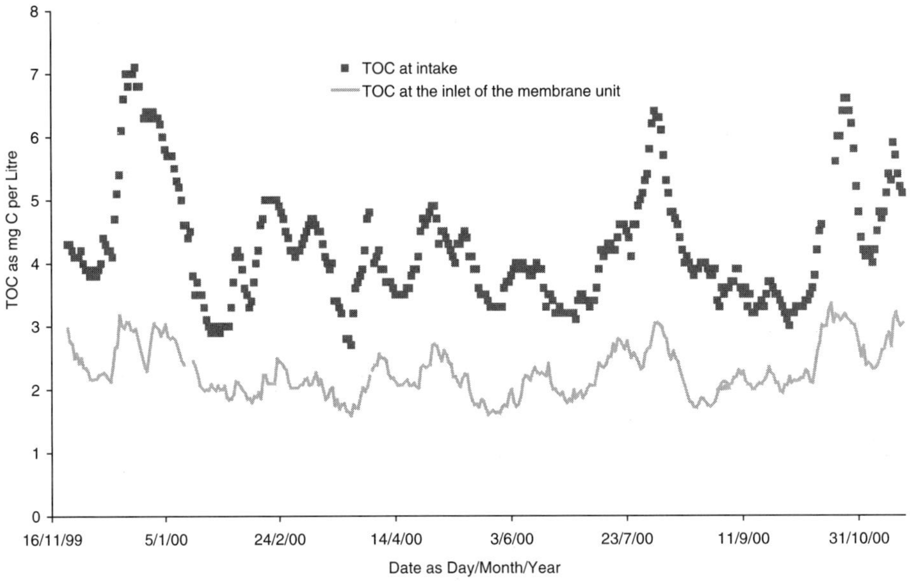

Figure 5 TOC (mg C.l^{-1}) variation in the water at the inlet of the plant and at the inlet of the membrane skids

quality of the water at the prefilter outlet is identical to that entering the prefilters. This was foreseeable given that the size of the bacteria is smaller than the filtration cartridge cutoff point.

Clogging Caused by Aluminium

The clogging of nanofiltration membranes can take several forms. The quality of the water supplying the membranes is dependent on the quality of the river water, which varies according to the season [8]. As membrane clogging is very closely linked to the quality of the supply water, the different seasons of the year lead to characteristic clogging levels.

The precipitation of calcium carbonates in the membrane concentrate depends on the pH of the supply water. Thanks to the

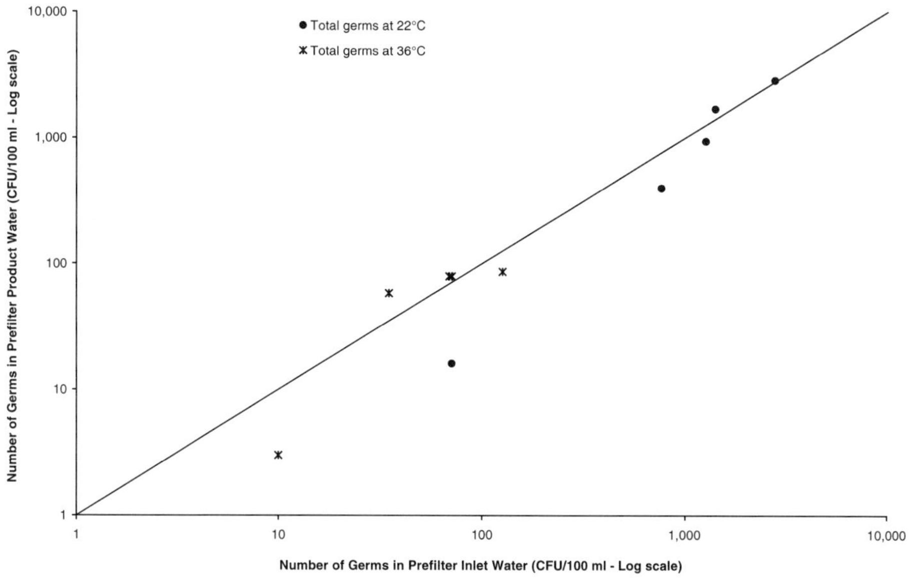

Figure 6 Comparison of microorganism quantities before and after prefiltration: total germs at 36°C and total germs at 22°C (expressed as UFC/100 ml)

use of an antiscalant*, the pH of the water entering the membranes can be lowered to a value of only 7 in order to prevent the precipitation of calcium carbonates in the concentrate. This leads to great savings in the quantity of sulphuric acid used. The usage of sulphuric acid is restricted in Méry-sur-Oise because it would increase the amount of sulphate ion in the water, thus leading to sulphate precipitates very difficult to remove. The usage of chlorhydric acid, which would not have presented this drawback, is not preferred for safety reasons.

For Méry-sur-Oise, the intention was to lower the pH at the inlet of the plant, as soon as the coagulation step. This presented

* Hypersperse AF200 by BetzDearborn at 2 g per m^3 of inlet water

the advantage of increasing the elimination of TOC and aluminium as from the pre-treatment stage.

During start-up, a malfunction occurred at the pre-treatment acidification stage during May and June 1999. This led to the pH being lowered to 7 by directly adding sulphuric acid at the membrane train intakes. This resulted in the operating membrane trains becoming rapidly clogged at the third stages.

Analyses carried out on the concentrates of the third stages revealed high aluminium concentrations. Figure 7 shows that this clogging might be caused by the presence of a large quantity of aluminium in the membrane concentrate. The loss of permeability over time increases alongside the amount of aluminium to be found in the concentrate.

Studies carried out on pilots revealed an optimal pH for coagulation with an aluminium salt. In accordance with the

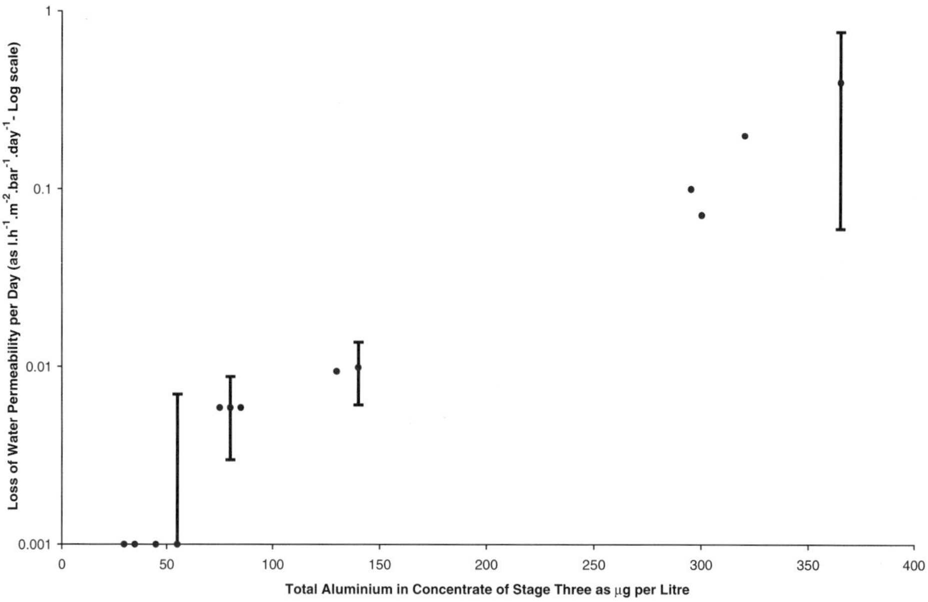

Figure 7 Loss of membrane permeability as a function of the quantity of aluminium in the membrane concentrate of third stages using a standardized water permeability (standardized at 25°C)

aluminium solubility curves, the minimum quantity of residual dissolved aluminium is obtained with a pH of between 6.8 and 7. In addition, the TOC reduction during pre-treatment also depends on the pH and is best provided within this pH range (6.8 to 7).

If coagulation is not carried out at an optimal pH, aluminium may be precipitated whenever acid is subsequently added. This is amplified in the membranes as, in addition to this precipitation, there is also the aluminium concentration factor on the concentrate side of the membranes.

MONITORING NANOFILTRATION MEMBRANES

In the plant, the membranes are monitored online. Flow rate, pressure and conductivity sensors installed on each skid allow a real-time and automatic calculation of membrane water permeabilities, the produced flows, the longitudinal head loss factors, etc., for each skid.

The values provided by the online sensors are transmitted to "machine" controllers. These controllers represent the first process control level and are used to run the machines and the processes. They manage all sensor measurements, alarm triggering and system faults. All membrane train calculations are carried out in real time. The upper control level, being level 2, is a central calculator supervising the entire facility. It receives information from the lower level (level 1) and can be used to print histories expressed in graphic form.

In addition, each membrane train also has an individual control system. This system manages the sensor measurements and calculations concerning the train. The 8 control systems of the 8 trains are supervised by a central fail-safe* control system (Figure 8). This central system manages the trains in relation to one

* A fail-safe control system is a dual system where every component is doubled in order to avoid any failure.

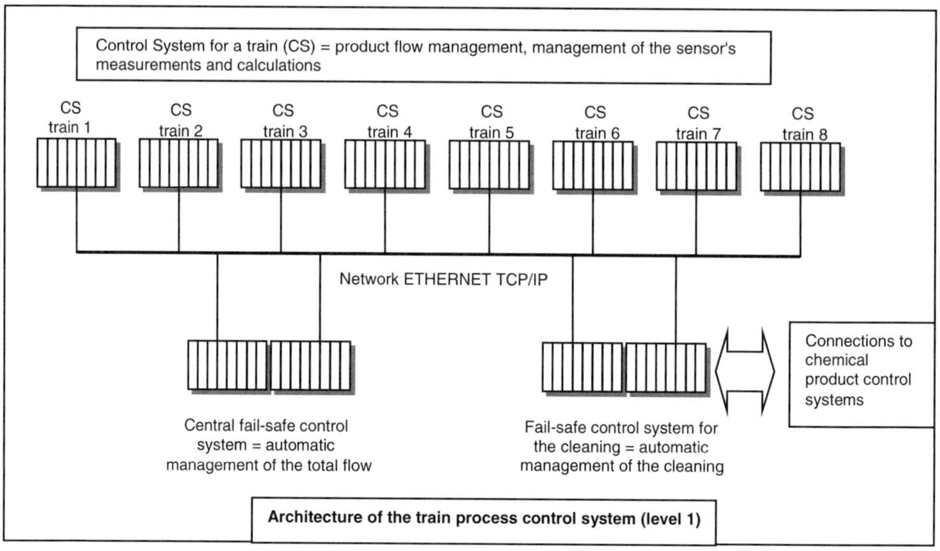

Figure 8 Simplified architecture of the membrane train process controls (level 1)

another and according to the total flow rate to be produced. For the cleaning, a fail-safe control system manages the 8 control systems of the 8 trains.

Membrane water permeabilities are standardized at 25°C (77°F) and calculated by taking into account the osmotic pressure to either side of the membrane, using conductivity meters installed in the permeate and concentrate of each skid. The longitudinal head losses are standardized relative to the temperature and the flow rate on the concentrate side. This leads to a calculation of a head loss factor. When the trains were first put onstream, longitudinal head loss curves were traced out according to the flow rate through each pressure vessel. The head loss factor corresponds to the increase of a longitudinal head loss caused by a potential clogging and also takes any flow rate fluctuations into consideration. Calculations are also standardized at 25°C.

The percentages of standardized permeability loss or increase in the head loss factor are calculated on the basis of each of these parameters to trigger the clogging thresholds.

Two thresholds are calculated for the water permeability. A loss percentage based on the initial condition of the stage following cleaning and a loss calculation based on the water permeability when the train was started up. These calculations both establish the clogging level of a membrane train during its production cycle and evaluate its evolution over time in terms of its operating capacity, as well as making it possible to estimate the efficiency of a cleaning operation. When one of the thresholds is reached on one of the trains, the train control system generates a cleaning order for the concerned train. A 25% standardized water permeability loss threshold or an increase in the standardized longitudinal head loss on one of the stages triggers a cleaning order.

To ensure this monitoring, it is very important that the measurements given by the sensors are reliable. Consequently, to validate the measurements, calculations are carried out that check the sensors against one another to see if there is a drift in one or more of the devices. Three types of balance are carried out on each membrane skid:

- For intake and outlet flows:

$$\frac{\text{Qinlet} - (\text{Qpermeate} + \text{Qconcentrate})}{\text{Qinlet}} = \text{x}\,\%$$

- For intake pressures and longitudinal head losses:

  P at skid entry – skid delta L = P at skid entry of the next stage

- For conductivity meters by carrying out balances based on measured flows:

$$\frac{\begin{array}{c}\text{Quantity of ions in inlet} -\\ (\text{quantity of ions in concentrate} +\\ \text{quantity of ions in permeate})\end{array}}{\text{Quantity of ions in inlet}} = \text{Y}\,\%$$

When the error percentages are too great (greater than 5% for flow rates and pressures, and greater than 20% for the conductivity meters) "sensor discordance alarms" are triggered, indicating that the operator needs to carry out maintenance works on one or more devices. Most of the time, there is less than 1% of errors recorded on the flow rate and pressure balances. This means that operators can have great confidence in the permeability and head losses values given by the system.

MEMBRANE CLEANING

Membrane cleaning is entirely automatic. Two hydraulic loops each have a 40 m^3 cleaning solution preparation tank, a variable-speed pump and a cartridge prefilter. The permeate is used to prepare the solutions. It is heated to 58°C (140°F) in a 60 m^3 tank.

Four chemical products are available on the site to clean the membranes. Two basic products: a detergent* and sodium hydroxide; an acid product that is citric acid and, last, a disinfectant formulated from acetic acid, peracetic acid and hydrogen peroxide. Because all these products are in liquid form, the preparation of the cleaning solutions can be automated.

There are three ways in which the cleaning solutions can be circulated through the hydraulic loops downstream from the membrane skids:

- A return to the corresponding 40 m^3 tank allowing the solutions to be recirculated in the skids (closed circuit recirculation sequences)
- Drainage to the sewers if the pH of the solution to be drained is between 6.5 and 8.5
- Drainage to the neutralisation tanks when the pH of the solution to be drained is lower than 6.5 or greater than 8.5

* Basic detergent: P3 Ultrasil 110 by Henkel

These three neutralisation tanks (150 m^3 each) allow the pH of the solutions to be neutralised before being discharged into the sewers.

The variable-speed pumps in the loops allow a range of flow rates to be covered. Two types of flows are used: a slow flow rate of 2.1 $m^3.h^{-1}$ per pressure vessel* (9.25 gpm per vessel) and a rapid flow rate of 7.8 $m^3.h^{-1}$ per pressure vessel (34 gpm per vessel). This permits a flow rate ranging from 60 $m^3.h^{-1}$ (264 gpm) to 420 $m^3.h^{-1}$ (1,850 gpm) that is then used to clean a 54-vessel or 28-vessel skid.

Each skid in each train can be individually connected to each of the two hydraulic loops. Because there are two cleaning loops, they can work in parallel on separated skids and have a maximum level of availability. For example, loop 1 can clean successively skid 1 then 3 in a train, while loop 2 cleans skid 2 then 4.

The membrane cleaning system allows the operator to choose the type of cleaning formula he prefers. The operator can define the following different parameters for each type of product:

- The preparation temperature of the solutions
- The quantities of product used
- The various soaking and recirculation sequences of the solutions in the membranes

The flow rates cannot be modified.

A soaking sequence corresponds to shutting down the pump and closing the skid valves so that the membranes can soak in the cleaning solution. A recirculation corresponds to operating the pump in closed circuit mode, in other words, the solution leaving the membrane skid (permeate and concentrate) is recirculated in the corresponding tank.

Water rinsing is carried out following each product sequence. These rinsing operations are necessary for safety reasons and

* Pressure vessel: housing 6 membranes in series

cannot be cancelled by the operator. These rinsings are carried out in two phases:

1. A slow flow rate phase (2.1 $m^3.h^{-1}$ per vessel) with drainage to the sewer or to the water neutralisation tank (depending on its pH)

2. A rapid flow rate recirculation phase in the tank (7.8 $m^3.h^{-1}$ per vessel).

The cycle times are adjustable for each of these rinsing phases. It is possible to extend the drainage to the sewer phase in the same way that it is possible to increase the rapid flow rate recirculation time for the membrane rinsing water.

As a result, the operator defines the work stages programme for two cleaning loops in such a way that the two units work in parallel with a common hot water management and, as a result, a maximum time gain. A standard cleaning formula, applicable in most cases, has been defined and programmed via the computerised supervision system. For the NF 200 B-400 membranes, the standard procedure is defined as follows:

- First product used: basic detergent P3 Ultrasil 110 at 0.3% at 30°C (86°F)

- Second product used: citric acid at 0.3% at 30°C

The procedure is as follows for each of these products:

1. Slow flow rate drainage of the volume of water contained in the membrane skid in such a way that it is replaced by the cleaning solution (plug flow)

2. Soaking for 30 minutes

3. Rapid flow rate recirculation for 15 minutes

4. Soaking for 30 minutes

5. Rapid flow rate recirculation for 30 minutes

Rinsing using permeate is then carried after each product cleaning operation, as follows:

1. Slow flow rate rinsing to the sewer or to the neutralisation tanks until a pH at the skid outlet of between 6.5 and 7.5 is obtained
2. Rapid flow rate closed circuit rinsing for 10 minutes

The time required to clean the three stages (4 skids) of a train is between 24 and 48 hours.

Cleaning requires a large quantity of hot water. The energy needed to clean the membranes (Figure 9) depends on the temperature of the water to be heated. This energy is used to heat the cleaning solutions.

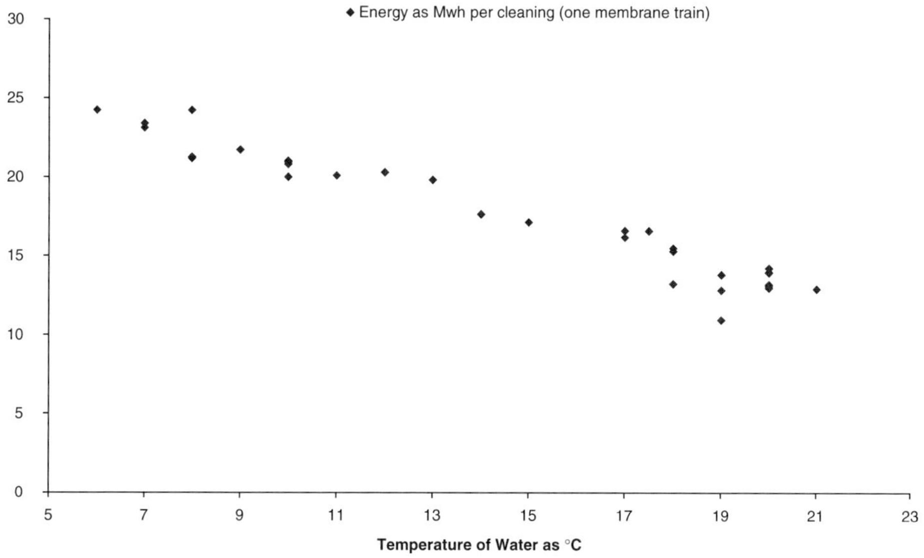

Figure 9 Energy needed to clean a membrane train in accordance with the water temperature.

EVOLUTION OF MEMBRANES AND CLEANING OPERATIONS

Figure 10 shows the evolution of a train's water permeability from August 1999 to December 2000.

When the trains were commissioned, the standardized water permeabilities were 7 l.h-1.m-2.bar-1 and fell to 6 l.h-1.m-2.bar-1 after several weeks of operation. However, permeability losses were visible after each cleaning operation (see cleaning numbers 1, 2, 3, 4 in Figure 10). No clogging was observed during the cycle, but, amazingly, cleaning seemed to affect the permeability of the membranes. It is possible that during the cleaning operations, intermediary rinsing (between the detergent and the acid cleaning operations) carried out using cold water did not fully rinse off the

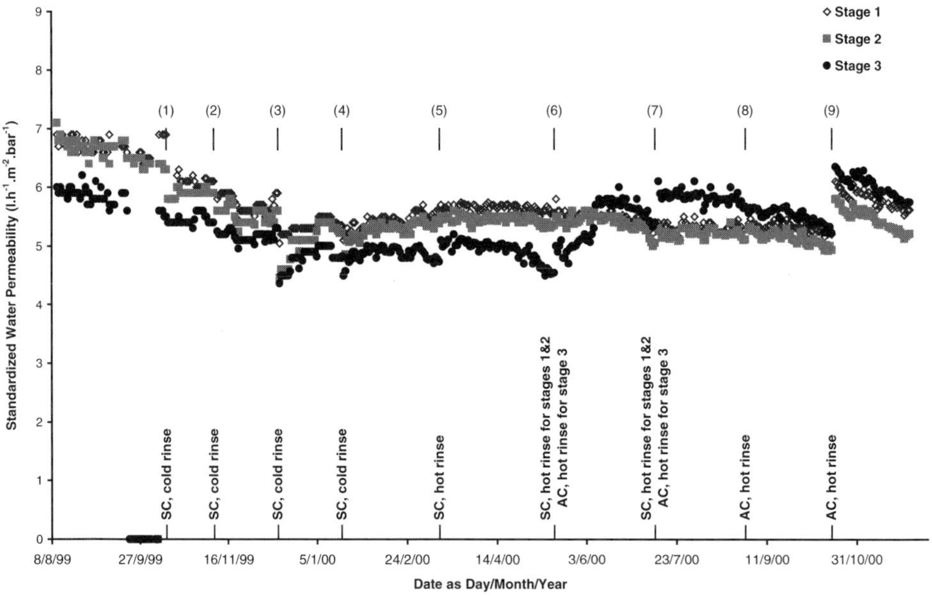

Figure 10 Monitoring of standardized water permeabilities for the three stages of train 1. Indication of the various cleaning operations carried out: SC: standard cleaning as described in paragraph 4; AC: adapted cleaning

residual detergent, which then precipitated when it came into contact with the acid. The river water is occasionally only at 1°C and, as a result, the efficiency of rinsing operations is diminished. Cleaning with hot rinsing was introduced as of February 2000.

Cleaning Tests on a Test Rig

To check the efficiency of hot rinsing directly following the product phases, cleaning trials on a test rig were carried out on six clogged NF 200 B-400 membranes. The standardized permeabilities were measured before and after each type of cleaning. Cleaning was carried out in compliance with the standard formulas applied on the big plant.

The first product used was the P3 Utrasil 110 detergent (at 0.3% and 30°C) followed by citric acid (at 0.3% and 30°C). Different types of rinsing were tested after each product:

a. Cold water rinsing (at 2°C) comprising its two phases (slow and rapid flow rates)

b. Cold water rinsing (at 2°C) followed by recirculation of water heated to 30°C

c. Hot water rinsing at 30°C comprising two phases (slow and rapid flow rates) directly following the products

Two trials were carried for each of the tests (A, B, C) and led to the results presented in Table 2.

Table 2 Efficiency under different types of rinsing expressed in % of increased (+) permeabilities. Kw = standardized water permeability.

	Cleaning Conditions		
(Kw before cleaning—Kw after cleaning)/Kw before cleaning expressed as percentage	Cold rinsing (A)	Cold then hot rinsing (B)	Hot rinsing (C)
Standardized water permeability	+1 and +3%	+ 5 and +9%	+12 and +15%

The test results confirm the inefficiency of the cleaning when it is carried out using cold rinsing following the product phases. For case (A), the increased permeability is not sufficiently high given what is expected from a cleaning operation. However, and all else being equal, it can be noted that cleaning using hot rinsing is more efficient in recovering the water permeability of the membranes.

Evolution of the Membranes

Regular monitoring of membrane standardized water permeability, as shown in Figure 10, allows us to verify the impact of changing the type of cleaning. In fact, the permeability losses related to the cold rinsing cleaning operations are no longer visible after cleaning operation number 4. Consequently, hot water rinsing is efficient on the big plant. Following cleaning operation number 5, with the permeability having fallen to around 5 $l.h^{-1}.m^{-2}.bar^{-1}$, more extensive cleaning operations were carried out [9]. The cleaning solution levels were adapted (cleaning N°6—onward). The soaking and recirculation times were increased. Figure 10 shows a progressive rise in water permeability after cleaning N°6. The last cleaning operation (number 9) saw the value increase to approximately 6 $l.h^{-1}.m^{-2}.bar^{-1}$. The expected effect was noted on all 7 other trains.

In addition, the importance of the water temperature on the energy expended by the high-pressure pumps can also be noted. Figure 11 shows the energy required to produce 1 m^3 of permeate for the first year of operation. This varies according to the temperature, and the lower the water temperature, the more energy is required from the high-pressure pumps to produce the same water flow rate. The design was set to run the membrane units at the same flux whatever the temperature or the degree of clogging (17 litres per hour per square metre of membrane—10 gfd). One should note that the actual operation of nanofiltration on such a scale leads to an affordable level of energy consumption given the water temperature in winter: the mean consumption energy is set

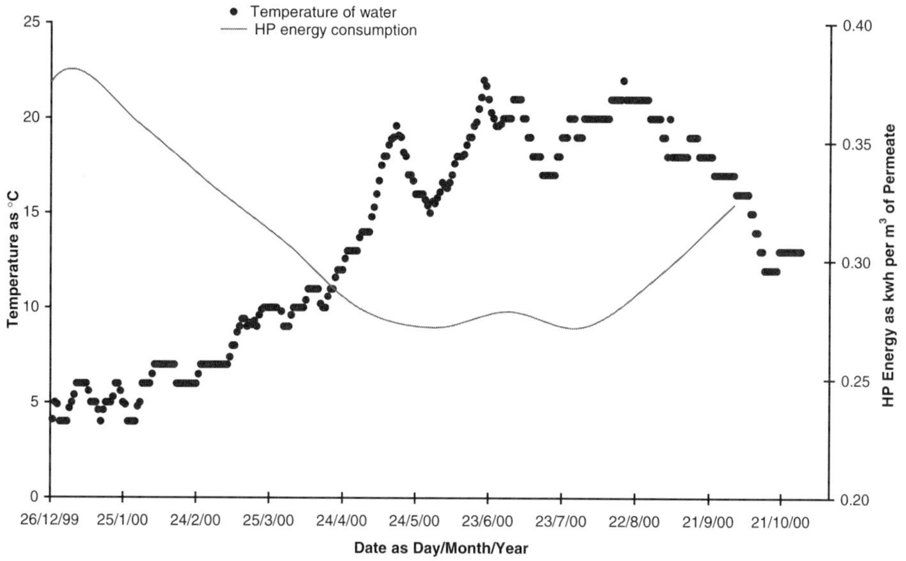

Figure 11 High-pressure pumping (HP) energy consumption in conjunction with variations in the water temperature

around 0.32 kWh/m^3 of permeate produced for this first year of operation.

QUALITY OF WATER PRODUCED BY THE NANOFILTRATION MEMBRANES

The Méry-sur-Oise nanofiltration membranes were specially engineered to treat water from the River Oise and, in particular, organic matter and pesticides. The goal was therefore to use membranes able to remove organic matter while permitting the passage of minerals and calcium. The TOC at the membrane intake can be as high as 3.5 mg.l^{-1}. During the first year of operation, the atrazine in the river has reached values of 620 nanog.l^{-1}.

The NF200 B-400 membrane permits the production of a water containing an average of 0.18 mgC/L TOC (Figure 12), and

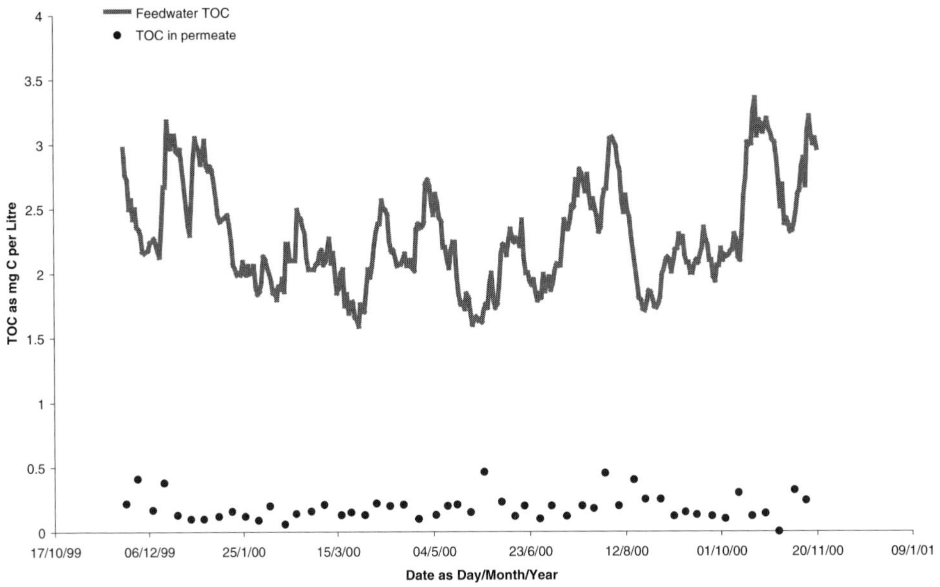

Figure 12 TOC content in feedwater and permeate (as mgC/L)

exempt from pesticides such as atrazine (concentration lower than the analysis detection limit of 50 μg.l^{-1}). It also permits, throughout the year, the production of a water containing an average 40 mg.l^{-1} of calcium.

CONCLUSION

The Méry-sur-Oise membrane facility has produced 140,000 m^3 of permeate per day since September 1999. Using the highly degraded and variable waters of the River Oise, this new facility allows the production of an exceptionally pure water. It contains very little organic matter and the biodegradable part of this organic matter is below the measurement detection limits. Thanks to a wise monitoring and control of the permeability (by means of an adapted pretreatment and cleanings), the consumption of energy required to perform nanofiltration of river water is kept to a minimum. The operation of the new facility is ensured by the on-site personnel assisted by a large number of automated control systems.

ACKNOWLEDGEMENTS

The authors thank Prof. Jan Schippers for his perspicacity and his invaluable assistance in setting up the ideal conditions for the running of the membranes.

REFERENCES

1. C. Ventresque, G. Bablon. The integrated nanofiltration system of the Méry-sur-Oise surface water treatment plant (37 mgd). *Desalination.* 113 (1997) 263–266.
2. K.M. Agekodo, B. Legube, P. Coté. Organics in NF Permeate. *J. AWWA*, May 1996 (88:5, 67).
3. A. Boireau, J. Cavard, G. Randon. Positive Action of Nanofiltration on Materials in Contact with Drinking Water. *Proc. IWSA. Corrosion Workshop*, Buenos Aires, Argentina 1996.
4. B. Legube, K. Agbekodo, P. Coté, M.M. Bourbigot. Removal of Organohalide Precursors by Nanofiltration. *Water Supply*, 1995 (13:2, 171).
5. C. Ventresque, V. Gisclon, G. Bablon, G. Chagneau. An outstanding feat of modern technology: the Méry-sur-Oise Nanofiltration Treatment Plant (340,000 m^3/d). *Conference on Membranes in Drinking and Industrial Water Production*, October 2000.
6. C. Ventresque, G. Turner, G. Bablon. Nanofiltration: from prototype to full scale. *J. AWWA*. October 1997 (65–76).
7. C. Démocrate, A. Plottu, G. Chéré, D. Gatel, P. Bonne, J. Cavard. New filtration solutions for high turbidity removal. *Conference IWA-ASPAC*, November 2000.
8. N. Her, G. Amy, C. Jarusutthirak. Seasonal variations of nanofiltration (NF) foulants: identification and control. *Conference on Membranes in Drinking and Industrial Water Production*, October 2000.

9. D. Paul. Obstacles to the effective chemical cleaning of a reverse osmosis unit. *Ultrapure Water.* October 1994 (33).

Part 6 Integrated Membrane Systems

Integration of Lime Softening and Ultrafiltration: A Powerful Water Quality Combination

Disinfection by Integrated Membrane Systems for Surface Water Treatment

Prediction of Full-Scale IMS Performance Using a Resistance Model and Laboratory Data

CHAPTER · 26

Integration of Lime Softening and Ultrafiltration: A Powerful Water Quality Combination

Steven J. Duranceau, PhD, PE
Jill A. Manning, PE
Robert K. Anderson, PE
Jackie Foster, EI
Boyle Engineering Corporation, Orlando, FL

H. John Losch, PE
Frank Kane
Craig Anderson
Florida Water Services Corporation, Marco Island, FL

BACKGROUND

In May 2000, the Marco Island Lime Softening Water Treatment Plant (LSWTP) was expanded from 5.0 to 6.67 million gallons per day (MGD) with the addition of a 1.67 MGD immersed membrane ultrafiltration (UF). The Marco Island LSWTP was previously limited to 5.0 MGD of production based on conventional sand filter capacity. This facility provides treatment of surface water from the Marco Lakes. Raw water from the Marco Lakes supply is pumped 9.5 miles to the Marco Island LSWTP. The raw water source contains elevated color (>40 cpu), total organic carbon (TOC) (>12 mg/L), UV-254 (0.181), hardness (>350 mg/L as $CaCO_3$), and moderate total dissolved solids (TDS) (>900 mg/L). Implementation of a full-scale 1.67 MGD UF process downstream of the Marco Island lime softening process demonstrated an average turbidity of 0.06 NTU versus an average sand-filtered finished water turbidity of 0.2 NTU.

INTRODUCTION

Marco Island, Florida, is located in Collier County, on the southwest coast of Florida. The island measures approximately 6 miles long by 4 miles wide and has experienced rapid population growth over the past several years. Marco Island is a popular tourist resort, having variable population changes, and is located off the southwest coast of Florida near Naples just west of the Everglades. Florida Water Services Corporation (Florida Water) owns and operates the Marco Island Water System (MIWS), which is composed of an ROWTP (6 MGD permeate capacity) that treats 10,500 mg/L brackish groundwater, and a lime-softening WTP (recently expanded to 6.67 MGD) that treats surface water on Marco Island.

Because of increased growth, increased system water demands, and more stringent regulations further defining the Safe Drinking Water Act (SDWA), Florida Water has been investigating treatment alternatives for expanding capacity and enhancing water quality within the MIWS. In May 2000, Florida Water re-rated the permitted capacity at the LSWTP by placing online a 1.67 MGD UF Demonstration unit supplied by Zenon Environmental Systems of Burlington, Ontario, Canada, and associated facilities at the Marco Island LSWTP. This chapter summarizes the results of the UF project that was implemented to evaluate the viability of a UF membrane process as a polishing step to conventional lime softening treatment of hard and highly natural organic and colored surface water supplies.

OVERVIEW OF ULTRAFILTRATION

Membrane processes that have applications in drinking water treatment include reverse osmosis (RO), nanofiltration (NF), ultrafiltration (UF) and microfiltration (MF), and electrodialysis reversal (EDR). Table 1 lists the regulated solutes controlled by membrane processes, including ultrafiltration. Note that combinations of membrane processes with other membrane or traditional

Table 1 Characteristics of membrane processes

			Regulated Solutes		
Process	Mechanism	Exclusion	Pathogens	Organics	Inorganics
EDR	C	0.0001 μ	None	None	All
RO	S, D	0.0001 μ	C, B, V	DBPs, SOCs	All
NF	S, D	0.001 μ	C, B, V	DBPs, SOCs	All
UF	S	0.001 μ	C, B, V	None	None
MF	S	0.01 μ	C, B	None	None

Mechanism: C=charge, S=sieving, D=diffusion
Pathogens: C=cysts, B=bacteria, V=viruses
Organics: DBPs=disinfection by-product precursors, SOCs=Synthetic Organic Compounds

treatment processes have become known as integrated membrane systems (IMSs). The coupling of MF and NF processes, or a combination of coagulation, sedimentation and filtration with a NF process, is an accepted example of IMSs. UF is a pressure-driven filtration process that utilizes hollow-fiber membranes capable of separating both insoluble and soluble materials from the treated water. Hollow-fiber membranes are either "inside-out" cross-flow membranes or "outside-in" transverse flow membranes. The applied pressures utilized with UF processes are much lower than feed pressures for NF or RO processes, and are referred to as the transmembrane pressures.

UF has been shown to be very effective for particle and turbidity removal [1, 2, 3]. Turbidity can be lowered to below 0.05 NTU on a consistent basis for a variable feedwater quality. Unlike MF, UF is capable of removing many viruses. Coliforms, bacteria, viruses, and cysts can be effectively removed from a water supply by UF. However, similarly to MF, UF does not effectively remove disinfection by-products or other dissolved substances from a water supply unless additional treatment (such as powdered activated carbon addition) is provided [4, 5, 6, 7].

UF membranes cover a wide range of molecular weight cut-offs (MWCOs) and pore sizes. UF membranes of 1,000 daltons may be employed for removal of colloidal organic materials from freshwater, while 10,000 dalton UF membranes are used for liquid/solid separation (that is, particle and microbial removal) [8]. Materials of construction for MF and UF membranes continue to improve as a result of recent advances in technology [9]. Improvements in fouling resistance and increased surface areas have been made. MF and UF modules can vary in terms of module dimensions, membrane material, pore size, capillary diameter, position (horizontal vs. vertical), inside-out and outside-in membranes, and air scour/chemical back-pulsing systems.

The US water quality requirements determine membrane selection. Many of the regulatory constraints for drinking water can be related to control of inorganic, organic or pathogenic solutes in the finished product. The specific application of membrane processes to drinking water applications is shown in simplified format in Table 2. The word "yes" indicates the membrane process can remove significant amounts of contaminate specified by the rule, and "no" indicates the membrane process cannot remove the regulated contaminate.

Existing Lime Softening Water Treatment Plant Description

Figure 1 presents the flow diagram of the MIWS showing the addition of the immersed UF membrane process. The Marco Island LSWTP is supplied by raw surface water withdrawn from the Marco Lakes water supply, which is located 9.5 miles from the island and on the adjacent mainland. Raw water is pumped to the Marco Island LSWTP, where it enters the WTP's solids contact unit (SCU). In the SCU, quick lime (CaO) is added in slurry form (average of 152 mg/L as CaO), in addition to alum (average of 10 mg/L), in order to aid settling. Chlorine (average of 8.6 mg/L), sulfuric acid, and ammonia (average of 1.7 mg/L) are added to the settled water in the effluent launder from the SCU.

The treated water (finished hardness averaging 160 mg/L as $CaCO_3$) travels through the overflow weir to the filters. Water passes through the filters (loading rate of 2.43 gpm/sf) to the

Table 2 Summary of membrane process applications for drinking water regulations

	Membrane Process				
US Regulation/Rule	EDR	RO	NF	UF	MF
SWTR/ESWTR	no	yes	yes	yes	yes
TCR	no	yes	yes	yes	yes
LCR	no	yes	yes	no	no
IOC	yes	yes	yes	no	no
SOC	no	yes	yes	yes	yes
Radionuclides	yes (-Rn)	yes (-Rn)	yes (-Rn)	no	no
DBPR	no	yes	yes	no	no
GWDR	no	yes	yes	yes	yes
Arsenic	yes	yes	yes	no	no
Sulfates	yes	yes	yes	no	no

SWTR: Surface Water Treatment Rule
ESWTR: Enhanced Surface Water Treatment Rule
TCR: Coliform Rule
LCR: Lead and Copper Rule
IOC: Inorganic Rule (Phases I, II, IIA, V)
SOC: Synthetic Organic Chemicals (Base Neutrals and Extractables)
DBPR: Disinfection By-Products Rule
GWDR: Groundwater Disinfection Rule

clearwell. Filtered water is treated with a phosphate inhibitor (1.9 mg/L) for distribution system corrosion control, and then is transferred to on-site ground storage tanks, or to the ROWTP's ground storage tanks, for blending. High-service pumps at both facilities are used to pump finished water from ground storage to the MIWS.

INTEGRATED LIME SOFTENING AND UF PILOT STUDY

In order to evaluate the effectiveness of integrating an immersed-membrane process into the Marco Island LSWTP, the pilot-scale ZeeWeed® hollow-fiber ultrafiltration (UF) membrane process

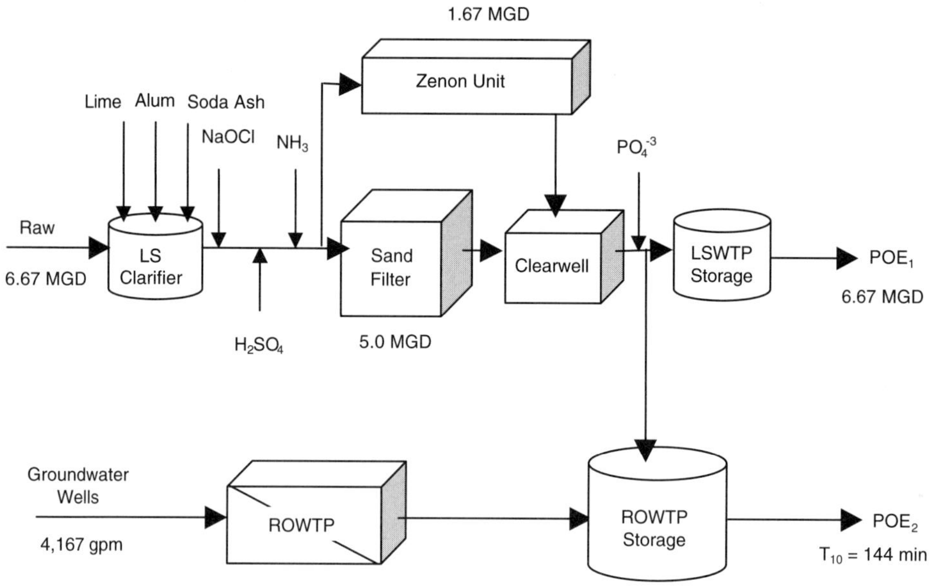

Figure 1 Marco Island Water System Treatment processes flow diagram

(Zenon Environmental, Burlington, Ontario, Canada) was placed online at the existing Marco Island LSWTP. The purpose of the pilot study was to evaluate using the existing LSWTP facilities followed by an immersed ultrafiltration (UF) membrane process.

The immersed-membrane pilot plant evaluation included three separate testing scenarios. The immersed membrane pilot plant was utilized to evaluate filtration of raw water, the softened weir overflow water, and the concentrated precipitant sludge from the SCU process blanket. This paper will focus on the evaluation of the weir overflow water from the SCU following pH adjustment with sulfuric acid, and disinfection with chloramine prior to filtration. Two UF pilot units were provided by Zenon during pilot testing. Each of the Zenon UF pilot units was configured with a ZeeWeed® module that consisted of hollow-fiber membranes mounted on a vertical frame, with permanent extraction from headers at the top and bottom of the assembly. Each unit consisted of one module to evaluate performance of a full-scale cassette. Each hollow fiber had a 0.085 micron absolute pore rating and 0.1 micron nominal pore rating. The Zenon UF pilot unit No. 1 provided approximately 461 ft^2 (based on OD) of active surface area, while pilot No. 2 provided 646 ft^2.

Evaluating the UF technology included submerging a ZeeWeed® module into a 185-gallon process tank filled with the pretreated (launder overflow) water. Filtration is accomplished through an "outside-in" flow pattern accomplished by pulling (vacuum) the pretreated water through the ZeeWeed® hollow fibers. The vacuum inside the fibers creates the transmembrane pressure differential for filtration. In addition, the Zenon technology uses external continuous aeration to scour the membrane fibers, induce mixing and maintain a high flux. Operating under suction imposes a practical limit to the transmembrane pressure and operating flux, which remain in the stable pressure-controlled region of the filtrate curve. These operating conditions result in less membrane fouling, reduced energy consumption, and a long membrane life.

The launder overflow water treatment evaluation was performed between November 8, 1999, and February 10, 2000, at the Marco Island LSWTP. Both pilot units were used for this evaluation. Pretreated supernatant weir overflow water collected in the launders was placed directly into the UF process pilot units for pilot testing of the treatment process. Each test scenario began after a chemical cleaning on each of the Zenon UF pilot units. Initial operating conditions for the Zenon pilot unit included a flux of 30 GFD and 95% recovery. The backwash frequency was maintained at 15-minute intervals with 15-second durations for the entire length of pilot testing. Filtrate or permeate flow was maintained at an average of 12 gpm for the duration of the test. The results of the pilot testing formulated the following: the Zenon UF process pilot was capable of treating lime-softened, launder overflow with favorable results; optimum operating conditions include a flux rate of 30 GFD and a recovery of 95 percent. At these conditions the transmembrane pressure averaged 5 in-Hg; bleach- and citric acid-based cleanings were able to restore the hollow fibers back to initial operating conditions; turbidity in the filtrate averaged 0.05 NTU; and additional TOC removal was minimal utilizing the UF testing units.

ULTRAFILTRATION PROCESS BASIS OF DESIGN

The basis of design for the UF full-scale demonstration unit (supplied by Zenon Environmental) was a maximum daily flow of 1.67 MGD and historical water quality. The full-scale UF unit consists of two transportable trailers. One trailer is provided to house the process equipment, and the second trailer contains the process tank and submerged ultrafiltration cassettes. The design criteria for the 1.67 MGD unit are presented in Table 3.

During normal operations softened water will flow by gravity from the SCU overflow weir to the UF process tank. The UF process tank includes twelve hollow-fiber membrane cassettes. Similar to the pilot unit, filtration was accomplished through an "outside-in" flow pattern by using the vacuum/permeate pumps to remove and transfer permeate to the LSWTP clearwell. A reject pump was utilized to remove process waste (concentrate) from the process tank. The reject is pumped to an on-site lift station that discharges to the on-site Marco Island wastewater treatment plant (WWTP).

Table 3 Ultrafiltration process design information

Design Parameter	Units	Quantity
Unit Recovery	%	95–99
Feed Rate	MGD	1.68–1.75
Reject flow	GPD	17,000–88,000
Membrane Module Area	sf	500
Number of Treatment Streams	–	1
Number of Membrane Modules	–	96
Number of Membrane Cassettes	–	12
Net Flux	GFD	30–35
Process Tank Volume	gallons	20,000

To maintain optimal operating conditions the UF process includes aeration and back-pulse cleaning with sodium hypochlorite. The UF hollow fibers are subject to a continuous air scour to induce mixing and to maintain membrane flux rates. Blower room air pressures typically run about 110 to 120 psi. The back-pulse cleaning consists of 5 mg/L sodium hypochlorite back-pulses for durations of 30 seconds every fifteen minutes. The back-pulse volume typically measures approximately 600 gallons. Cleaning of the membrane cassettes is accomplished with clean-in-place (CIP) chemical cleanings. The chemical feed systems available for the CIP chemical cleanings include sodium hypochlorite, sodium bisulphite, citric acid, and sodium hydroxide feed systems.

START-UP RESULTS

On May 18, 2000, the unit was placed online, and the following start-up testing information was collected over a one-week period. Bacteriological evaluations were conducted on the UF equipment, and results indicated the absence of coliform. Table 4 presents the results of the start-up testing phase. Back-pulse conditions were established at an average back-pulse duration of 45 seconds with a corresponding permeate back-pulse interval of every 15 minutes. Problems encountered during the initiation of treatment included

Table 4 Immersed 1.67 MGD UF membrane process start-up data

Parameter	Before Back-Pulse	After Back-Pulse
Transmembrane Pressure (psi)	1.1	3.2*
Permeate Flow (gpm)	790	1030
Permeate Turbidity (NTU)	0.061	0.063
2–5 μ Permeate Particle Counts (counts/mL)	1.8	2.4
5–10 μ Permeate Particle Counts (counts/mL)	0.51	0.41

* During back-pulse conditions.

air compressor failures, but were resolved by installing a new monitor system. Also, the plant was off-line because work was performed by operations on the turbine pumps at the Marco Lakes raw water site.

OPERATIONAL RESULTS

The unit was taken off-line after start-up testing and optimization was completed; the plant was placed into service in June 2000. Following the start-up tests, the UF plant was then placed into service, and operation assessments were conducted from June through August 2000. Table 5 presents the results of the initial operation phase. On July 8, back-pulse conditions were altered and reestablished at an average duration of 30 seconds with a corresponding 15-minute operating time between each back-pulse event. The UF unit ran continuously in August with minimal reported problems.

Table 5 Turbidity and particle count operating data

Parameter	Before Back-Pulse	After Back-Pulse
Transmembrane Pressure (psi)	–2.0 to –2.5	–1.2 to –2.1
Permeate Flow (gpm)	1,090 to 1,150	880 to 1,050
Permeate Turbidity (NTU)	0.06 to 0.10	0.07 to 0.14
2–5 μ Permeate Particle Counts (counts/mL)	5 to 18	0 to 12
5–10 μ Permeate Particle Counts (counts/mL)	3 to 14	0 to 4
10–15 μ Permeate Particle Counts (counts/mL)	0 to 8	0 to 2

COMPARISON OF TURBIDITY BETWEEN SAND FILTERS AND ULTRAFILTRATION

During the first full month of steady-state operation, the ultrafiltration process produced an average turbidity of 0.04 NTU. The sand filters in that same time period produced an average of 0.1 NTU. The permeate clearwell, representing the mixture of UF and sand-filtered water, averaged 0.08 NTU. Particle count data are only available for the UF system so that comparisons could not be made.

COSTS

This project was implemented in such a way that the project would allow for a lease-purchase agreement for the immersed membrane process component of the work. Costs associated with engineering, legal, administration, planning, contractor mobilization, permitting and site development improvements approximated $0.5 million. The lease cost for the membrane process equipment was established at $17,000 per month for one year, at which time a purchase agreement would be approved if the performance of the equipment met the engineer's design criteria. The total installed UF equipment cost for the 1.67 MGD expansion at time of purchase is approximately equivalent to $0.68 per gallon per day filtrate water. Operating costs are preliminary, but are estimated to run at $23,800/yr for power, $1,600/yr for back-pulse and recovery-cleaning chemicals, and $4,500 for general maintenance (such as oil and gaskets for pumps, blowers, valves and instruments). Costs are not available for residuals management or for operating labor. Recommended manufacturer labor requirements are estimated at one person for a maximum of 2 hours per day.

SUMMARY

The primary impetus for Florida Water to implement an expansion of the Marco Island LSWTP using an immersed UF membrane process has been a direct result of the more stringent regulations and the impacts of increased population growth on the island. Membrane technology has become a popular alternative treatment technology for drinking water treatment. It has been demonstrated that UF is capable of meeting the requirements of the Surface Water Treatment Rule for Giardia (3 logs removal), and is capable of meeting future log removal conditions of recently promulgated legislation.

The implementation of UF for polishing of lime-softened surface water has been demonstrated on a high-hardness, high-color, high-TOC surface water. Pilot testing demonstrated an average turbidity of 0.05 NTU for the ultrafiltration permeate. Implementation of a full-scale 1.67 MGD UF process downstream of the Marco Island lime softening process demonstrated an average turbidity of 0.04 NTU versus an average lime-softened, sand-filtered finished water turbidity of 0.1 NTU. The UF demonstration unit has demonstrated that innovative technology can be utilized for filtration following lime softening. The expansion occurred within a footprint approximately 40 percent smaller than that for traditional sand filtration.

ACKNOWLEDGEMENTS

This chapter could not have been completed without the combined support and contributions of several individuals. Special thanks go to Brian Matthews and the Marco Island operations staff of Florida Water Services. Also, the efforts of Jodi Cumin and Graham Best of Zenon Environmental are appreciated. And last, the support of Anke Backer, Robbie Gonzalez, Orinthia Thomas and Will Lovins at Boyle Engineering has been greatly appreciated.

REFERENCES

1. Taylor, J.S., S.J. Duranceau, W.M. Barret and J.F. Goigel. Assessment of Potable Water Membrane Applications and Research Needs. AWWA Research Foundation and AWWA, Denver, Colorado (1989).
2. Reiss, C.R. and J.S. Taylor. Membrane Pretreatment of a Surface Water. In Proceedings of the American Water Works Association Membrane Technology in the Water Industry Conference. AWWA, Denver, Colorado (1991).
3. Jacangelo, J.G., J.M. Laine, K.E. Carns, E.W. Cummings, and J. Mallevialle. Jour. AWWA 81 (11): 97–106 (1991).
4. Taylor, J.S., L.A. Mulford, S.J. Duranceau, and W.M. Barrett. Cost and Performance of a Membrane Pilot Plant. *Journal AWWA* 81 (11): 52–60 (1990).
5. Taylor, J.S., C.R. Reiss, P.S. Jones, K.E. Morris, T.L. Lyn, D.K. Smith, L.A. Mulford, and S.J. Duranceau. Reduction of Disinfection By-Product Precursors by Nanofiltration. Final Report, Cooperative Agreement No. CR-815288, EPA/600/SR-92/023. EPA Risk Reduction Environmental Laboratories, Cincinnati, Ohio, April (1992).
6. Mallevialle, J., P.E.Oedendaal, and M.R. Weisner. Water Treatment Membrane Processes. McGraw-Hill, New York (1996).
7. Duranceau, S.J., K.A. Thorner, and P.A. McAleese. The Olivenhain Water Storage Project and Integrated Membrane Systems. In National Water Research Institute Microfiltration II Conference Proceedings. AWWA, San Diego, California, November 12–13, 83–94 (1998).
8. Jacangelo, J.G., S. Adham, and J.M. Laine. Membrane Filtration for Microbial Removal. AWWA Research Foundation and AWWA, Denver, Colorado (1997).
9. Duranceau, S.J., and K.A. Thorner. Engineering and Construction Considerations for Integrated Membrane Systems. In Proceedings of the AWWA 1999 Engineering and Construction Conference. AWWA, Orlando, Florida, March 14–17 (1999).

CHAPTER · 27

Disinfection by Integrated Membrane Systems for Surface Water Treatment

Joop C. Kruithof and Jan C. Schippers
Kiwa N.V. Research and Consultancy
Nieuwegein, The Netherlands

Jan Peter van der Hoek
Amsterdam Water Supply
Amsterdam, The Netherlands

Peer C. Kamp
N.V. PWN Water Supply Company North Holland
Bloemendaal, The Netherlands

BACKGROUND

Within the framework of a project cofunded by AWWARF and USEPA two promising integrated membrane systems (IMSs) are identified for disinfection of surface water:

- Advanced physical/chemical treatment including coagulation, sedimentation, filtration (CSF), biological activated carbon filtration and slow sand filtration followed by reverse osmosis. Investigation carried out by Irvine Ranch Water District and Amsterdam Water Supply (AWS).
- Microfiltration and ultrafiltration followed by ultra-low-pressure reverse osmosis. Investigation carried out by American Water Works Service Company Inc. and NV PWN Water Supply Company North Holland (PWN).

This chapter will focus on the research carried out by the Dutch utilities in a cooperative research effort with Kiwa.

At Amsterdam Water Supply spiking with MS-2 phages in the slow sand filtrate showed a removal capacity of 3.4 log units for reverse osmosis. Removal of spores of Sulphite Reducing *Clostridia* proved to be complete. *Giardia* and *Cryptosporidium* removal should be equal to this removal efficiency. For direct cell counts and heterotrophic plate counts a lower efficiency was found. Probably this is due to regrowth on the permeate side of the reverse osmosis installation.

At NV PWN Water Supply Company North Holland, indicator organisms, *E. coli* and spores of sulphite-reducing *Clostridia* present in CSF treated Yssel Lake water were reduced after ultrafiltration to values below the detection limit. Generally ultrafiltration showed a complete (> 5 log) removal of MS-2 phages. After an introduced leakage of 0.2% removal was 2.7 log units only. Ultra-low-pressure reverse osmosis showed a high (4.7–4.9 log) although no complete removal.

At PWN integrity of the ultrafiltration is measured by a pressure hold test and monitored by particle counting. Ultra-low-pressure reverse osmosis integrity is measured by vacuum testing and in situ conductivity and monitored by particle counting and sulphate removal.

The perspectives of IMSs for disinfection purposes are very promising. Integrity measurements will be a critical issue.

INTRODUCTION

Membrane processes such as microfiltration, ultrafiltration, nanofiltration and reverse osmosis are potentially applicable to remove a broad range of contaminants from surface water. Microfiltration and ultrafiltration are applicable to remove particles, colloids and microorganisms. Nanofiltration and reverse osmosis can additionally remove DBP-precursors (colour), synthetic organic chemicals such as pesticides, hardness constituents and salts. However, membrane processes, especially nanofiltration and reverse osmosis, are susceptible to fouling. Pretreatment steps, needed to restrict fouling, may include coagulation, sedimentation, filtration (CSF), ozonation followed by activated carbon filtration, slow sand filtration and soil passage. Additional treatment steps can be microfiltration and ultrafiltration. (Advanced) pretreatment processes combined with conventional membrane systems (CMS) such as nanofiltration and reverse osmosis are integrated membrane systems (IMS).

At Leiduin a study is executed in close cooperation between the Amsterdam Water Supply and Kiwa. The feedwater to this plant originates from the River Rhine and is pretreated by coagulation, sedimentation and rapid sand filtration at the Water Transport Company Rijn Kennemerland (WRK I, II) plant at Nieuwegein. The raw water has a relatively high turbidity and is microbiologically and chemically contaminated. At Leiduin a demonstration installation is available. The reference treatment of this installation is composed of slow sand filtration and reverse osmosis. An alternative scheme is ozonation, biological activated carbon filtration, slow sand filtration and reverse osmosis.

At Andijk a study is executed in close cooperation between the Water Supply Company of North Holland (PWN) and Kiwa. The raw water, pretreated Yssel Lake water, has a DOC content of around 3 mg/l, is turbid and is microbiologically and chemically polluted. At the Andijk plant the suitability of reverse osmosis with cellulose acetate membranes has been investigated for many years. In these studies surface water pretreated with coagulation, sedimentation and filtration has been additionally pretreated by rapid filtration as a polishing step. Recently the application of ultrafiltration and reverse osmosis with composite membranes has been pursued. During this research effort pretreatment of reverse osmosis was focused on ultrafiltration.

The primary objectives of both integrated membrane system projects are:

- Fouling and productivity aspects
- Removal of microbials including membrane integrity
- DBP precursor removal

Additional objectives are:

- Removal of SOCs such as pesticides
- Removal of inorganic constituents
- Biological stability
- Corrosivity

This chapter will focus on the removal of microbials and membrane integrity.

EXPERIMENTAL SETUP

At Leiduin (AWS) the reference treatment scheme is composed of CSF, slow sand filtration and reverse osmosis. An alternative treatment scheme is CSF, ozonation, biological activated carbon filtration, slow sand filtration and reverse osmosis. As feedwater, pretreated River Rhine water is used. The average water composition is summarized in Table 1. The RO-blocks of both treatment schemes are identical. They are constructed as three staged plants with 4, 2 and 1 pressure vessels in respectively first, second and third stage. The pressure vessels contain six 4 × 40–inch Fluid Systems 4821 ULP membrane elements. The plant is operated at a capacity of 7.8 m^3/h with an average flux of 25 $L/m^2/hr$.

At Andijk (PWN) the reference treatment scheme is composed of microstraining, CSF, ultrafiltration and reverse osmosis. Alternative treatment schemes are microstraining, CSF and reverse osmosis or microstraining, ultrafiltration and reverse osmosis. As feedwater, pretreated Yssel Lake water is used. The average water composition is summarized in Table 2. The water shows large seasonal temperature variations. In extreme years the temperature may vary between 0° and 250° C. The demonstration-scale ultrafiltration plant consists of 3 units. One unit has a capacity of approximately 45 m^3/h and produces the feedwater for the reverse osmosis units. The other two units, 15 m^3/h each, are used to optimize the process and operating conditions. For this research project the ultrafiltration plant is loaded with X-flow hollow-fiber polyethersulphone elements.

The demonstration-scale reverse osmosis plant consists of 3 units with a product capacity of 7–9 m^3/h each. Recently the plants were retrofitted into two staged systems with seven elements per pressure vessel, due to the introduction of ultra-low-pressure reverse osmosis membranes. Data will be presented for Hydranautics 4040-UHY-ESPA membranes.

DISINFECTION AND INTEGRITY AWS DATA

Removal of Microorganisms

Cryptosporidium *Oocyst and* Giardia *Cyst Removal*

To determine the concentration of *Cryptosporidium* and *Giardia* the pretreated River Rhine water was sampled at the intake of the treatment plant of Leiduin. Table 3 shows that *Cryptosporidium* oocysts and *Giardia* cysts were found in 33% of the samples. Comparison with data of the concentration of *Cryptosporidium* oocysts and *Giardia* cysts in the water before pretreatment indicates that coagulation, sedimentation and rapid sand filtration resulted in a 1–2 log removal of both organisms. Because the numbers of *Cryptosporidium* oocysts and *Giardia* cysts in the CSF treated water are very low, the removal of (oo)cysts by either SSF, RO or O3, BAC, SSF, RO could hardly be measured.

Fecal Indicator Organism Removal

Table 4 gives an overview of the number of colony-forming units of fecal indicator bacteria present in the pretreated River Rhine water. The average of all individually measured fecal indicator bacteria was below 27 cfu/100 ml. Data on the number of colony-forming units of total coliforms (n = 8), thermotolerant coliforms (n = 8), *E. coli* (n = 6) after SSF (reference scheme) indicate a removal efficiency in the range of 1.6–1.7 log units. The maximum numbers of colony-forming units found after SSF were near the detection limit. Table 5 shows the data on the number of SSRC in pretreated River Rhine water and after both the reference and alternative scheme. Already, prior to RO the reference scheme (SSF) gives a reduction of 0.8 log units, while the alternative treatment (O3, BAC, SSF) gives a reduction of 2.2 log units, resulting in very low numbers.

Because of the low numbers of indicator bacteria and spores of sulphite reducing *Clostridia* in the slow sand filtrate it wasn't possible to accurately measure the removal with reverse osmosis. Therefore a dosing test with MS-2 phages was performed to monitor membrane integrity.

Membrane Integrity Monitoring

MS-2 Phages

Spiking with MS-2 phages was performed at the water before RO (Scheme O3/BACF/-SSF/RO). Control samples before spiking taken from the RO feed and RO permeate showed that natural phage concentrations are too low to influence the spiking experiment. The concentration of dosed MS-2 phages in the RO feed was 2.1×10^5 per ml. Based on this concentration and the average concentration in the RO permeate (7.3 × 101 per ml), a removal capacity of 3.4 log units was calculated.

Heterotrophic Plate Counts

The presented data show a log removal of MS-2 phages of 3.4 log units. Because of their larger size the removal of other micro-organisms should be at least equal. However, the reduction of the number of colonies found on heterotrophic plate counts was much lower. This indicates that regrowth plays a part in the permeate of the reverse osmosis system, probably caused by materials in contact with the water.

PWN DATA

Removal of Microorganisms

Cryptosporidium *Oocyst and* Giardia *Cyst Removal*

To determine the concentration of *Cryptosporidium* and *Giardia* in the source water of the Andijk treatment plant, water of the Yssel Lake was sampled at the intake of the reservoir. Since reservoir storage and microstraining may affect the concentration of protozoa, additional samples were taken of the water leaving the microstrainers. Because of the low values in the feedwater, no samples were taken in the treatment system. Samples of approximately 20–400 liters were processed and analyzed according to the method of LeChevallier et al., with minor modifications. The average recovery of the method (n = 3) was 22.9% and 4.2% for *Giardia* and 4.3% and 2.7% for *Cryptosporidium* in samples of the Yssel Lake and after microstraining, respectively.

Table 6 shows the results of the monitoring program from September 1996 to July 1997. In the Yssel Lake, at the abstraction point, *Giardia* cysts were found in 91% of the samples and *Cryptosporidium* oocysts in 64% of the samples. After reservoir storage and microstraining, the percentage of positive findings was reduced to 83% for *Giardia* cysts and 50% for *Cryptosporidium* oocysts. The (uncorrected) average concentration of *Giardia* in the Yssel Lake is 1.1 cyst per liter, with a maximum of 5.2 cysts per liter. Reservoir storage and microstraining reduce the average concentration with 87% (= 0.9 logs).

The average *Cryptosporidium* concentration in both the Yssel Lake and after microstraining is two- to threefold lower than the *Giardia* concentration at these sites. The reduction in the *Cryptosporidium* concentration during reservoir storage and microstraining is 91% (1.0 log), similar to the reduction of the *Giardia* concentration.

Fecal Indicator Organism Removal

Figure 1 gives an overview of the reduction of microorganisms present in the raw water. The data show, amongst others, that several tens to hundreds of *E. coli* and *Clostridia* per milliliter are present in the raw water of the Yssel Lake. After pretreatment by WRK III the number is reduced to 1/100 ml, which is just above the standards for drinking water. After treatment with UF, *E. coli* and *Clostridia* are absent in the water.

Based on these data, it is concluded that the combination UF-ULPRO is a reliable disinfection barrier, when integrity of the system can be guaranteed. Despite some leakages in the UF system, the overall UF-ULPRO process showed a log reduction of around 4 based on particle numbers (> 0.05 μm), indicating that membrane integrity monitoring is a critical issue.

Membrane Integrity Monitoring

Vacuum Testing

Before the elements were loaded into the ULPRO installation, a vacuum test was applied. Figure 2 shows the results of the vacuum test of the elements in the ULPRO installation. When this test was carried out, the 20 kPa/min criterion was still used. Only 1 out of 42 elements tested had a too high vacuum decrease rate of 22 kPa/min. Nevertheless this element was loaded in the system.

Conductivity Measurement

After loading of the elements, an in situ test with conductivity of the pressure vessels was carried out. The results of stage 1 of the installation are shown in Figure 3, showing a gradually increasing conductivity. From these data it can be concluded that the installation had no leaks at the moment it was tested.

MS-2 Phage Spiking

To determine the reduction factors of the ultrafiltration and ULPRO the feedwater to both installations was spiked with MS-2 phages.

Table 7 summarizes the results. UF after WRK III pretreatment shows a relatively low log reduction for MS-2 phages. The log reduction of 2.7, while a value around 5 was expected, was caused by defects in the UF membranes. The ULPRO and also the UF on surface water treatment show much higher log reductions. These results show that integrity monitoring is absolutely necessary if the membranes are used as a disinfection barrier, without application of an additional chemical disinfection or soil passage. In addition to vacuum testing and in situ conductivity measurements, particle counting (> 2 μm) and sulphate removal are used for integrity control.

EVALUATION

Within the framework of a project cofunded by AWWARF and USEPA, some very promising IMSs were identified for surface water treatment. This chapter describes the disinfection aspects of the pilot research carried out by Amsterdam Water Supply, NV PWN Water Supply Company of North Holland and Kiwa.

At Water Treatment Station "Cornelis Biemond," water from the River Rhine is treated by coagulation, sedimentation and rapid filtration. The treated water is transported to AWS's pilot facilities at Leiduin. At the pilot plant the water is treated by either slow sand filtration or ozonation, biological activated carbon filtration and slow sand filtration followed by ultra-low-pressure reverse osmosis.

Coagulation, sedimentation, and filtration give a 1–2 log *Cryptosporidium* oocyst and *Giardia* cyst removal, resulting in very low numbers. Fecal indicator organisms present in the CSF treated River Rhine water are removed to very low numbers after the slow sand filtration step.

To determine the elimination capacity of the reverse osmosis system the feed was spiked with MS-2 phages. The system showed a high (3.4 log) although no complete removal. In the reverse osmosis permeate some heterotrophic bacteria were found, indicating the presence of regrowth.

At Water Treatment Station "Prinses Juliana," water from the Yssel Lake is treated by coagulation, sedimentation and upflow filtration. The treated water is transported to PWN's pilot facilities at Andijk. At the pilot plant the water is post-treated with ultrafiltration followed by ultra-low-pressure reverse osmosis.

After microstraining, the numbers of *Cryptosporidium* oocysts and *Giardia* cysts were very low. Indicator organisms *E. coli* and spores of sulphite-reducing *Clostridia*, present in conventionally treated Yssel Lake water, were reduced by ultrafiltration to values below the detection limit.

To determine the elimination capacity of both membrane systems the feeds were spiked with MS-2 phages. Generally ultrafiltration showed a complete (> 5 log) removal. A damaged ultrafiltration element with 0.2% leakage showed a 2.7 log removal only. Ultra-low-pressure reverse osmosis showed a very high (4.7–4.9 log) although no complete removal.

These results show that for disinfection purposes, integrity measurements and monitoring of both membrane steps are essential. Ultrafiltration integrity is measured by a pressure hold test and monitored by particle counting. Ultra-low-pressure reverse osmosis integrity is measured by vacuum testing and monitored by conductivity measurement, particle counting and sulphate measurement. Both IMSs based on pretreatment by slow sand filtration or ultrafiltration followed by ultra-low-pressure reverse osmosis are ideally suited for control of microbials.

ACKNOWLEDGEMENTS

This research was funded by the American Water Works Association Research Foundation and the U.S. Environmental Protection Agency, Drinking Water Research Division. Part of the research was conducted within the framework of the research program of the Netherlands Water Supply Companies in a cooperative effort by the Amsterdam Water Supply (GWA), NV PWN Water Supply Company of North Holland (PWN) and Kiwa NV Research and Consultancy.

The authors want to thank the numerous staff members of the water supply companies and Kiwa for their cooperation in this study.

REFERENCES

1. J.A.M. van Paassen, J.C. Kruithof, S.M. Bakker and F. Schoonenberg Kegel: *Desalination* 118 (1998) 239–248.
2. J.P. van der Hoek, P.A.C. Bonné, E.A.M. van Soest and A. Graveland: *Proc., IWSA World Congress 1997*, Madrid. SS1-11.
3. M.M. Nederlof, J.C. Kruithof, J.A.M.H. Hofman, M. de Koning, J.P. van der Hoek and P.A.C. Bonné: *Desalination* 119 (1998) 263–273.
4. P.C. Kamp: *Proc., Membrane Technology Conference*, New Orleans 1997.
5. J.C. Kruithof, J.C. Schippers, P.C. Kamp, H.C. Folmer and J.A.M.H. Hofman: *Desalination* 117 (1998) 37–48.
6. S.F.E. Boerlage, M.D. Kennedy, P.A.C. Bonné, G.Galjaard and J.C. Schippers: *Proc., Membrane Technology Conference*, New Orleans 1997, 979–999.
7. S.F.E. Boerlage, M.D. Kennedy, G.J. Witkamp, J.P. van der Hoek and J.C. Schippers: *Proc., Membrane Technology Conference*, New Orleans 1997, 745–759.
8. H.S. Vrouwenvelder, J.A.M. van Paassen, H.C. Folmer, J.A.M.H. Hofman, M.M. Nederlof and D. van der Kooy: *Desalination* 118 (1998) 157–166.

Table 1 Characteristics of the pretreated River Rhine water (period 1992–1995)

Parameter	Range	Parameter	Range
Ca^{2+} (mg/l)	75–85	DOC (mg C/l)	2–3
SO_4^{2-} (mg/l)	50–70	AOC (µg/l)	10
HCO_3^- (mg/l)	150–170	AOX (µg Ac C eq/l)	15–20
Ba^{2+} (µg/l)	60–90	MFI (s/l^2)	1.5–3
pH	7.6–8.0	Temp. (°C)	0–25

Table 2 Average water composition of WRK III water

Cations			Anions			Others		
Calcium	mg/l	90	HCO_3^-	mg/l	249	Silica	mg/l	19.3
Magnesium	mg/l	14	Sulfate	mg/l	140	pH	–	7.8
Sodium	mg/l	110	Chloride	mg/l	200	Temp	°C	2–22
Potassium	mg/l	7	Nitrate	mg/l	11	DOC	mg/l	3
Barium	mg/l	0.07	Fluoride	mg/l	0.05	TDS	mg/l	640
Strontium	mg/l	0.6						

Table 3 *Cryptosporidium* oocysts and *Giardia* cysts in the source water of the Leiduin pilot plant

	Cryptosporidium	*Giardia*
No. of samples	12	12
% positive	33	33
Average* (n/l)	0.0015	0.0049
Minimum (n/l)	< 0.001	< 0.001
Maximum (n/l)	0.012	0.023
Average recovery (%)	4.0	14.3
Average corrected for recovery (n/l)	0.04	0.04

* Determined as total number of (oo)cysts counted in all samples/total volume analyzed from all samples.

Table 5 Fecal indicator bacteria in pretreated River Rhine water (WRK I, II)

	Total Coliforms *n/100 ml*	Thermotolerant Coliforms *n/100 ml*	*E. coli* *n/100 ml*	SSRC *n/100 ml*
No. of samples	11	16	9	16
% positive	91	81	89	100
Average*	10.9	8.4	12.0	3.3
Minimum	< 0.1	< 0.1	< 0.1	0.3
Maximum	27	27	27	10

* Determined as total number counted in all samples/total volume analyzed from all samples.

Table 4 Spores of sulphite-reducing *Clostridia* in pretreated River Rhine water and after both the reference and alternative pretreatment scheme

	Pretreated River Rhine Water	SSF	O3/BACF/SSF
No. of samples	16	16	17
% positive	100	69	18
Average* (n/100 ml)	3.3	0.50	0.02
Minimum (n/100 ml)	0.3	< 0.1	< 0.1
Maximum (n/100 ml)	10	2.4	0.2
Log removal **	–	0.8	2.2

* Determined as total number counted in all samples/total volume analyzed from all samples, and not determined.

** Log removal related to pretreated River Rhine water.

Table 6 *Cryptosporidium* and *Giardia* in the source water of Andijk treatment plant

	IJssellake		After Microstrainers	
Location	*Giardia*	*Crypto.*	*Giardia*	*Crypto.*
No. of samples	11	11	12	12
% positive	91	64	83	50
Average* (n/l)	1.1	0.51	0.14	0.05
Minimum (n/l)	< 0.13	< 0.05	< 0.04	< 0.02
Maximum (n/l)	5.2	2.6	0.43	0.12

* Determined as total number of (oo)cysts counted in all samples / total volume analyzed from all samples. The data are not corrected for the recovery of the detection method.

Table 7 Results of spiking of feedwater with MS-2 phages and the calculated log-reduction

	Feedwater (*pfu/ml*)	Product Water (*pfu/ml*)	Log Reduction
UF after WRK III pretreatment	3,400	6.73±1.43	2.7
ULPRO	6,200	0.81±0.17	4.6
UF on surface water	18,000	0.25±0.10	4.9

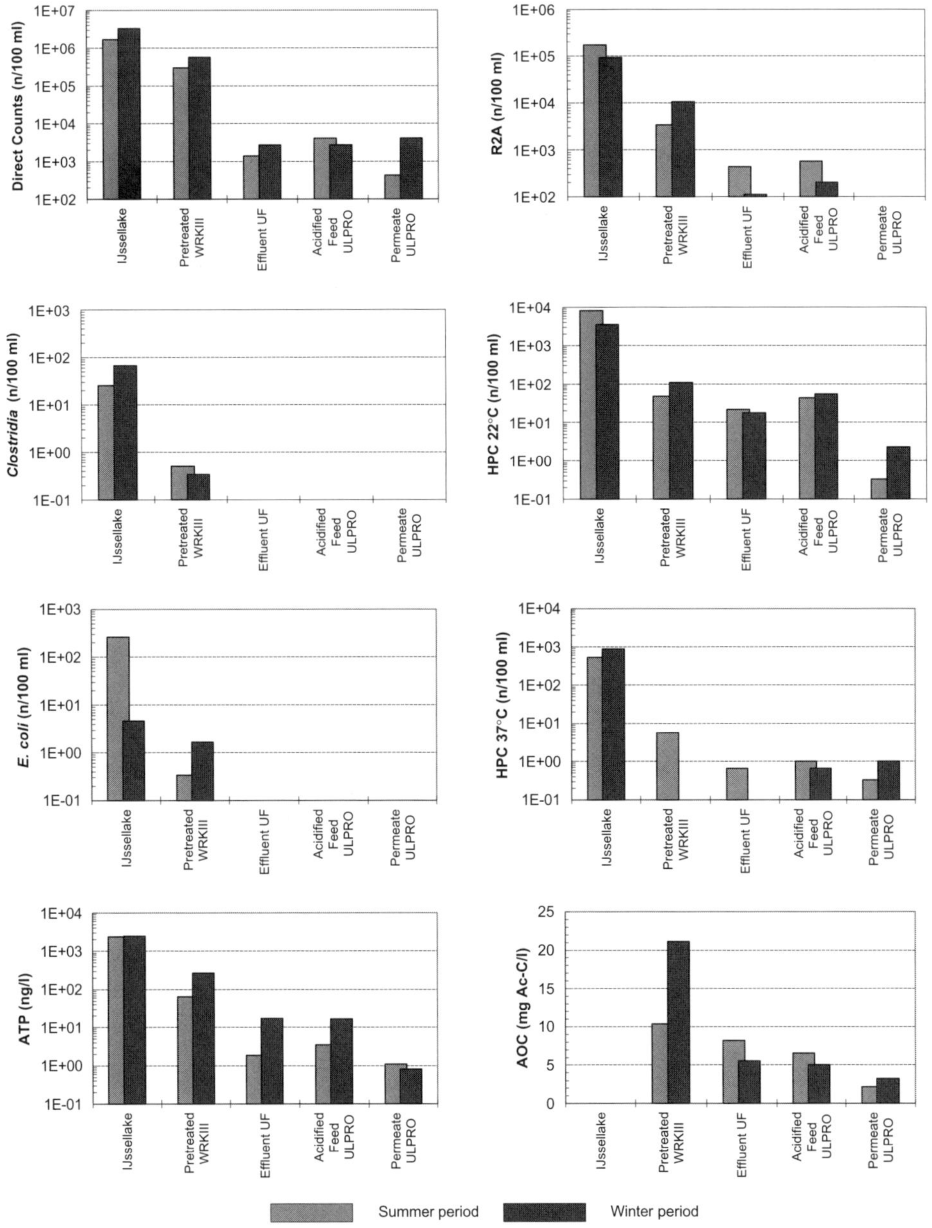

Figure 1 Reduction of microorganisms by UF-ULPRO treatment

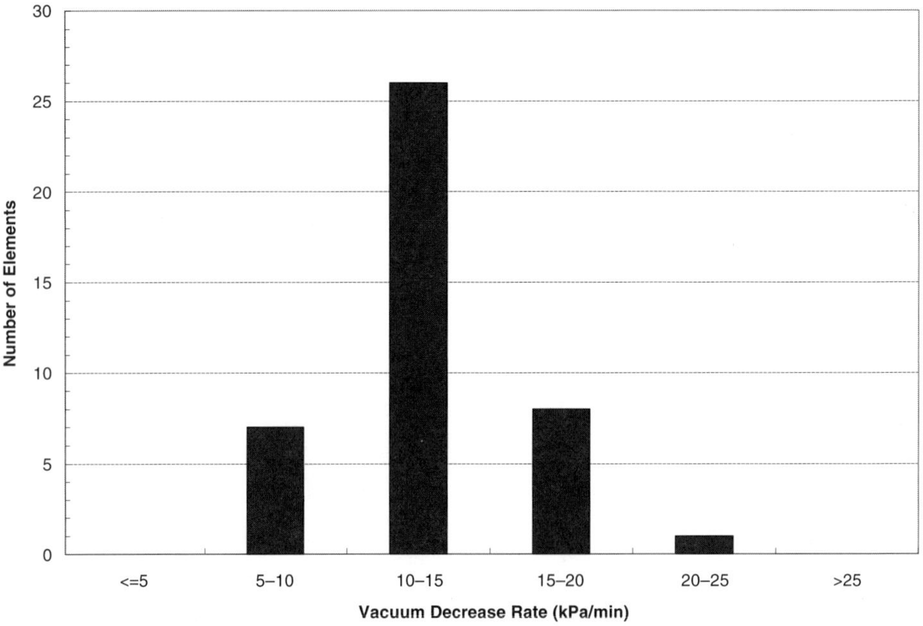

Figure 2 Frequency distribution of the vacuum test of the Hydranautics 4040-UHY-ESPA membrane in the ULPRO installation

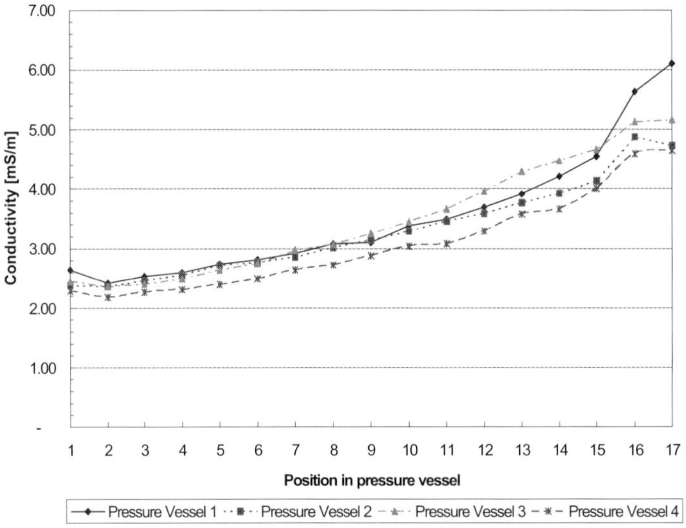

Figure 3 In situ measurement of conductivity in permeate tubes of the first-stage pressure vessels of the ULPRO installation

CHAPTER · 28

Prediction of Full-Scale IMS Performance Using a Resistance Model and Laboratory Data

William A. Lovins, Ph.D.
Boyle Engineering Corp., Orlando, FL

James S. Taylor, Ph.D., P.E.
University of Central Florida, Orlando, FL

C. Robert Reiss, P.E.
Reiss Environmental, Winter Park, FL

BACKGROUND

Integrated membrane system (IMS) productivity and water quality were evaluated using a newly developed resistance model. Models were developed using laboratory data and verified using independently collected field data. These results demonstrated that field-scale IMS operation could be simulated using field-scale hydraulic and water quality data with models developed from laboratory-scale systems. Laboratory productivity was modeled using a modified form of Darcy's Law. This model was based on hydraulic resistance due to cumulative foulant mass loading. Additive resistance was assumed and related to accumulation of organic, biological, and particle foulants. The model also accounted for decreasing membrane film resistance resulting from combined chlorine loading. Statistically significant models were obtained for flat-sheet tests conducted with composite thin-film (CTF) nanofilters (NFs). For one of these nanofilters (LFC1), the model indicated fouling due to organic, biological and particle accumulation. The model also indicated degradation due to combined chlorine. Model verification was performed using pilot- and full-scale data. The LFC1 flat-sheet model using laboratory data was able to simulate field-scale IMS productivity decline.

INTRODUCTION AND OBJECTIVES

Drinking water utilities are in need of operational and water quality information to assess membrane system cost and performance. Information on the treatment of surface water with membrane systems is especially limited because few full-scale systems exist and a limited number of pilot studies have been conducted. By using bench- or pilot-scale evaluations, the rates of productivity decline, forms of fouling and required cleaning procedures can be assessed. Laboratory testing provides an attractive alternative to pilot testing given potential capital and operational cost savings, although limitations associated with laboratory tests must be recognized, especially when evaluating membrane fouling. Membrane productivity is commonly modeled using empirical curve fitting procedures. This is useful for projecting productivity decline rates but lacks a mechanistic basis. Productivity models are needed that relate fouling to water quality and hydraulic operating conditions.

While studies have been published showing comparable results in water quality and productivity between flat-sheet and membrane single-element and multi-element pilot units (Allgeier et al., 1995, 1997; Gusses et al., 1997; Lovins, 1996 and Speth et al., 1997), pilot testing is still crucial for verification of laboratory data in addition to evaluation of impacts associated with variable raw water quality. Difficulties presented by the use of batch-recycle systems presented in the literature include difficulty in maintaining steady-state test conditions (i.e., changes in water constituent mass due to foulant deposition and batch volume loss); limitations in simulating concentration effects due to product water recovery using systems operated at low recovery; and variability of membrane sample (DiGiano, 1996; Reiss et al., 1997; Allgeier and Summers 1995; Gusses et al., 1999).

Objectives of this chapter include: (1) evaluation of limitations associated with laboratory-scale IMS testing, (2) evaluation of pretreatment requirements for sustained treatment of surface water by nanofiltration, (3) validation of laboratory-scale testing, and (4) development of measures to improve correlation of laboratory test results to field- or full-scale performance.

METHODS AND MATERIALS

Membrane Films

Composite thin-film (CTF) and cellulose acetate (CA) nanofilters (NFs) were tested in the laboratory as batch-recycle systems using flat-sheet and single-element systems and as continuous-feed single-pass systems in the field. Laboratory- and field-scale IMS evaluations were conducted using three different membranes. Characteristics of these membrane films are summarized in Table 1. These membranes were obtained in both flat-sheet and single-element forms as needed and were used to assess productivity decline for high-fouling surface waters. Batch volumes were 15 gallons for flat-sheet tests and 30 gallons for single-element tests. Laboratory testing with pretreated feedwaters was preceded by continuous operation with distilled and deionized water at representative flux and recovery to reduce apparent fouling caused by membrane compaction.

Raw Waters and Advanced Pretreatments

The Hillsborough River utilized by the Tampa Water Department (located in Tampa, Florida) and the Mississippi River utilized by the Illinois-American surface water treatment plant (located in East St. Louis, Illinois, American Water Works Service Company) served as the raw water supplies presented in this communication. Overall, Hillsborough River water is classified as a highly organic and moderate-turbidity water supply, whereas Mississippi River water is classified as moderate organic water with high particulate content. Select pretreatments were used to vary organic, biological and particle mass loading. As shown in Table 2, pretreatments included cartridge filtration, microfiltration, ultrafiltration, coagulation sedimentation, sand filtration, and acid and scale inhibitor chemical addition. Combined chlorine, predominantly in the form of monochloramines, was used to inhibit biofouling. Pretreatment, water and membrane film combinations tested are summarized in Table 3. In Table 4, raw and pretreated water quality is summarized for Hillsborough and Mississippi River waters. The parameters listed reflect constituents used to assess fouling potential and reduction by pretreatment. These diverse water supplies provided a unique opportunity to investigate NF fouling potentials and associated pretreatment requirements by water.

Productivity Modeling

Permeate flux, F_w, in a pressure-driven membrane process at a given time can be described by Darcy's Law. Shown as Equation 1, Darcy's Law relates flux to pressure, viscosity and total hydraulic resistance, R_T.

$$F_w = \frac{1}{A_m}\frac{dV}{dt} = \frac{\Delta P}{\mu R_T} \qquad (1)$$

where:

F_w = water flux (L/t)
A_m = effective membrane area (L^2)
V = cumulative permeate volume (L^3)
t = time of filtration
ΔP = net driving pressure (M/Lt^2)
R_T = the combined resistance to flow due to the membrane film and the diffuse fouling layer (1/L)

Changes in membrane resistance due to foulant accumulation are complex. Exact methods for predicting productivity profiles in natural waters are not available. Darcy's Law provides a fundamental basis for meeting this objective. One approach is to adapt Darcy's Law for use with cumulative mass loading of contaminants responsible for fouling. In this manner, an incremental approach using combined water quality and hydraulic information was used. The mass of foulant (M) transported to the membrane film/foulant layer was approximated using water flux (F_w), foulant concentration (C) and time (t) for a given period of operation ($M=F_wCt$). Unlike microfiltration, which is commonly operated in dead-end mode, complete foulant accumulation in a nanofiltration system would not be expected. Physical and chemical effects may preclude complete deposition, including but not limited to lateral transport across the surface (i.e., sloughing), back diffusion away from the surface, hydrophobic/hydrophilic interactions, and charge repulsion. Foulant resistance, R_f, representing total hydraulic resistance due to the diffuse fouling layer, is expressed in Equation 2. The K terms represent fitting parameters, which acted as sticking factors specific to foulant type and were evaluated empirically.

$$R_F = \sum R_i = \sum K_i F_w C_i t \qquad (2)$$

where:

R_F	=	resistance due to foulant accumulation (1/L)
R_i	=	resistance due to a specific foulant (1/L)
K	=	proportionality coefficient (L/M)
C_i	=	specific foulant concentration (M/L^3)
F_w	=	water flux (L/t)
t	=	time

Membrane resistance was calculated directly. By estimating a reference membrane resistance, i.e., prior to foulant deposition, changes in resistance due to fouling were tracked. Membrane resistance is represented by Equation 3, using initial pressure (ΔP_o) and initial flux (F_{wo}) values. Units used for R_m were ft^2-d-psi/gal-cp.

$$R_m = \frac{\Delta P_o}{\mu F_{wo}} \qquad (3)$$

Empirical relationships were used to adjust viscosity for variations in temperature (*Handbook of Chemistry and Physics*, 1990).

Adding the general form of Equation 2 to Darcy's Law results in Equation 4.

$$MTC_w = \frac{1}{\dfrac{\Delta P_o}{F_{wo}} + \mu\phi F_w Ct} \tag{4}$$

By rearranging Equation 4, it can be shown that the resistance model follows second-order kinetics.

$$\frac{1}{MTC_w} = \frac{\Delta P_o}{F_{wo}} + \mu\phi F_w t = \frac{1}{MTC_w} + A_2 t \tag{5}$$

where:

$\Delta P_o/F_{wo}$	=	the reciprocal of MTC_w at time zero
$\mu\phi F_w Ct$	=	A_2

This reinforces results shown by Tram (1997) suggesting models of this form are likely to provide a more mechanistic basis for predicting fouling.

Given the choice of pretreatments utilized in these evaluations, the potential existed for evaluating organic, biological and particle fouling. Mass loading for these potential foulants was based upon NPDOC, HPC and particle count concentrations, respectively. While UV-254 and turbidity may be used as surrogates for organic and particle fouling, preference was given to NPDOC and particles. The potential for chemical oxidation of membrane films by combined chlorine was recognized. Consequently, combined chlorine mass loading was considered. Membranes also degrade as natural waters are passed through them, i.e., irreversible fouling or "wear-and-tear." This was accounted for by incorporating cumulative water mass loading into the model. Substituting Equations 2 and 3 into Equation 1, with the appropriate solutes, yields the semi-empirical model shown as Equation 6.

$$MTC_w = \frac{1}{\mu R_T}$$

$$= \frac{1}{\mu\left[A \cdot R_m + \frac{Q_t}{A_m}\begin{pmatrix} K_a(\text{NPDOC}) + K_b(\text{HPC}) \\ + K_c(\text{Particles}) \\ + K_{d_1}(\text{Total Cl}_2) \\ + K_{d_2}(\text{H}_2\text{O}) \end{pmatrix}\right]} \tag{6}$$

This equation provided a basis for modeling membrane productivity over time under conditions of varying flux and recovery. The *A* coefficient does not have physical meaning related to fouling. This term was added to enable use of initial membrane resistances for each test sequence. Without this coefficient, R_m would be solved for by the nonlinear regression algorithm, which was not desired. It is noted that HPC and particle count concentrations do not provide a measure of cumulative mass per se; however, cumulative cfu and particles are analogous to mass. Mass is used to describe accrual of each foulant from this point forward. The magnitude and sign of these coefficients indicate the extent and effect upon resistance to flow. For instance, positive coefficients indicate deposition by a particular foulant, whereas negative coefficients suggest membrane degradation or loosening. Combined chlorine is an oxidant with the potential to degrade a membrane film, especially polyamide-based membranes, in which case a negative coefficient would result. A negative coefficient could also result for HPC loading, but would apply to cellulose acetate membranes rather than composite thin-film membranes.

RESULTS AND DISCUSSION

Foulant Reduction by Advanced Pretreatments

The pretreatment IMS unit operations were effective for reduction of foulants. As shown in Table 4, coagulation-sedimentation-filtration (CSF) reduced organic, particle and biological foulants. Organic removal by pretreatment varied by water. For Tampa water, 61 to 83 percent NPDOC removal and 58 to 93 percent UV-254 removal were observed for CSF. At East St. Louis, 20-percent NPDOC removal by CSF was reported, although raw water NPDOC was significantly lower at East St. Louis.

CSF and microfiltration (MF) pretreatments achieved enhanced removal of particulates. For Tampa water, particle log removal was typically above 1.4-log and ranging as high as 2.2-log for MF or coagulation-sedimentation-microfiltration/sand filtration (CS-MF/CSF) pretreatments. All pretreatments were capable of reducing turbidity to below 1 NTU. Turbidity following MF or CS-MF/CSF pretreatment averaged 0.2 NTU whereas turbidity following cartridge filtration (CaF) averaged 0.7 NTU.

A semi-log relationship between chloramine dose and HPC levels following disinfection was indicated (not shown). The minimum chloramine dose for continuous maximum deactivation of HPCs for the Tampa water is equal to or greater than 5 mg/L as Cl_2 and capable of 4 to 5 log HPC deactivation.

Laboratory Productivity Evaluations

IMS testing with Tampa Water was performed using cellulose acetate (CA) and composite thin-film NFs, denoted as CALP and LFC1, respectively. As will be shown, water quality, mode of operation, treatment scale, and membrane type had a pronounced impact on membrane productivity decline rates, also referred to as projected run time. Projected run time between chemical cleaning was based on a 15 percent decline rate from an assumed 0.15 gsfd/psi initial MTC_w value.

CALP Laboratory Testing

CALP flat-sheet and single-element productivity profiles are shown in Figures 1 and 2. Target water flux was 14 gsfd for all laboratory experiments; however, system recoveries differed by apparatus, 1 percent for flat-sheet and 85 percent for single-element. The productivity profiles for CALP testing show each test sequence by pretreatment or IMS. For flat-sheet and single-element tests CS-MF-NH_2Cl, MF-NH_2Cl, CaF-NH_2Cl and CS-MF pretreatments were evaluated (note: NH_2Cl denotes combined chlorine pretreatment). A compact version of the commercially available US Filter/Memcor MF was used to perform MF in the laboratory.

Inspection of Figures 1 and 2 reveals relatively linear MTC_w profiles for laboratory tests. Rates of MTC_w decline were estimated using linear regression trends generated in Microsoft Excel™. The rates of productivity decline and projected run times are shown to vary by pretreatment and flat-sheet and single-element. For all tests, the CALP membrane projected run times ranged from 4 days to infinity (i.e., no fouling). Comparison of flat-sheet and single-element cleaning frequencies/run times indicates fouling was more pronounced in flat-sheet tests versus the single-element. Differences are expected in light of notable differences in water quality shown in Table 4; however, differences in membrane surface area per batch volume are also recognized.

Enhanced NF operation corresponded to tests including coagulation, microfiltration and combined chlorine pretreatments, CS-MF-NH_2Cl. Fouling was effectively controlled for these systems with operational run times between chemical cleaning approaching performance typical of low-fouling NF groundwater applications—151 days and infinity (i.e., no fouling) for flat-sheet and single-element tests, respectively.

NPDOC, UV-254, turbidity and particles were significantly removed by the advanced pretreatment(s). For instance, pretreated levels for these parameters during testing were less than 3.9 mg/L NPDOC, 0.12 cm^{-1} UV-254, and 414 particles/ml for both flat-sheet and single-element tests. Single-element MTC_w was sustained using CSF-MF pretreatment without combined chlorine addition with a projected 3-month cleaning frequency. Flat-sheet evaluation with CS-MF pretreatment resulted in a projected 18-day cleaning frequency. Comparison of flat-sheet NF feedwater quality indicates significant differences among pretreated HPC levels, 8,600 cfu/ml without combined chlorine (CS-MF) and 4 cfu/ml with combined chlorine (CS-MF-NH_2Cl). With comparable NPDOC and UV-254, biological fouling is implied for the flat-sheet experiment. Particle counts were actually greater during the lesser-fouling CSF-NH_2Cl evaluation, 65 versus 414 #/ml.

For tests with MF-NH_2Cl and CaF-NH_2Cl pretreatment, higher fouling generally corresponded to higher foulant concentrations, e.g., organics, particulates and microbes. CaF-NH_2Cl pretreatment represented flat-sheet testing, with the least amount of pretreatment having 18.5 mg/L NPDOC, 0.82 cm^{-1} UV absorbance, 22,400 cfu/ml and 0.95 NTU, resulting in a projected 4-day run time. Replacement of cartridge filtration (CaF) with MF reduced turbidity to 0.20 NTU and HPCs to 2,630 cfu/ml, improving performance to a 13-day run time. In contrast, single-element results indicated less fouling for CaF-NH_2Cl pretreatment over MF-NH_2Cl. Inspection of water quality for these tests would suggest greater particle and biological fouling potential for the MF-NH_2Cl single-element experiment, although combined chlorine was twice the amount in the CaF-NH_2Cl test of that applied during MF-NH_2Cl operation. Operation with increased batch volume and extended run time may have led to more insightful single-element results. Consequently, insight into fouling mechanisms by single-element tests is uncertain. Overall, flat-sheet tests were more meaningful. Flat-sheet pretreatments rank in descending order of projected run time as CS-MF-NH_2Cl > CS-MF > MF-NH_2Cl > CaF-NH_2Cl.

LFC1 Laboratory Testing

LFC1 flat-sheet and single-element productivity profiles are shown in Figures 3 and 4. Laboratory LFC1 productivity profiles were predominantly linear, although some curvilinear trends were observed. A notable aspect of these profiles was the increasing trend shown for CS-MF-NH_2Cl testing. Target hydraulic conditions are the same as discussed for the CALP membrane film.

For laboratory evaluations, CS-MF-NH_2Cl, CS-MF, MF-NH_2Cl, MF, CaF-NH_2Cl and CaF pretreatments were considered (Tables 2 and 3). Comparison of the LFC1 operational profiles to those for the CALP NF indicate the CALP film was generally more resistant to fouling, having significantly lower rates of MTC_w decline. Differences in performance may be attributed to differences in physical and chemical properties. LFC1 was not as smooth and was more negatively charged relative to the CALP NF film, indicating that choice of membrane film, composite thin-film versus cellulose acetate, has significant performance implications related to fouling.

Projected run times by pretreatment are summarized in Figures 3 and 4. Projected operational run times for LFC1 flat-sheet and single-element operations ranged from 1.6 days to infinity and 5 days to infinity, respectively. Like that of the CALP NF, LFC1 fouling was generally reduced with increasing level of pretreatment. The "highest" level of pretreatment corresponded to CS-MF-NH_2Cl pretreatment, which provided significant removal of organic, particulate and biological fouling agents. Operation with this feedwater indicated fouling was controlled, i.e., infinite run time; however, the LFC1 NF became more productive with time of operation, indicated by an increasing MTC_w trend. This is thought to indicate combined chlorine tolerance limitations for this film. Glater et al. (1983 and 1994) have investigated membrane sensitivity to chlorinated disinfectants. From his investigation, reaction of chlorine with a membrane film may cause structural damage leading to decreased solute rejection and eventual catastrophic failure. The long-term impacts of combined chlorine exposure are undetermined in these tests; however, this topic was the focus of an independent evaluation recently conducted by Beverly (2000).

General comparison of LFC1 flat-sheet and single-element run times indicated flat-sheet and single-element tests do not correlate for prediction of run time using productivity decline rates. The data show flat-sheet experiments experienced significantly higher rates of decline relative to single-element tests. Differences in water quality and foulant mass are the suspected sources of reduced single-element fouling.

Comparing CS-MF pretreatment with and without combined chlorine for flat-sheet and single-element tests would suggest a biological fouling component. Heterotrophic plate count analysis of biofilms extracted from flat-sheet specimens following testing confirmed, to varying degrees, biological activity (not shown). Unfortunately, threshold HPC densities at which biological fouling is manifested are not documented. However, Vrouwenvelder et al. (1998), using direct cell counts, report cell densities of $108/cm^2$ in applications where biological fouling has been observed. LFC1 flat-sheet operation was optimized using CS-MF-NH_2Cl pretreatment, as fouling was not experienced. CS-MF testing resulted in a 4.3-day run time. Even though fouling was not indicated during CS-MF-NH_2Cl testing, it is possible that organic and particle deposition did occur but was masked by increases in productivity due to interaction of combined chlorine with the membrane film.

Organic fouling is indicated by comparison of trials with and without coagulation, particularly flat-sheet evaluations. MF-NH_2Cl pretreatment provided enhanced particulate removal and reduced HPC levels, allowing fouling principally due to organics to be assessed. CS-MF pretreatment resulted in a 4.3-day run time whereas MF with and without combined chlorine resulted in 2- and 1.7-day run times, respectively. The rate of fouling decreased with decreasing NPDOC feed concentration (18.8 versus ~3.8 mg/L C). Despite differences in water quality, cleaning frequencies for CaF pretreatment were comparable to MF pretreatments. The similarity in decline rate and higher particle counts in CaF pretreated water further support organic fouling. Given only slight differences in projected run time for CaF and CaF-NH_2Cl pretreatment, 1.6 to 2 days, organic fouling may have outweighed particle and biological fouling. Flat-sheet turbidity and particle counts following CaF and CaF-NH_2Cl were significantly higher than the other pretreatments. Flat-sheet CaF and CaF-NH_2Cl turbidities ranged from 0.75 to 0.80

NTU and particles ranged from 1,952 to 4,728 #/ml. Turbidity during single-element CaF and CaF-NH_2Cl pretreatment ranged from 0.58 to 1.0 NTU. Particles during these tests ranged from 2,526 to 4,872 particles/ml. The associated rates of fouling ranged from 6 to 1.6 days.

Combined chlorine had a relatively insignificant impact upon performance with CaF and CaF-NH_2Cl experiments, which would seemingly contradict the previous conclusion regarding biological fouling during CS-MF operation. However, organic and biological fouling mechanisms may outweigh one another depending on pretreated feedwater. Biological fouling seems to be more apparent for CS-MF when organics were lowered by coagulation. Organic fouling appeared to prevail over biological and particle fouling when organics were high, exemplified by projected run times for MF and CaF pretreatments with and without combined chlorine. Therefore a single dominant fouling mechanism for the LFC1 NF is not certain, although organic fouling and biological fouling seem to overshadow particle fouling. Refined efforts to uncover fouling mechanisms were addressed by productivity modeling.

Productivity Modeling

Using nonlinear regression, model coefficients were developed (SPSS 8.0™). Cumulative mass loading was estimated from combined hydraulic and water quality monitoring data. When possible, volume-weighted water quality concentrations were used to account for changes in bulk concentration resulting from deposition and/or batch volume losses. Insignificant independent variables were eliminated one at a time until a model resulted in only statistically significant coefficients (95% confidence interval). Model diagnostics included parameter confidence intervals, coefficient of determination (r^2), correlation matrices and plots of predicted versus actual profiles.

Without a definitive basis for modeling NF mass loading, three model types, differing by choice of mass loading, were considered. Model types included: (1) cumulative mass loading calculated by the product of feedwater concentration (C_f), permeate flow rate (Q_p) and time (t); (2) product of feedwater concentration (C_f), feed flow rate (Q_f) and time (t); and (3) product of average feed channel concentration (blended and concentrate, C_b), blended feed flow rate (Q_b) and time (t). Following numerous iterations, models based on the first approach, $M=C_f Q_p t$, were found to be the most meaningful and consistent with actual operational data. Accordingly, preference was given to this approach. Note: Only approaches (1) and (2) were applied to flat-sheet data as concentrate recycle was not used and feed-side concentration gradient was negligible due to low recovery operation (~1 percent). Flat-sheet and single-element data were modeled separately; however, models were evaluated for either flat-sheet or single-element experiments as one complete data set.

LFC1 NF Productivity Modeling

For the LFC1 data set, the model shown as Equation 7 resulted. All terms except for water mass loading (K_{d2}) were significant. This model indicated fouling due to organic, biological, and particle mass loading, along with chemical degradation by combined chlorine exposure. The coefficient of determination, r^2, was 0.815 for this model. Predicted and actual productivity profiles are shown in Figure 5. Predicted and actual productivity profiles were in good agreement for five of the six test sequences. The predicted productivity profile for MF-NH_2Cl pretreatment, the second test series, indicated less fouling than actually observed. The slight increasing trend shown in the first test sequence, CSF-MF-NH_2Cl pretreatment, was predicted by the model—negative K_{d1} coefficient, hence reduced film resistance due to chemical degradation by combined chlorine mass loading.

$$MTC_w = \frac{1}{\mu R_T}$$

$$= \frac{1}{\mu\left[0.97R_m + \frac{Q_p t}{A_m}\begin{pmatrix} 0.184(\text{NPDOC}) \\ + 3.99 \times 10^{-11}(\text{HPC}) \\ + 7.70 \times 10^{-10}(\text{Particles}) \\ - 0.335(\text{Total Cl}_2) \end{pmatrix}\right]} \tag{7}$$

where:

MTC_w	=	water mass transfer coefficient (gsfd/psi-d)
μ	=	viscosity (cp)
Q_p	=	permeate flow rate (gal/day)
t	=	cumulative run time (days)
A_m	=	membrane area (ft^2)
R_m	=	initial membrane resistance (ft^2-day-psi/gal-cp)
NPDOC	=	cumulative mass loading NPDOC (g/ft^2)
HPC	=	cumulative HPC loading (cfu/ft^2)
Particles	=	cumulative particle loading ($\#/ft^2$)
Total Cl_2	=	cumulative total chlorine mass loading (g/ft^2)

Modeling results for single-element testing are far less compelling than those for flat-sheet. Single-element models did not indicate significant or consistent relationships between membrane productivity and cumulative mass loading for any of the associated mechanisms. Here again, the single-element evaluations would have likely benefited from larger batch volume and increased run time, which were impractical at the laboratory scale.

CALP NF Productivity Modeling

Resistance modeling was conducted in likewise fashion for CALP flat-sheet and single-element data sets. For the flat-sheet model evaluated with all coefficients, all fouling and degradation terms were insignificant. Following successive iterations, a model indicated fouling due to biological mass loading or HPCs was realized. However, r^2 was a mere 0.204, indicating the model was able to account for only 20 percent of the productivity variation. Interestingly, organic adsorption was not significant. Recall, organic adsorption was significant for the LFC1 membrane, a hydrophobic film. The CALP film is hydrophobic and as such would be thought to have a greater affinity for organic adsorption. Like that for LFC1, surface charge is negative for CALP. While the hydrophobic conditions seemingly favor adsorption onto the CALP NF, surface charge may offset adsorption, especially if the dissolved organic constituents are negatively charged.

Model results for CALP single-element testing were in contrast to flat-sheet results. Single-element models indicated particle fouling and chemical degradation due to combined chlorine. While it is not clear why flat-sheet tests indicated biological fouling and single-element tests indicated particle fouling, credence was given to the flat-sheet result on the basis of foulant mass limitations per membrane area for CALP single-element tests.

Model Verification

LFC1 productivity models developed using laboratory data were used to predict productivity profiles for IMSs tested in the field, a staged pilot membrane array (Hillsborough River) and a single-element membrane system (Mississippi River water). This approach served to validate the laboratory-based approach by using independently collected field data with the associated model.

Hillsborough River Verification

IMS testing at Tampa included evaluation of CALP and LFC1 NFs, the same NFs evaluated for laboratory tests, and pretreatments similar to those used for laboratory tests. Predicted pilot productivity results for the LFC1 pilot operated at Tampa are shown with actual MTC_w profiles in Figure 6.

In general, the predictive model corroborates actual performance, especially during the first two test sequences, Memcor™

microfiltration (MMF) and MMF-NH_2Cl pretreatments, respectively. The model does not appear to consistently over- or under-predict actual conditions, indicating error may be in large part due to random error in either hydraulic data or water quality data. For two of the six test intervals, model predictions do not agree with actual profiles, Zenon microfiltration followed by combined chlorine addition (ZMF-NH_2Cl, hours 251 to 497) and CSF-NH_2Cl (hours 1,034 to 1,332). In these experiments productivity profiles are in conflict, showing opposing trends. However, while not perfect for all cases, the model reflected general trends observed in five of seven cases.

The predictive model was also evaluated using the upper and lower 95 percent nonlinear regression coefficients confidence interval limits. In most cases, the actual profiles are bound by the upper and lower coefficient predictions, reinforcing the practicality of the model.

Mississippi River Verification

Unlike Hillsborough River water, the Mississippi River is a highly turbid, relatively low-organic raw water. A single test interval was modeled for two separate membranes operated in parallel. Test intervals correspond to roughly 40-day operating periods at 55 percent recovery and 10 gsfd flux. Predicted and actual data are shown in Figures 7 and 8 for LFC1 and DK-C NFs, respectively.

Model results are in relatively good agreement with the actual LFC1 MTC_w profile shown in Figure 7. The data set is largely bound by the upper and lower prediction intervals using the upper and lower values of the 95 percent confidence level for each coefficient. The predicted profile for another CTF NF, DK-C, shown in Figure 8, was also in good agreement.

From a practical standpoint, the model was capable of predicting productivity decline for operation with a water fundamentally different from that used to develop the model, an important underlying question. However, while the model was also in good agreement when used to predict productivity of another composite thin-film NF, DK-C, this would not suggest the LFC1 model can be applied to any nanofilter regardless of type. Performance predictions of a fundamentally different film would be inappropriate, i.e., prediction of cellulose acetate NF performance. While the model adequately predicted performance of both membranes, the films were apparently similar enough to show conformity. However, if a completely mechanistic model were available that accounted for physical and chemical characteristics of the membrane surface and membrane foulants (e.g., charge, hydrophobicity, roughness, etc.), extension to different film and water combinations would be better served.

CONCLUSIONS

- Laboratory IMS pretreatment unit operations effectively reduced NF fouling mechanisms, with NF fouling generally decreasing with corresponding decreases in foulant concentration or cumulative mass loading.
- Coagulation-sedimentation-filtration followed by combined chlorine provided significant increases in projected run time between chemical cleanings. The highest rates of productivity decline were generally observed for the least amount of pretreatment, cartridge filtration.
- Cellulose acetate and composite thin-film NFs differed significantly in terms of resistance to fouling, with enhanced performance provided by cellulose acetate NF over composite thin-film NF.
- Laboratory-based flat-sheet and single-element projected runtimes were not correlated, as single-element productivity decline rates were generally much lower than analogous tests conducted with flat-sheet cell test equipment. Differences are attributed to significantly less foulant mass available per membrane area in single-element tests versus flat-sheet tests.

- The results showed that laboratory tests could be used to evaluate IMS performance. Predicted productivity profiles using a series resistance model were in good agreement with actual data.
- Single-element testing in batch-recycle mode was a poor test method for quantifying membrane fouling at laboratory scale. Consequently, flat-sheet tests are strongly favored over single-element tests in that flat-sheet tests more closely approach a continuously operated (nonbatch) pilot- or full-scale system.
- In order to achieve comparable loading, the single-element system would require several thousand gallons of feedwater. Thus, single-element testing was shown to be a poor candidate for assessing productivity decline in the laboratory due to practical limitations.
- Flat-sheet productivity profiles were generally in agreement with relative performance rankings observed in the field.
- Membrane productivity modeling with a semi-empirical model based on Darcy's Law and cumulative foulant mass loading was useful for assessment of membrane fouling. The approach used was capable of overcoming water quality and hydraulic differences between laboratory and field evaluations. Furthermore, the model served to elucidate membrane fouling mechanisms and quantification of individual contributions by mechanism.
- Flat-sheet models were more insightful than single-element models. In most cases, single-element models did not converge to meaningful results and in most cases were statistically insignificant. Difficulties in modeling single-element data were attributed to reduced single-element fouling (i.e., foulant mass per membrane area limitations) and increased variability.

- Flat-sheet LFC1 productivity modeling with Tampa water indicated organic, biological and particle fouling. Degradation due to total chlorine mass loading was also statistically significant. Overall, predicted productivity profiles reasonably corroborated actual results.
- Laboratory testing was validated using resistance models developed from flat-sheet experiments. The results demonstrate that field productivity can be predicted by applying pilot-scale hydraulic and water quality data to resistance models developed from laboratory tests.
- Despite relatively good agreement, cleaning frequencies determined directly from laboratory-based models are still regarded as limited, although laboratory-based evaluations may provide a cost-effective means of screening membrane pretreatment combinations for testing at pilot scale and desktop cost estimating.

RECOMMENDATIONS

- Prior to design, construction and operation of a full-scale integrated membrane system, membrane films and pretreatments should be screened to optimize performance for fouling control and product water quality. While useful performance data can be developed in the laboratory, pilot testing with a staged array provides the best available approach for simulating full-scale performance and should be used to verify laboratory data.
- Extraction of flat-sheet membrane specimens from spiral-wound single-element membranes may reduce differences observed between initial productivity.
- Evaluation of membrane fouling with batch-recycle operated systems may present significant foulant mass limitation depending on membrane area, flux and batch volume.

For this reason nonbatch testing is preferred. The rate of fouling as a function of batch volume would be useful. This could be used to establish threshold criteria for batch-recycle systems as a function of membrane area, flux, recovery, and length of operation.

- Differences in flow regime between flat-sheet cell systems and spiral-wound membrane configurations could be offset by using flat-sheet systems designed to simulate a differential element that is representative of a spiral-wound membrane (e.g., RBSMT with transverse flow, concentrate recycle for increased recovery and use of feed channel spacers to increase turbulence).
- When comparing fouling observed during laboratory tests to fouling at pilot or full scale, differences in system hydraulics and water quality should be quantified. The resistance model developed in this chapter provided a reasonable means for meeting this objective.
- Isolation and evaluation of organic, biological and particle foulants in natural waters are limited, as pretreatment processes are not entirely solute-specific. Therefore, models used to project full-scale performance should be based on multiple evaluations that include significant foulant variation by category and combinations thereof.
- Additional testing is needed to evaluate interaction among foulants. The resistance model as developed assumes additive resistance for organic, biological and particle fouling. If these foulants interact, which is possible, the effect may not be additive. Also, it stands to reason that as fouling occurs, membrane characteristics may change as well, both physical and chemical (e.g., surface roughness, hydrophobicity and charge, to name a few).
- In the absence of a model that accounts for intrinsic physical and chemical characteristics of membrane films and foulants, use of a resistance model with cumulative

foulant mass loading is regarded as membrane- and water-specific. For conservative measure, model projections should only be applied to a source water and membrane combination from which the model was developed.

Table 1 Select manufacturer specifications for NF membrane systems employed during testing

Parameter	Membrane System			
Commercial designation	Fluid Systems ROGA CALP	Hydranautics LFC1[1]	Hydranautics PVD1	Desal DK-C
Cut-off rating	300–500[2] Da	100–200[2] Da	150–300[2] Da	150–300[2] Da
Configuration	Flat-sheet/ Spiral-wound	Flat-sheet/ Spiral-wound	Flat-sheet/ Spiral-wound	Flat-sheet/ Spiral-wound
Material	Cellulose acetate (CA)	Composite thin-film (CTF)	Composite thin-film (CTF)	Composite thin-film (CTF)

[1] Prototype membrane at the time of testing.
[2] Vendor-based molecular weight cut-off rating.

Table 2 NF advanced pretreatment summary by process

Pretreatment Process	Foulant Impacted: Particles	Organics	Biological	Scaling	Comment
Cartridge filtration (CaF)	√–[1]				Used to simulate conventional NF pretreatment (5 μm nominal pore size).
Microfiltration (MF/MMF/ZMF)	√+[2]				Used for reduction of fouling potential due to particles.
Ultrafiltration	√+[2]		√		Used for reduction of fouling potential due to particles and bulk microbial removal.
Coagulation and sedimentation (CS)		√			Used for TOC removal for reduction of fouling potential due to organic adsorption.
Sand filtration	√				Used for reduction of fouling potential due to particles.
Acid/scale inhibitor				√	H_2SO_4 and commercially available scale inhibitor were employed for scaling control.
Combined chlorine (NH_2Cl)			√+[2]		Biofouling potential was reduced by application of combined chlorine. Monochloramine was the predominant combined chlorine specie.

[1]Minus denotes limited foulant removal capability.
[2]Denotes enhanced foulant removal capability.

Table 3 Test summary by NF test matrix, raw water, NF membrane and test scale

	Tampa						East St. Louis		
NF/Pretreated Feedwaters	Flat-sheet	Single-element	Pilot-plant	Flat-sheet	Single-element	Pilot-plant	Single-element	Single-element	Single-element
Nanofilter	CALP	CALP	CALP	LFC1	LFC1	LFC1	LFC1	CALP	DK-C
$CSF\text{-}NH_2Cl$[1]	√	√	√	√	√	√	√	√	√
CSF[2]	√	√	√	√	√	√			
CaF[3]				√	√				
$CaF\text{-}NH_2Cl$[4]	√	√		√	√				
MF[5]	√			√	√	√			
$MF\text{-}NH_2Cl$[6]	√	√	√	√	√	√			
$UF\text{-}NH_2Cl$[7]								√	

[1] Coagulation-sedimentation-(sand) filtration with combined chlorine, also denoted by $CS\text{-}MF\text{-}NH_2Cl$ or $C/ZMF\text{-}NH_2Cl$ for Tampa testing and CS-UF or ILC-UF for East St. Louis testing.

[2] Coagulation-sedimentation-filtration without combined chlorine, also denoted by CS-MF or C/ZMF.

[3] Cartridge filtration without combined chlorine.

[4] Cartridge filtration with combined chlorine.

[5] Microfiltration without combined chlorine, also denoted by MMF and ZMF.

[6] Microfiltration with combined chlorine, also denoted by $MMF\text{-}NH_2Cl$ and $ZMF\text{-}NH_2Cl$.

[7] Ultrafiltration with combined chlorine.

Table 4 Average raw and pretreated water quality summary

			NPDOC (*mg/L C*)		UV-254 (*cm-1*)		HPC (*cfu/ml*)		Turbidity (*NTU*)		Particles (*#/ml*)		Cl_2[1] (*ppm*)
Scale	Water	Pretreatment	Raw	Pre.[2]	Raw	Pre.[2]	Raw	Pre.[2]	Raw	Pre.[2]	Raw	Pre.[2]	Pre.[2]
Lab Flat-sheet	Tampa	CaF	18.8	19.5	0.74	0.784	140,000	237,000	2.4	0.80	12,500	1,952	0
		MF	18.8	18.6	0.74	0.702	140,000	131,000	2.4	0.12	12,500	2,344	0
		CS-MF	18.8	3.8	0.74	0.078	140,000	61,300	2.4	0.12	12,500	72	0
		CaF- NH_2Cl	18.8	19.8	0.74	0.810	140,000	11,300	2.4	0.88	12,500	2,498	9.4
		MF-NH_2Cl	18.8	18.3	0.74	0.762	140,000	1,315	2.4	0.30	12,500	409	3.6
		CS-MF-NH_2Cl	18.8	3.3	0.74	0.090	140,000	3	2.4	0.43	12,500	316	5.2
Lab Single-element	Tampa	CaF	9.9	9.9	0.27	0.240	106,000	56,500	10.1	0.50	7,600	719	0
		MF	9.9	9.9	0.27	0.250	106,000	6,690	10.1	0.33	7,600	112	0
		CS-MF	9.9	3.9	0.27	0.062	106,000	20,600	10.1	0.34	7,600	220	0
		CaF- NH_2Cl	9.9	9.9	0.27	0.230	106,000	58	10.1	0.70	7,600	2,260	7.6
		MF-NH_2Cl	9.9	9.9	0.27	0.211	106,000	30	10.1	0.10	7,600	149	4.7
		CS-MF-NH_2Cl	9.9	3.9	0.27	0.113	106,000	52	10.1	0.20	7,600	212	5.6
Field Pilot	Tampa	MF	17.0	17.1	0.77	0.800	ns[3]	5,780	1.83	0.12	3,992	528	0
		CS-MF	10.6	1.8	0.40	0.031	ns	2,400	3.76	0.13	26,951	232	0
		MF-NH_2Cl	7.8	7.3	0.33	0.283	ns	16,113	1.94	0.13	9,624	387	0.8
		CS-MF-NH_2Cl	21.1	4.6	1.31	0.098	ns	301	1.91	0.11	8,383	253	4.0
Field Single-Element	East St. Louis	CSF	4.4	3.5	0.13	0.058	ns	144	109	0.08	471,000	82	<1
		CS-UF	4.9	3.9	0.19	0.139	ns	ns	118	0.07	892,000	18	1–5
		ILC-UF	4.3	3.5	0.14	0.076	ns	ns	59	0.08	294,000	48	1–5
		UF	4.3	4.2	0.13	0.122	ns	ns	116	0.08	238,000	27	1–5

[1] Average feed combined chlorine concentration.
[2] Water quality value following pretreatment of raw water.
[3] Not sampled.

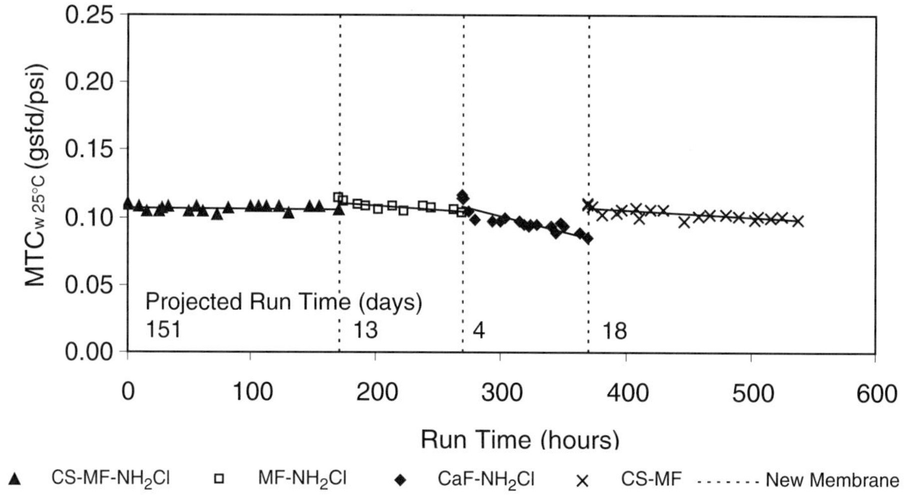

Figure 1 CALP flat-sheet productivity profiles for varying pretreatment—13 gsfd, 1% recovery

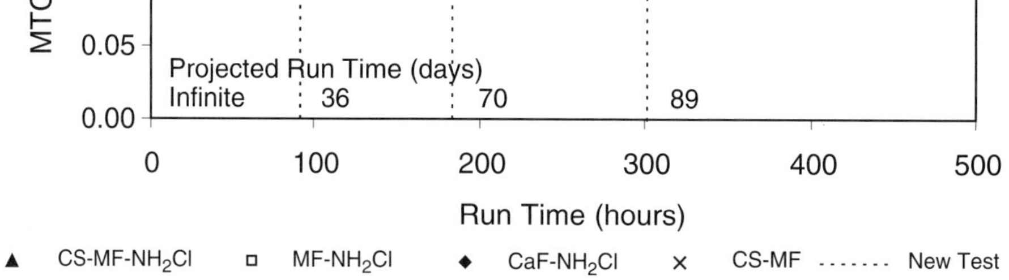

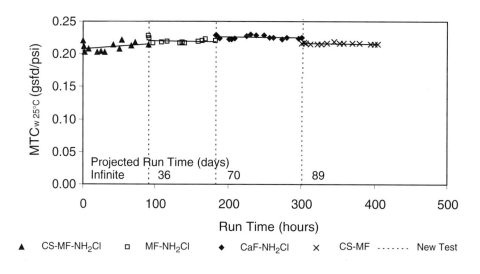

Figure 2 CALP single-element productivity profiles for varying pretreatment—14 gsfd, 86% recovery

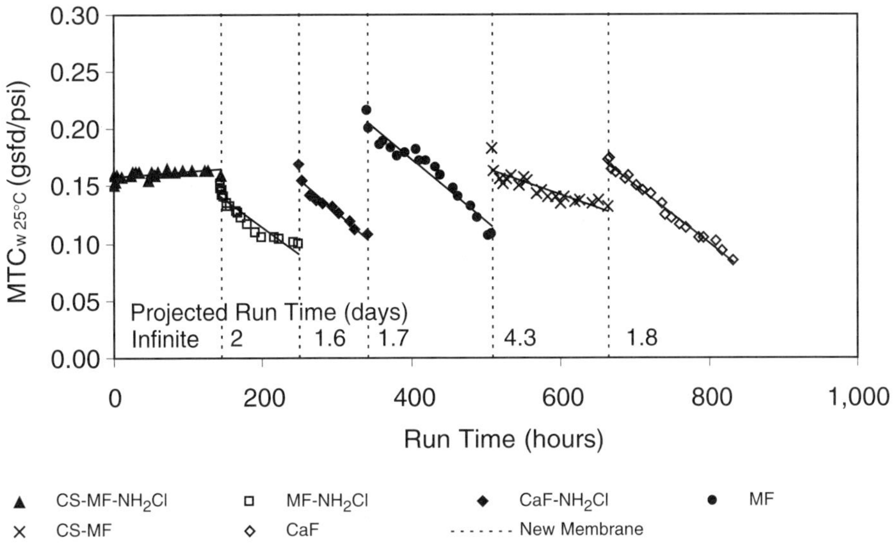

Figure 3 LFC1 flat-sheet productivity profiles for varying pretreatment—14 gsfd, 1% recovery

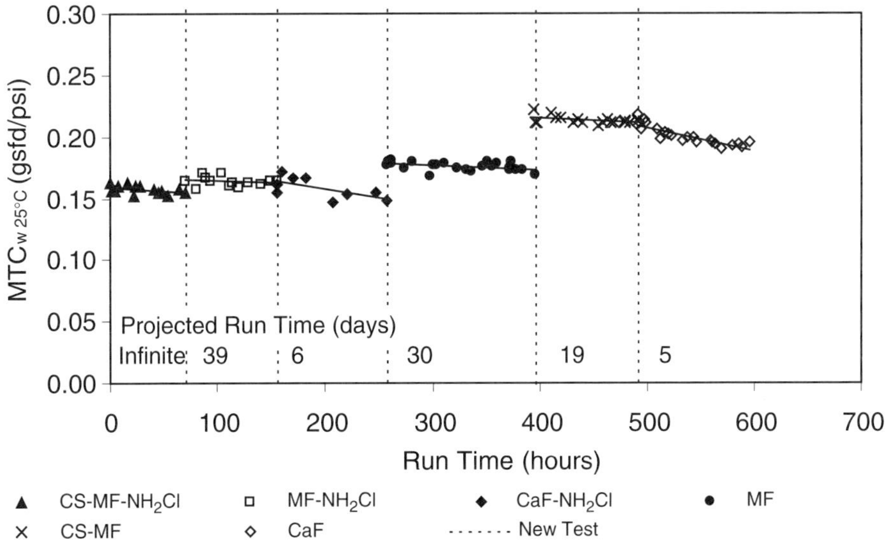

Figure 4 LFC1 single-element productivity profiles for varying pretreatment—13 gsfd, 80% recovery

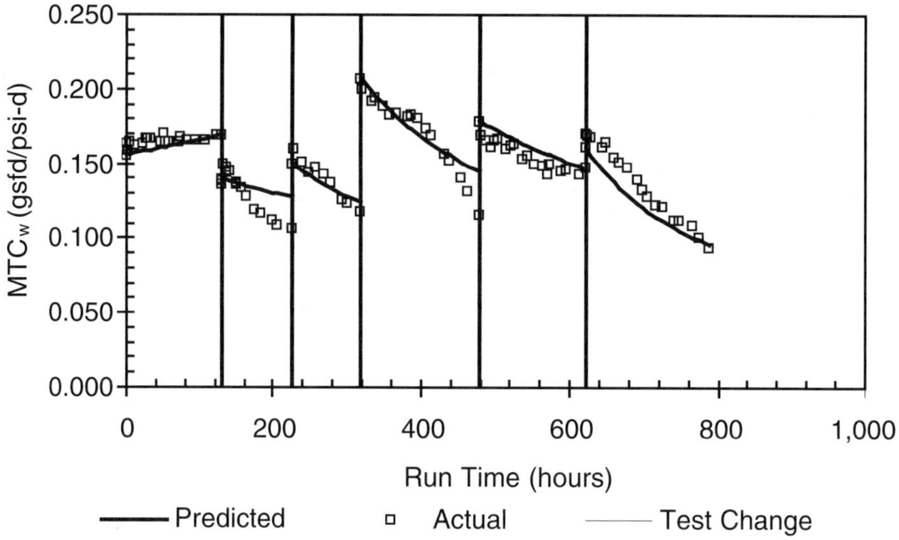

Figure 5 Predicted laboratory productivity profiles—LFC1 flat-sheet, cumulative NPDOC, HPC, particle and NH_2Cl loading

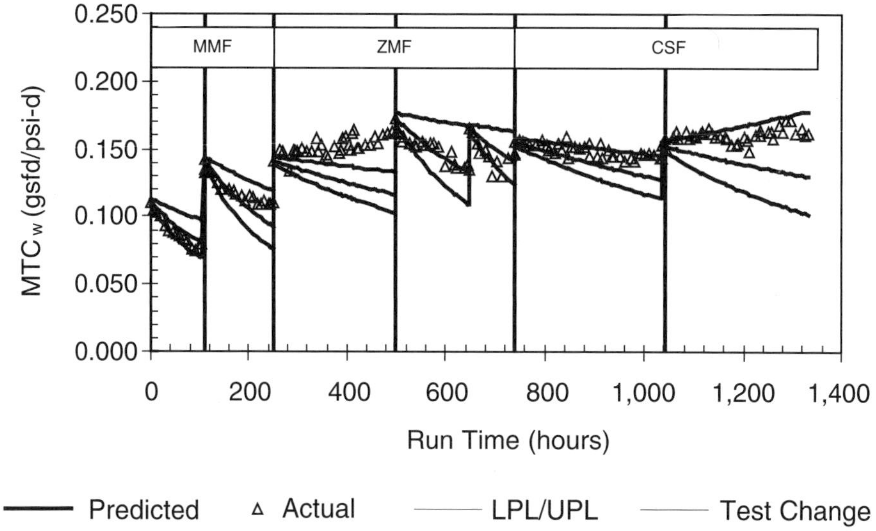

Figure 6 Tampa LFC1 pilot performance—flat-sheet model verification, cumulative NPDOC, HPC, particle and NH_2Cl loading

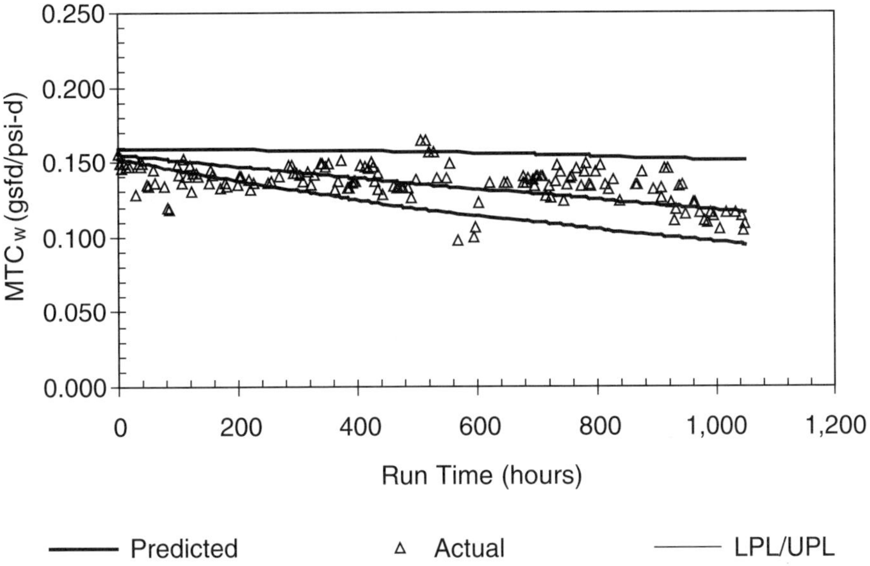

Figure 7 East St. Louis pilot performance—flat-sheet model verification, cumulative NPDOC, HPC, particle and NH_2Cl loading

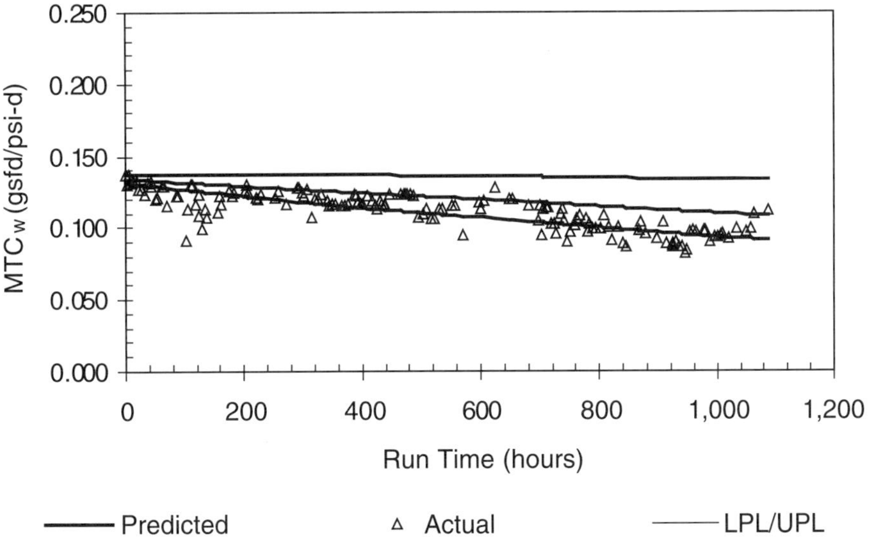

Figure 8 East St. Louis pilot performance—flat-sheet model verification, cumulative NPDOC, HPC, particle and NH_2Cl loading

ACKNOWLEDGEMENTS

The study was completed while William Lovins was a Ph.D. candidate at the University of Central Florida. The information presented originates from tasks required for completion of research sponsored by the American Water Works Association Research Foundation (AWWARF)—*Integrated, multi-objective membrane systems for control of microbials and DBP precursors* (AWWARF Project Number 264). The authors are grateful for the support provided by the project advisory committee and the analytical skills of the Environmental Systems Engineering Institute, particularly Charles Norris, Maria Pia Real-Robert, David Ingram, Christophe Robert, Luke Mulford and Hola Sfeir.

REFERENCES

1. Allgeier, S.C and R.S. Summers. Evaluating NF for DBP control with the RBSMT. *J. AWWA* Vol. 87 No. 3 (1995).
2. Allgeier, S.C.; A.M. Gusses; T.F. Speth; and R.S. Summers. Verification of the rapid bench-scale membrane test for use in the ICR. *Proceedings Membrane Technology Conference*, New Orleans, LA (1997).
3. Beverly, S. *Effects of chlorine and chloramine on membrane properties.* Master's Thesis, University of Central Florida, Orlando, FL (2000).
4. DiGiano, F.A. Evaluation of natural organic matter fouling in bench-scale, batch recycle tests of nanofiltration. *Presented at the 1996 AWWA GAC, Membranes and the ICR: A Workshop on Bench-scale and Pilot-scale Evaluations*, Cincinnati, OH.
5. Glater, J.; M.R. Zachariah; S.B. Mcray; and J.W. McCutchan. Reverse osmosis membrane sensitivity to ozone and halogen disinfectants. *Desalination 48* (1983).
6. Glater, J.; S.K. Hong; and M. Elimilech. The search for a chlorine tolerant reverse osmosis membrane. *Desalination* 95 (1994).

7. Gusses, A.; T.F. Speth; S.C. Allgeier; and R.S. Summers. Evaluation of surface water pretreatment processes using the rapid bench scale membrane test. *Proceedings Membrane Technology Conference*, New Orleans, LA (1997).
8. Gusses, A.M.; S.C. Allgeier; T.F. Speth; and R.S. Summers. Impact of membrane sample variability on the performance of the rapid bench-scale membrane test. *Proceedings Membrane Technology Conference*, Long Beach, CA (1999).
9. Lovins, W.A. *Modeling mass-transfer in single element bench units.* Master's Thesis, University of Central Florida (1996).
10. Reiss, C.R.; C. Robert; and J.S. Taylor. Multi-contaminant removal by integrated membrane systems. *Proceedings Membrane Technology Conference*, New Orleans, LA (1997).
11. Speth, T.F.; A.M. Gusses; S.C. Allgeier; R.S. Summers; J.W. Swertfeger; and B.W. MacLeod. Predicting pilot-scale membrane performance with bench-scale studies. *Proceedings Membrane Technology Conference*, New Orleans, LA (1997).
12. Tram, Q.M. *Modeling membrane mass transfer in a full-scale membrane water treatment plant.* Master's Thesis, University of Central Florida (1997).
13. Vrouwenvelder, J.S.; J.A.M. van Paasen; H.C. Folmer; J.A.M.H. Hofman; M.M. Nederlof; and D. van der Kooij. Biofouling of membranes for drinking water production. *Desalination*, Vol. 118 (1998).

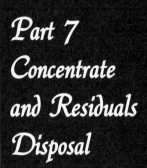

Part 7 Concentrate and Residuals Disposal

Survey of Membrane Concentrate Reuse and Disposal

Options for Treatment and Disposal of Residuals Produced by Membrane Processes in the Reclamation of Municipal Wastewater

A Methodology for Calculating Actual Dilution of a Membrane Concentrate Discharge to Tidal Receiving Waters

CHAPTER · 29

Survey of Membrane Concentrate Reuse and Disposal

Eric Kenna, Graduate Research Assistant

Amy K. Zander, Associate Professor
Department of Civil and Environmental Engineering
Clarkson University, Potsdam, NY

INTRODUCTION

In recent years the demand for potable water has put a significant strain on the United States' freshwater supplies. There are a number of factors that make it increasingly more difficult for freshwater supplies to meet the world's potable water needs. The United States' freshwater supplies are finite and are distributed unevenly in surface and groundwater sources. The already enormous demand for water is growing at a rapid rate and may eventually outgrow supplies. This continually growing demand for water resources along with the increasingly stringent standards being placed on drinking water treatment has turned attention to the use and development of technologies that can produce potable water from non-potable waters, sea waters and brackish waters. Additionally there remains the issue of meeting the water demands without effecting a negative impact on the environment or humans.

One such technology capable of producing potable water from non-potable sources as well as meeting the projected future regulations is the use of pressure-driven membrane processes for drinking water treatment. Reverse osmosis (RO) is referred to as the best available technology (BAT) for four radionuclides and 13 of the inorganics due for regulation under the Safe Drinking Water Act. Nanofiltration (NF) can be used for removing hardness or dissolved organic compounds when TDS removal is not a main treatment objective. Ultrafiltration (UF) is not effective at removing dissolved salts but can be useful in the removal of particulate matter, colloids and some high-molecular-weight, dissolved organic compounds. Microfiltration (MF) applications are used mainly for particulate removal, and more specifically removal of bacteria and protozoa in some cases.

The most glaring barrier to the implementation of membrane processes in water treatment is the issue surrounding the disposal and reuse of the concentrate stream produced, which is a natural by-product of these processes. Two of the primary obstacles, from a regulatory standpoint, surrounding membrane concentrate disposal are the classification of membrane concentrate and the perceived toxicity of concentrate and subsequent required testing procedures. The American Desalting Association has concluded that this classification of concentrate as an industrial waste is the result of a simple process of elimination rather than the result of any sound fact or judgement [1]. This classification lends itself to a lengthy and often expensive permitting process for water plant operators discharging a concentrate, not to mention the perception it implants in the public's eye as to the nature of concentrate streams and the pollutants they contain. In reality membrane concentrates contain simply the impurities that were contained in the raw water, albeit at more concentrated levels. There are little to no new substances added to the water by the drinking water facilities, as there would most likely be in an industrial production process.

There exists a "Catch-22" of sorts within the issue of regulating membrane concentrate disposal. New, more stringent drinking water standards are one reason for the growing use and need for such technologies as membrane processes. However, one of the major barriers preventing the implementation of membrane technologies is the regulatory restriction surrounding the disposal and reuse of the concentrate.

The overall objective of this project is to collect and analyze existing information on membrane concentrate reuse in water treatment processes and compile a complete report on this subject, in order to allow informed decisions to be made regarding these issues. Decisions regarding how to deal with concentrate must be made in the design stages of a membrane plant. It is in these early stages that the final report from this project is intended to be of aid to engineers/designers in making informed decisions regarding concentrate disposal and reuse.

Discharge of membrane concentrate to a receiving body of water is a fairly common method of disposal. Whether that body is a marine water (ocean discharge) or a non-marine surface water (lake, river, etc.), a National Pollutant Discharge Elimination System (NPDES) permit must be obtained. This permit is authorized under the Clean Water Act and establishes in-stream water quality criteria and standards. Another disposal option, and perhaps the most popular, is to discharge the concentrate to the local wastewater treatment plant via the sewer system or a direct pipeline. This may be the most popular method simply because it is often the easiest as well as the least expensive. This option typically requires only some type of local permit or agreement with the local POTW (Publically Operated Treatment Works). Deep well injection (DWI) is another option for concentrate disposal. This method usually requires a permit for well operation and underground injection with the appropriate state agency. DWI is typically a costly method. The costs are incurred in the construction of the well and the extensive monitoring that is required. Evaporation ponds are a viable disposal option for those geographic regions with warm arid climates. For the most part there are no specific permits for the use of evaporation ponds.

Land application is a disposal option that can be considered reuse. Both soil conditioning and irrigation fall under the category of land application. These are both applications that do not require the use of potable water and for which concentrate streams can successfully fulfill the need for water. There are no permits required or regulations written specifically to address land applications. Another possible method of reuse for which membrane concentrate would be suitable is industrial cooling water. Again there is no need to waste potable water in industrial processes where it is not needed. Source water recharge is also a possible reuse method for concentrate streams. There are several other possible reuse alternatives for membrane concentrates within the water industry. These include saline-tolerant turf operations, fire water, aesthetic fountains and lagoons, and toilet flush water. All of these are instances in which it is not necessary to waste valuable potable water supplies.

METHODS AND MATERIALS

Several procedures were utilized in this project in order to meet the previously stated objectives. The first was a literature review, which is ongoing throughout the course of the project. This included a search of conventional library indexes as well as the use of Internet search engines.

A survey of membrane plant operators in the United States of America (U.S.) was conducted. A mailing list was compiled from several sources. The Water:\Stats Database (1998) [4] was consulted, as well as the earlier AWWARF study *Membrane Concentrate Disposal* [3]. Membrane equipment suppliers were contacted in an effort to include newer plants, especially the growing number of MF, UF and NF systems. Any plants uncovered in the literature review have also been included. The survey was written to allow the operators to provide as much information as possible while consuming little time. The survey includes the following information:

- Treatment plant type (RO, NF, UF or MF) and specific treatment goals
- Source water
- Concentrate characteristics such as pH, TDS, radionuclides, organic content, hardness, and any toxicity measurement available

- Current and proposed concentrate discharge methods
- Current and proposed concentrate testing methods
- Perceived and real barriers to concentrate reuse

A similar survey was distributed to the appropriate regulatory personnel in each state. The list of regulatory personnel was compiled from the list in Mickley et al. (1993) and the 1998 Conservation Directory. Information gathered from the survey of regulatory personnel included:

- Indication of the regulatory process necessary for concentrate reuse options
- Current and proposed test methods for membrane concentrate
- Regulatory position on concentrate reuse as determined by concentrate characteristics
- Regulatory position towards concentrate reuse based on current regulatory standards

A second round of survey mailings and follow-up phone calls was used in attempt to collect as much data, in the form of returned surveys, as possible.

Data were sorted by a number of factors. The data from the membrane plant surveys were sorted by type of membrane process, source water characteristics (i.e., groundwater, surface water or ocean), disposal/reuse practices, concentrate characteristics, and geographical region or state. The data collected from the regulatory survey will be sorted by geographical region. A complete record of regulatory concerns is being compiled and each will be addressed to the extent possible.

RESULTS AND DISCUSSION: PLANT SURVEYS

A total of 240 membrane plants/utilities were surveyed, including 41 that were in the planning stages in the 1993 Mickley et al. report. To date, 92 responses have been received, of which 20 stated that membrane processes were no longer being used at that facility, and two are used for water reclamation purposes and not drinking water production. Therefore, the data presented and discussed in this section will be based upon 70 partially or fully completed surveys. The data are broken down by type of membrane process used, by source water, by concentrate discharge method and by EPA region. Dividing the results by membrane process may give some insight into any existing relationships between membrane process and target contaminants, concentrate characteristics or disposal methods, as displayed in Table 1. The use of the word *processes* is needed due to the fact that six plants use integrated membrane systems composed of two different membrane processes. As a result, from the 70 plants that returned surveys, data have been collected on 76 processes. The results are discussed in a similar fashion. Separating the data by source water will provide a look at which membrane processes are typically used to treat certain types of source water, as well as the target contaminants of different sources. Dividing by concentrate discharge method allows the authors to examine what permits are required for different discharge methods. Finally, breaking down the data by EPA region gives a better picture of what areas of the United States are using which membrane processes and which disposal methods are allowed in different areas.

Survey Results by Process

Microfiltration (MF) processes are practiced by 21 of the utilities surveyed. Nine of the 21 are located in California, three in Hawaii and a total of 14 in EPA Region 9. Surface water sources are by far the most common and are used by 19 of the utilities that practice MF. MF targets the removal of particulate matter and large organic compounds and therefore is best suited for use in treating surface water, as these compounds are much more likely to be found in surface water than in a groundwater. In fact only two of the plants practicing MF used a groundwater source.

The overwhelming reason for the MF processes, as specified by the operators and shown in Table 1, was turbidity. This could be expected, as one of the main targets of MF is particulate matter. Five of the operators also specified protozoa/bacteria removal as a reason for the plant.

There were a number of concentrate discharge methods reported by plants using MF processes. These are shown in Table 2, an overview of concentrate and cleaning chemical disposal for membrane processes in the U.S. The most common method reported was sewer discharge, practiced by 10 of these processes.

Table 1 Overview of survey of membrane processes in the United States

Membrane Process	Percent of Respondents	Surface Water Source	Groundwater Source	Reason for Plant
MF (N = 21)	28	90%	10%	Turbidity Protozoa/ Bacteria
UF (N = 6)	8	33%	67%	Turbidity Organic compounds Iron/ Manganese
NF (N = 7)	9	0%	100%	Hardness Dissolved solids Organic compounds
RO (N = 42)	55	12% (5% ocean, 7% fresh)	86%*	Dissolved solids Radionuclides Water reclamation
Total Processes (N = 76) Total Plants (N = 70)	100	34%	65%	–

*One RO plant operator neglected to report the water source.

This method seems to be a commonly selected option. Three of the MF processes have local or regional permits with the receiving POTW and one of the processes has a permit with the appropriate state agency. The next most common discharge option practiced is land application. Concentrate from MF application is mostly wash water and basically amounts to "dirty" water or water with high turbidity. This is supported by the fact that turbidity was the most common concentrate characteristic reported and was listed by eight operators. Land application is practiced by seven of the MF

Table 2 Surveyed concentrate and cleaning chemical disposal for membrane processes in U.S.

Process	Concentrate Disposition	Cleaning Chemicals Disposition
MF	Sewer – 48%	Sewer – 52%
(N = 21)	Land Application – 33%	Mixed with Concentrate – 38%
	Irrigation – 10%	Irrigation following Neutralization – 5%
	Percolation – 13%	Surface Water – 5%
	Sludge Ponds – 10%	
	Surface Water – 19%	
UF	Land Application – 50%	Mixed with Concentrate to Land – 50%
(N = 6)	Sewer – 33%	Mixed with Concentrate to Sewer – 33%
	Surface Water – 17%	Recycle to Plant – 17%
NF	Deep Well Injection – 43%	Deep Well Injection – 43%
(N = 7)	Sewer – 43%	Sewer – 43%
	Ocean – 14%	Ocean – 14%
RO	Surface Discharge – 60 %	Sewer – 54%
(N = 42)	Fresh – 48%	Mixed with Concentrate – 17%
	Ocean - 12%	Evaporation Pond – 5%
	Sewer – 17%	No answer given – 24%
	Deep Well Injection – 12%	
	Land Application – 9%	
	No answer – 2%	

processes, two for the purposes of irrigation, three for percolation/ infiltration and two using sludge ponds. Of these seven, four operators indicated that they did not need a permit, including the three processes in Hawaii, two of which use sludge ponds. The three processes using percolation/ infiltration have permits through the respective state health department. The health department likely gets involved in land application due to the concentrated levels of protozoa and/or bacteria, which may exist in MF concentrates. The fear of spreading these on the land comes through the fear of subsequent groundwater and surface water contamination and resulting impacts on human health. Five of the MF processes discharge their concentrate to a surface water. Two of the operators indicated that the plant has an NPDES permit and one has a permit with the Department of Health in its respective state. One plant practices DWI and has a state injection well permit with the appropriate state agency.

Cleaning chemicals generally included citric acid and caustic soda. Some processes use a pre-made cleaning solution provided by the membrane manufacturer, for example Memclean®. Cleaning wastes were either discharged to the sewer (Table 2) or mixed with the concentrate and then discharged accordingly. Also, one process discharges cleaning waste to surface water and another uses it for irrigation after the waste has been neutralized. As will be seen these are the primary options for dealing with cleaning waste.

The only reuse exhibited by plants practicing MF processes were those using land application for concentrate discharge. This can also be seen in Table 3, which breaks down reuse practices in more detail by membrane process. Infiltration can be seen as reuse in the form of groundwater recharge, as well as the irrigation applications. These applications involved the fewest permits and one would seem to be a very viable option for concentrate disposal for these types of processes. Those processes not practicing reuse mentioned politics, in the form of the permitting process or public acceptance, and cost as their main barriers to reuse applications.

Ultrafiltration (UF) processes are used by six of the plants that responded to the survey. There was no noticeable pattern or preference as to location of these processes. Four of the UF processes use a groundwater source and two a surface water source. The reasons for the plant were somewhat varied (Table 1). Turbidity and organic compounds were listed by one of the processes, which uses a surface water source, as would be expected; however, they were also listed by both processes using a groundwater source. Treatment of surface water for turbidity is much more common than that of groundwater. The other plant using a surface water source listed Surface Water Treatment Rule compliance as the reason for the plant, likely meaning turbidity as well. The remaining

Table 3 Surveyed concentrate reuse practices

Process	Potential Practices	Actual Practices	Percent Practicing	Barriers to Reuse
MF (N = 21)	Wetlands Dust Control	Land Application Infiltration	33%	Cost Permitting
UF (N = 6)	Groundwater Recharge Industrial Water Wetlands Dust Control	Land Application	50%	Cost Turbidity Permitting
NF (N = 7)	Groundwater Recharge Industrial Water Toilet Water Aesthetic Fountains	None	0%	Cost Permitting
RO (N = 42)	Fire Water Industrial Water Toilet Water Aesthetic Fountains Saline-Tolerant Turf Operations	Irrigation	10%	Dissolved Solids Cost Permitting
Total (N = 76)			18%	

processes using a groundwater source stated Disinfection By-Products/ Precursors (DBP/Ps) and iron and manganese removal, respectively, as the reasons for the plant. Iron and manganese are more typical targets in groundwater treatment than turbidity would be.

Operators of the UF plants did not give a lot of information on concentrate characteristics. Turbidity and calcium precipitates were noted by one of the processes treating groundwater. Turbidity was the reason given for this particular plant and as such was expected to be present in the concentrate. Calcium limestone particulate would most likely be a result of the bedrock surrounding the groundwater source. Also, as would be expected, the plant listing Fe and Mn removal as the reason for the plant listed these two metals as the main concentrate characteristics.

The six UF processes practice three methods of concentrate disposal (Table 2). Land application is the most common, being practiced by three of the processes. In two cases the plant had an NPDES permit for its discharge and in a third a state health department permit was required. The use of NPDES regulations to oversee land applications is not uncommon and is a result of the lack of regulations that are written specifically for land application procedures. In the case of these three processes the two with NPDES permits were using a groundwater source and the plant using a surface water source was subject to the health department permit. The health department's involvement again may be a result of the fear that treatment of surface water by UF processes can result in high concentrations of disease vectors, i.e., protozoa and/or bacteria. Two of the UF processes practice sewer discharge with their concentrate and both have a local permit with the receiving POTW. The remaining plant discharges its concentrate to a surface water and has a permit through the county Department of Environmental Protection office, in Florida.

Cleaning chemicals used are very similar to those of MF processes (Table 2). Sulfuric or citric acid is commonly used in conjunction with caustic soda. Again some processes used premade solutions or detergents. Cleaning waste discharge methods consisted of two processes practicing sewer discharge, three processes mixing with the concentrate for discharge and one plant recycling the waste to the head of the plant after neutralization. In two of the cases in which the cleaning waste is mixed with concentrate, the concentrate is discharged through land application.

There was only one reuse application reported and that was irrigation in one of the cases in which the concentrate is disposed of through land application. The main barriers to reuse options, as mentioned by operators, were cost and concentrate quality, specifically turbidity in this particular case (Table 3).

Seven of the processes that have thus far returned the survey practice **nanofiltration** (NF). All seven of these are located in Florida and all seven of the processes use a groundwater source (Table 1). This is no coincidence as NF is best suited for removal of hardness, some total dissolved solids (TDS) and some small organic compounds. These are the contaminants most often associated with groundwater. Also, the presence of ample groundwater supplies in Florida lends itself nicely to the application of NF processes. As would be expected, five of the seven processes listed hardness as one of the reasons for the plant and three listed TDS. In addition, other reasons indicated were DBP/Ps and Trihalomethane Formation Potential (THMFP), each by three processes respectively.

Six of the seven processes reported information on their concentrate characteristics, and all six listed TDS levels, specifically sulfate and calcium in some cases, as a main component. Conductivity levels were also noted by two of the processes. As NF membranes do reject a large part of the TDS, it is expected that the concentrate exhibit these high levels. Also, hardness in water is often measured as calcium hardness, and when this is a treatment target of NF it can be expected that the calcium levels in the concentrate will be quite high. Three of the six processes practice DWI (Table 2) and all three have an operational permit for the injection well through the state. Three of the processes discharge the concentrate to the sewer and the only permits listed were NPDES permits for the receiving POTW in each case. One of the processes does discharge its concentrate to the ocean and has a state industrial waste permit for this practice. This classification of membrane concentrate as an industrial waste is one of the major hang-ups in the regulation of concentrate disposal. The use of an industrial waste permit to regulate ocean disposal is most likely the result of the lack of a specific regulation or set of regulations to deal with the subject.

The common cleaning waste in NF processes as reported by the respondents was citric acid. Caustic soda and detergents were listed as well. Three of the processes discharge the cleaning waste to the sewer and four mix it with their concentrate for discharge. Of the four that mix the cleaning waste with their concentrate three practice DWI and the forth is the ocean discharge plant. The presence of this cleaning waste in the concentrate may also be a factor in the issuance of an industrial waste permit for the discharge.

In two of the cases in which the concentrate is discharged to a POTW, the effluent from that facility is used for irrigation purposes. One of the processes practicing DWI uses some wastewater effluent from its POTW to dilute the concentrate before disposal. The main barriers to reuse, as stated by the operators of processes not practicing any reuse, were cost and permitting (Table 3); each was listed by two processes. Also, two of these plants expressed an interest in the practice of spray irrigation should they be allowed to do so.

Reverse Osmosis (RO) is by far the most common membrane process being used in the drinking water industry today. Forty-two of the utilities that returned surveys use RO processes, nearly 50% of those utilities that returned surveys. The main geographical area that jumps right out of the data is Florida with 21 RO plants that returned the survey. California also had eight RO processes report back with data.

The overwhelming reason for the use of RO processes is TDS removal, as was reported by 20 of the processes, with chlorides listed specifically in 16 instances. Radionuclides were listed as a reason for the plant by seven facilities, and four plant operators listed water reclamation, as well. The overwhelming source water for the RO processes was groundwater. Thirty-six use a groundwater source (Table 1), three processes indicated a surface water source and two use ocean water. One plant did not report. The use of RO systems to treat groundwater for TDS removal should not come as a surprise. RO applications' main use in the water industry is for removal of ionic species, such as TDS, and these contaminants are much more likely to be found in groundwater than in surface water.

The main characteristics of RO concentrate were TDS and conductivity levels, reported by 16 and five processes respectively. As RO concentrate contains the contaminants from the source water in higher concentrations, the TDS levels can be extremely high in the concentrates, sometimes up to five times as concentrated as the raw water, depending on the recovery rate of the system. Twenty of the processes dispose of their concentrate through discharging to a surface water, and this was the most common disposal method reported (Table 2). In nine cases the plant holds an NPDES permit for its discharge, and in six other cases a similar state permit is held for discharging in this manner. In three cases of surface water discharge an industrial waste permit is required. This again is due to the classification of membrane concentrate, and in particular brines from RO processes treating for high TDS, as an industrial waste. Seven of the processes discharge to their local POTW, with five holding a state permit for their discharge, one an NPDES permit and one a local permit with the receiving POTW, as seems to be a common method of regulation for this option. The waste discharge from the water plant is monitored through the maintenance of the POTW's compliance with its effluent permits. Five of the processes practice ocean discharge of their concentrate. Three of these hold NPDES permits, and three hold additional local permits, and the remaining plant has an industrial waste permit for its discharge. Another five of the processes use DWI to dispose of their concentrate. All five hold well operation permits with the appropriate agency within their respective state. This method can be quite costly when well construction and monitoring costs are all considered. Finally, four of the RO processes dispose of their concentrate stream through land application processes. Three of these hold state permits and one is not required to have a permit for this disposal method.

There are a variety of cleaning chemicals added by plants using RO processes. Some of the more common selections are citric acid, caustic soda, and sodium tripolyphosphate, as well as other detergents and surfactants. Twenty-three of the RO processes dispose of their cleaning waste through sewer discharge and nine by mixing it with their concentrate prior to discharge. There does not seem to be any significant tendency to mix cleaning waste with concentrate for any particular concentrate disposal practice.

The only reuse practice employed by the RO processes that responded is irrigation, and four plants practice this in some form. However, an additional eight plants expressed an interest in this practice if they were allowed to do so by the governing regulatory bodies. The main barriers to reuse practice, as indicated by plant operators and shown in Table 3, were concentrate quality, usually high TDS and chloride levels, and cost. Cost would include any capital costs, permitting costs and monitoring costs accrued through the implementation of a reuse practice.

Survey Results by Source Water

A **groundwater** source is used by 49 of the processes that returned a survey. Twenty-four of these processes are located in Florida. Groundwater seems to be the main drinking water supply in Florida, and hence the widespread use of RO systems to treat for TDS removal. Thirty-six of the 44 processes use RO systems to treat the water. Six processes use NF, another four use UF processes and one plant even uses MF to treat the water. This plant treats the groundwater for turbidity removal and thus the use of an MF system. As has already been stated, one of the main contaminants in groundwater is TDS levels and this was listed by 21 of the plant operators as the reason for treatment. More specifically, 13 of the processes indicated chlorides as a reason for treatment.

The main concentrate characteristics listed by the operators included high TDS or conductivity levels, listed by 23 processes. Eight processes also listed sulfate specifically as a concentrate characteristic. Ten of the processes use sewer discharge to dispose of their concentrate. Six hold a state permit, three a local permit with the receiving POTW, and one listed an NPDES permit, which may be held by the receiving POTW. Eight processes practice DWI with their concentrate and all eight hold state injection well operation permits with the appropriate state agency in each case. Land application of concentrate is practiced by six of the processes, with two holding an NPDES permit and three a state permit for the discharge. Four processes discharge to ocean water, three under an NPDES permit and one with an industrial waste permit. In addition to the NPDES permit one plant also holds a local discharge permit and another a state discharge permit. The most common disposal method reported by processes using a groundwater source is surface discharge. Nineteen processes practice surface discharge, with 10 holding an NPDES permit, four a state permit and two an industrial waste permit. The remaining processes did not complete this section of the survey. There does not seem to be any preferred disposal method for concentrate produced from groundwater treatment. There was a large percentage of processes using surface water discharge, but this may be simply due to the location of the plant in close proximity to a suitable surface water, and the variety of disposal options used is quite large as well.

A variety of acids and bases were listed as cleaning chemicals used in groundwater applications. As before citric acid, hydrochloric acid and caustic soda were popular selections. Twenty-four of the processes discharge their cleaning waste to the sewer and 13 mix it with the concentrate for discharge purposes. In no instance in which the concentrate was discharged to a surface water was the cleaning waste mixed with the concentrate. In all surface water discharge cases the cleaning waste is discharged to the sewer. Cleaning wastes were at times mixed with the concentrate in all the other concentrate disposal methods, i.e., DWI, ocean discharge and land application.

Five processes reported the practice of reuse methods, four in the form of irrigation and in one case, in which an evaporation pond is used to dispose of concentrate, the sediments dredged from the pond are used in roadwork. Twelve operators expressed an interest in reuse alternatives, should they be permitted to use such practices. Nine would like to use irrigation practices and three would like to mix the concentrate with an existing reuse application for wastewater treatment plant effluent. The main barriers reported by the operators were concentrate quality, listed in 16 cases; cost, in five cases; and regulatory or permitting problems, in four cases.

Twenty-four of the processes make use of a **surface water** source. The majority of these are located in the western U.S., including 10 in California and three in Hawaii. The dominant membrane application used is MF, which is used in 17 of the 22 processes. Three processes use UF and three also use RO processes, one of which also has an MF system and was included in the 17 such processes mentioned above. The reasons for the processes, as given by plant operators, were turbidity (15 processes), protozoa/bacteria removal (five processes) and removal of DBPs (three processes). All of these are common contaminants found in surface water and they are also common targets of MF processes, which is supported by the data, which show MF systems to be the preferred membrane application in treatment of surface waters.

The main concentrate characteristics reported were turbidity, in seven cases, and conductivity and color, each in two cases. Among the processes treating surface water that returned a survey, there were only three concentrate disposal methods used. Nine of the processes use land applications to dispose of the concentrate stream. One of these holds an NPDES permit and four hold state permits, three with the Department of Health. Land discharge seems to be common among processes treating surface water, and doesn't meet much regulatory resistance. This may be more a reflection of the areas of the United States in which surface water sources are more common rather than a relation to the type of source water. Eight of the processes discharge to the sewer with four holding local permits with the receiving POTW and four holding state permits for their discharge. Five of the processes discharge the concentrate back to a surface water. Surprisingly, only one of these indicated that it holds an NPDES permit for its surface discharge. Three of the processes hold a state permit.

Cleaning chemicals that are being used at these water treatment facilities include citric acid, caustic soda, sodium hydroxide and various cleaning solutions. Eight processes discharge their cleaning waste to the sewer, 10 mix it with the concentrate from the treatment process and one plant uses the cleaning waste for irrigation after it has been neutralized. In five cases in which the cleaning waste is mixed with the concentrate, the concentrate discharge method being used is land application. Again land application of wastes from membrane processes treating surface water seems to be an acceptable alternative.

A few of these processes are practicing reuse methods. Three processes use the land discharges as irrigation, and one plant uses its concentrate for a combination of fire hydrant water, toilet water, and industrial cooling water. The plant practicing this combination of reuse applications is an RO plant, and therefore its concentrate stream would have the necessary volume to be applied for such uses. The plant that operates both an RO system and an MF system expressed the desire to recycle concentrate directly for use as potable water. Cost was listed as a barrier to reuse by two processes, regulatory issues by another and concentrate quality by another.

Only two of the processes that returned the survey use an **ocean water** source. Ocean water is the last resort in terms of sources of drinking water. The high levels of chlorides make it extremely difficult to treat marine waters to meet drinking water standards. The only pressure-driven membrane process used in such desalination applications is RO, and both of the processes referred to herein use RO systems. The obvious reason for treatment is chloride levels in the source water, and both processes that responded to the survey indicated this as well.

Predictably, the main concentrate characteristics listed by both processes are high chloride levels. One of the processes disposes of its concentrate through DWI and has permits with the Florida

Department of Environmental Protection and the South Florida Waste Management District. The second plant disposes of its concentrate through land discharge. It holds a permit with the California Regional Water Quality Control Board.

Citric acid and detergent are cleaning chemicals used by both processes. The Florida plant disposes of its cleaning waste through DWI, as it does with its membrane concentrate. The California plant disposes of the cleaning waste through sewer discharge, following neutralization. Neither plant gave any information on reuse methods or possible barriers to such methods.

Survey Results by Concentrate Discharge Method

The discussion in this section is based on the number of plants that returned surveys rather than the number of processes. In the cases in which a plant has more than one process, of which there are six covered here, the concentrate from both is discharged together as one stream. Therefore this section is based on data returned by 70 plants.

Fourteen plants practice **land discharge** of their membrane concentrate. Broken down by membrane process, there are eight MF plants, three RO plants and three UF plants using land discharge for disposal of the concentrate. Five plants are located in California, while Florida has four and Hawaii three. Land applications seem to be the method of choice in Hawaii and no permits are required for any of the three plants.

Eight plants use a surface water source, five a groundwater and one uses ocean water. In connection with the eight surface water plants, turbidity was listed as a reason for the plant in seven cases. Four operators also listed turbidity as a concentrate characteristic, while six noted TDS or conductivity levels. Six plants hold state permits for their discharge, and two an NPDES permit. However, six of the plants are not required to hold any permits for land discharge. If land discharge of concentrate is an acceptable alternative for a membrane plant, the permitting processes involved, if any, seem to be fairly simple. Land discharge also seems to be an acceptable, and common, discharge method for MF processes treating surface water supplies.

Five of the plants using land discharge specified irrigation practices as the land application being used. This is the most common reuse method being used by respondents to the survey, and the only method being used with any kind of regularity. Several plants, which are not allowed to use their concentrate for irrigation purposes at this time, expressed an interest in doing so should they be granted the right in the future. Barriers to reuse, as listed by plant operators practicing land discharge of their concentrate, were chlorides in one case, cost in one case, and regulatory restrictions in another case.

Nine plants that returned the survey practice DWI of membrane concentrate and seven of these are located in Florida. Florida seems to be the only state that regularly practices DWI of membrane concentrates. Eight of the plants use a groundwater source, which again is the most common drinking water source in Florida, and the remaining plant uses ocean water.

Four plants listed TDS removal as the reason for the plant and three listed THMFP. This being the case, six of the plants use RO systems and three use NF systems. This fits together perfectly, as RO and NF are the primary membrane processes used in treating water for the removal of TDS. Concentrate characteristics reported by these plants include high TDS levels, conductivity and gross alpha levels. The idea behind DWI of these concentrates is to keep them from degrading surface and groundwater. However, in many cases, such as in Florida, the groundwater already contains high levels of TDS and conductivity.

None of the plants practice any direct reuse with their concentrate streams. Barriers to such practices were given as concentrate quality by three plants, regulatory and permitting issues by three plants and cost issue by two plants. However, four operators indicated that they would like to use irrigation systems and two others related an interest in blending their concentrate with reclaimed wastewater for reuse purposes. It would seem, especially in the case of Florida, that irrigation would be a viable reuse option for concentrate. With this practice the concentrate would eventually work its way back into the groundwater from which it came, and would likely lose some of the chlorides and other TDS components to sorption onto soil in the process.

Five of the RO plants that responded to the survey practice **ocean discharge**. California and Florida are each home to three of the six plants. Four plants listed TDS as the reason for the plants; one listed the need for a supplemental water source and one listed water reclamation. Five of the six employ a groundwater source, and the plant operating for water reclamation purposes uses secondary sewage as its source.

The main concentrate characteristics reported were TDS in three cases, radium in two cases, and sulfate in two cases. Four of the plants hold NPDES permits for their discharge, and additionally, three of those also hold a permit with the appropriate state agency as well. One plant has an industrial waste permit and one did not list any permits. In cases of ocean discharge it is likely that the NPDES permit would contain information regarding the sizing of mixing zones, which define the allowable contaminant levels at predetermined distances from the discharge point.

None of the plants indicated the practice of any reuse methods. Concentrate quality and cost were reasons mentioned for the lack of reuse. One did express an interest in an energy recovery system, usually a turbine of some type, which would be turned by the high-pressure concentrate stream produced by RO processes.

A large number of membrane plants simply discharge their concentrate to the **sewer**. In the case of this survey, 17 of the plants that responded practice this method of concentrate disposal. Ten plants use MF processes, seven RO, two UF and two NF. Similarly, 10 plants listed turbidity as the reason for the plant and seven listed TDS removal. Five plants also listed protozoa and/or bacteria removal as a reason for the plant. It isn't likely that there is a correlation between sewer disposal and any one type of membrane process. The number of MF and RO systems practicing this disposal method is most likely a reflection of the fact that these two systems are the more popular membrane processes in drinking water treatment. There didn't seem to be any relationship to source water either, as nine plants use surface water, eight groundwater and one secondary sewage.

Sewer discharge of concentrate is often the easiest option available to operators. The permitting process often simply entails a local agreement with the receiving POTW or an NPDES for the POTW and nothing for the water plant itself. Five of the plants reporting here hold local permits with their POTW, and in six cases either the water plant or the POTW holds an NPDES. Another five of the plants are required to hold a state permit as well. Sewer lines are often readily accessible to the water plant, making this disposal plan very easy to implement as well.

When dealing with cleaning waste the two methods used are generally sewer disposal and mixing with concentrate for disposal. In the case of these plants either method will lead to the cleaning waste eventually ending up in the sewer, as that is where the concentrate is disposed of. One of the plants hauls the cleaning waste off-site for disposal and in the remaining 17, the cleaning waste ends up in the sewer either directly or with the concentrate.

Reuse methods in the case of plants discharging to the sewer would likely involve the effluent from the POTW rather than the concentrate directly. In two cases irrigation is practiced and in one instance water is recycled. The main barriers to reuse practices were cost in two cases, concentrate quality in one case, and politics in another case.

Another commonly used disposal option is **surface discharge**. Twenty-five of the plants that responded to the survey use surface discharge. There does seem to be a link to RO processes for this discharge method, as 20 of the plants use RO systems. Nineteen of the plants use a groundwater source and this is most likely the reason for the use of the RO systems. This is supported by the fact that TDS, chlorides and radium were listed as the main target contaminants, which are more common in groundwater and easily removed by RO processes. One could infer a link between groundwater sources and surface discharge from these data but this is not likely. As discussed previously, plants using groundwater sources used a number of disposal methods to deal with their concentrate, including surface discharge, DWI, sewer discharge, ocean discharge and land application. The link may not be between surface discharge and RO processes, but may in fact be to the area of the United States that the plant is in. Fourteen of these plants are in the southeastern corner of the United States, including 12 in the state of Florida. It happens that RO systems are the most common membrane process used in this area, based on their source water, which is often groundwater, and their treatment needs. As compared to other states, Florida has a large number of membrane plants and as a result may have a better developed regulatory scheme to deal with concentrate disposal. This may be another reason for the wider acceptance of surface discharge in this area.

The most common concentrate characteristic listed by operators was TDS levels. Ten of the plants hold an NPDES permit for surface discharge and nine have a state permit with the appropriate state agency. In only two cases was cleaning waste mixed with the concentrate for disposal to a surface water.

One plant uses its concentrate irrigation and one plant reported the use of its concentrate as fire water, industrial cooling water and toilet water. Three more plants expressed an interest in irrigation practices. Concentrate quality was the main barrier to reuse methods and was listed by 10 of the plants as such.

Survey Results by EPA Region

The two EPA regions with the most membrane processes are Regions 4 and 9. Region 4 is located in the southeast part of the U.S. and includes Florida, South Carolina and Tennessee. Region 9 is located on the West Coast of the U.S. and includes California, Arizona and Hawaii. These two regions will be discussed in the following section. A breakdown of processes used in all the EPA regions is shown in Table 4.

EPA Region 4 has 33 processes in 30 plants (Table 4). Florida has 27; there are two in South Carolina and one in Tennessee. Twenty-four processes use RO systems, seven use NF and two use

Table 4 Surveyed processes by USEPA region

EPA Region	MF	UF	NF	RO	Total
1 (CT, ME, MA, NH, RI, VT)	–	1	–	–	1
2 (NJ, NY, Puerto Rico, Virgin Islands)	–	1	–	–	1
3 (DE, MD, PA, VA, WV, DC)	2	–	–	–	2
4 (AL, FL, GA, KY, MS, NC, SC, TN)	–	2	7	2	33
5 (IL, IN, MI, MN, OH, WI)	4	–	–	–	4
6 (AR, LA, NM, OK, TX)	–	–	–	1	1
7 (IA, KS, MO, NE)	–	–	–	4	4
8 (CO, MT, ND, SD, UT, WY)	1	–	–	2	3
9 (AZ, CA, HI, NV, Guam, American Samoa)	15	1	–	10	26
10 (AK, ID, OR, WA)	–	–	–	1	1
Totals	21	6	7	42	76

UF. Three of these plants are integrated membrane systems with NF and RO. There is certainly a preference for RO systems in this area of the U.S. and it is likely due to the almost exclusive use of groundwater sources. Of these 30 processes, 29 use a groundwater source and the remaining plant uses ocean water, which is treated by RO processes as well.

As would be predicted by the use of RO systems the main reasons given for the processes were TDS, specifically chlorides, and radium. As a result these contaminants were the main characteristics of the concentrate as well. There isn't a particular disposal method dominating this region, as all of the methods discussed above are used in more than one case. Surface discharge is the most common, currently used by 13 processes, and all of these processes hold either an NPDES permit or an equivalent permit with the appropriate state agency. The common use of surface discharge in this area was discussed previously. Seven processes in Florida use DWI. Florida seems to be the only state that uses this method with any regularity. In all seven cases the disposal is permitted through the Underground Injection Control program, overseen by the Florida Department of Environmental Regulation.

Two processes use their concentrate for irrigation purposes and two blend it with wastewater treatment plant effluent for reuse purposes. Seven more processes expressed interest in irrigation uses and three in similar blending practices. The overwhelming barrier listed was concentrate quality, which was indicated in 13 cases. Four processes also listed cost factors and two listed regulatory and/or permitting problems.

There are 26 membrane processes operating in **EPA Region 9** in a total of 24 plants (Table 4). Seventeen of these are in California, three in Arizona, three in Hawaii, and one in Nevada. Ten plants use RO processes, with two of these also using UF in an integrated process. Thirteen plants use MF exclusively, and one uses UF. Surface water seems to be the source of choice, with 16 processes using this. Additionally, seven processes treat groundwater; two treat secondary sewage and another ocean water. As would be expected by the use of surface water sources, turbidity was the main reason given for the processes and was thus a main concentrate characteristic.

Each of the concentrate disposal methods discussed is used in at least one case in this region. Sewer discharge is a common choice and is practiced by nine plants. Of these nine, six hold state permits and three local permits with the receiving POTW. Land application of concentrate is popular in this region and is used by nine processes as well. It is particularly common in California and Hawaii. Commonly MF and UF concentrates are handled this way and this is likely due to the fact that these concentrates commonly have small volumes and contain mainly particulate matter. In Hawaii none of the processes practicing land applications hold any permits for their disposal. In California a state permit is usually required and it is often through the State Department of Health. The Department of Health likely gets involved over the concern for concentrated bacteria and other microorganisms that may be present in UF and MF concentrates.

This region of the U.S. also disposes of its cleaning waste in a number of ways. In addition to the common practices of sewer discharge and mixing with the concentrate, two processes use their cleaning waste for irrigation, one discharges to surface water, one hauls it off-site and one plant recycles its cleaning waste.

Three processes use their concentrate for irrigation and one blends the concentrate for recycling purposes. Processes not practicing reuse at this time did indicate an interest in reuse in the forms of irrigation, energy recovery and recycling directly to potable water. The barriers to reuse were listed as concentrate quality in three cases, regulatory issues in two cases and "politics" in another case.

ONGOING WORK

As completed surveys continue to come in they will be added to the above results and analyzed in the same fashion. Also, efforts continue to gather information on the regulations involved in handling of membrane concentrate in each state. The appropriate regulations are being identified and copies are being obtained. Once this is done each regulation will be analyzed and interpreted to the extent possible within the scope of this project.

CONCLUSIONS

Based on preliminary results, it seems that cost and regulatory issues are the biggest barriers preventing reuse of concentrate produced by MF, UF, and NF processes. RO concentrates, on the other hand, seem to run into problems with concentrate quality (mainly TDS) limits. A trend can be seen toward the use of MF systems to treat surface water in western parts of the U.S. Another such trend can be seen in the use of RO and NF systems to treat groundwater in the southeastern U.S.

ACKNOWLEDGEMENTS

This survey is funded by the American Water Works Association Research Foundation under Project Number 498-97. The Project Officer is Frank Blaha.

REFERENCES

1. American Desalting Association (1995). *Concentrate from Membrane Desalting: A Logical Approach to Its Disposal. A White Paper.* American Desalting Association, Sacramento, CA.
2. *Conservation Directory,* 43rd edition (1998). R.E. Gordon, editor. National Wildlife Federation, Vienna, VA.
3. Mickley, M., Hamilton, R., Gallegos, L. and Truesdall, J. (1993). *Membrane Concentrate Disposal,* AWWA Research Foundation, Denver, CO.
4. Water:\Stats The Water Utility Database (1998). American Water Works Association, Denver, CO.

CHAPTER · 30

Options for Treatment and Disposal of Residuals Produced by Membrane Processes in the Reclamation of Municipal Wastewater

Mehul V. Patel, Engineer
Orange County Water District, Fountain Valley, CA

Greg Leslie, Senior Engineer
Orange County Water District, Fountain Valley, CA

John Yanguba, Graduate Student
California State University Long Beach, Long Beach, CA

Massoud Pirbazari, Professor
University of Southern California, Los Angeles, CA

Ilknur Ersever, Graduate Student
University of Southern California, Los Angeles, CA

INTRODUCTION

Water utilities in the southwestern United States are expanding the use of reclaimed water to supplement local water supplies. Recently, considerable effort has been made to evaluate and optimize membrane systems such as microfiltration (MF) and ultrafiltration (UF) for the removal of suspended solids and reverse osmosis (RO) for the removal of salts in reclaimed water applications in both northern [1] and southern [2] California.

Very few studies, however, have focused on the management of the residuals produced by membranes during the processing of wastewater. Often this can be attributed to the fact that the volume of either the backwash, generated by the MF/UF process, or the RO concentrate (brine) was relatively small and could easily be returned to the sanitary sewer. Currently the combined volume of residuals produced by all of the membrane plants in the southwestern United States using MF or UF upstream of RO amounts to only 2.5 million gallons per day of backwash and 2.1 million gallons per day of reverse osmosis concentrate [3]. However, capacity of reclamation plants in California and Arizona is expected to reach 130 mgd by the year 2020 [3]. As much as 14 million gallons per day (mgd) of combined backwash and brine would be produced if these facilities operate at 85% recovery for both membrane processes. Moreover, the concentrations of suspended solids in the MF backwash and salts and dissolved nutrients in the brine are approximately 8.5 times the influent concentration. Consequently disposal of membrane residuals would be expensive if disposed in the sewer and deleterious to the environment if disposed in receiving waters.

The following considers two options for the management of membrane residuals; these include the use of a second-stage MF process or conventional settling to reduce the volume of backwash and the use of a biological nitrification/denitrification process to remove nitrogenous compounds from the RO concentrate. Reducing the volume of MF backwash will significantly decrease sewer handling fees while the removal of nitrogenous compounds (particularly ammonia) will obviate some of the toxicity issues associated with the discharge of the brine into surface receiving waters. Options to manage the residuals produced by the MF/UF process are to reduce the volume of backwash water either by the two-stage microfiltration process or by conventional settling. The question posed in the MF backwash study was to determine if the higher solids loading on the second-stage membrane surface increased the rate of membrane fouling. The question posed by settling studies was if the high solids content of the backwash water could be greatly reduced by conventional settling techniques. The second option to manage the residuals produced by membrane processes is to reduce the toxicity of RO concentrate. The reduction of toxicity in the form of ammonia nitrogen can be achieved through biological nitrification/denitrification. Nitrification is the process by which ammonia is reduced to nitrite and finally to nitrate by bacteria through the reduction of ammonia oxygen demand. Denitrification is the process by which nitrate is converted to nitrogen gas by bacteria in an environment devoid of oxygen. A more comprehensive explanation of the nitrification/denitrification process can be found elsewhere [4]. The question posed in the brine nutrient removal study was to determine if the levels of TDS in the brine inhibited the rate of biological nitrification and denitrification.

MATERIALS & METHODS

Reclamation studies were conducted on oxidized and clarified (secondary) effluent produced at the Orange County Sanitation Districts (OCSD) plant No. 1 located in Fountain Valley, California. The secondary effluent contained 1,100 ppm total dissolved solids, 10–11 mgL^{-1} total organic carbon, 5–15mgL^{-1} biological oxygen demand, 5 mgL^{-1} suspended solids, 15–20 mgL^{-1} ammonia and approximately 10^6 coliforms per 100 ml (Table 1). Secondary effluent turbidity was typically 2 NTU; however, it was not uncommon to have turbidity excursions of >15 NTU.

Backwash Production

Clarified secondary effluent containing a 3 mgL^{-1} combined chlorine residual was processed using a Memcor 6M10C microfilter consisting of six Memcor M10 modules (233 ft^2 /module) operated in dead-end mode. Permeate was collected on the lumen side of the fibers at a constant flow rate. The fibers were backwashed with air followed by a feed flush every 18 to 20 minutes. The duration of the feed flush step was 20 seconds; 80 gallons of feed were used per backwash at a rate of 40 gallons per minute per module. Spent backwash was collected in a holding tank (120 gallons). Feed and filtrate pressure filtrate flow were recorded every 12 hours.

A comprehensive description of the Memcor Continuous Microfiltration (CMF) system and the air backwash may be found elsewhere [5].

Reverse Osmosis Concentrate Production

Microfiltered secondary effluent was demineralized in an RO system configured in four-stage array with two (4") pressure vessels in stages one and two and one (4") pressure vessel in stages three and four. Each pressure vessel housed three 4" × 40" polyamide thin film composite (Fluid System TFCL HR) membrane elements for a total membrane area of 1,440 ft^2. The system operated at a flux of 10.4 gallons per foot per day (gfd) for a total production of 23,000 gallons per day (gpd). The system operated at a constant feed flow of 13.8 gpm, or produced permeate at 10.4 gpm concentrate for a recovery of 75%; total residual flow (RO concentrate) was 3.4 gpm. Precipitation of inorganic salts in the third and fourth vessels was prevented by adjusting the pH of the microfiltered effluent to 7.4 through the addition of sulfuric acid (93% H_2SO_4) and a scale inhibitor (hypersperse 2000, Argo Scientific) at a dose of 3.0 mgL^{-1}. The membranes were cleaned with a 2% solution of

Table 1 Generation of microfilter backwash residuals

Parameter	Process Stream		
	Secondary Effluent (1)	Microfiltrate (2)	MF Backwash (3)
TSS (mgL^{-1})	5	< 1	25
TOC (mgL^{-1})	10.6	9.4	22.4
SiO_2 (mgL^{-1})	22.5	21.4	22.5
Turbidity (NTU)	2–5	0.1	15
Coliform (cfu/100 ml)	170,000	< 1	4.9×10^6
Fecal coli (cfu/100 ml)	11,000	< 1	–
Indig. Phage (PFU/mL)	38	< 1	1.2×10^2

NOTE: Numbers in parentheses correspond to labels on Figures 1, 2, and 3.

sodium tripolyphosphate and sodium dodecylbenzene sulphonate at pH 7.5. Total feed pressure; differential pressure across each stage; feed, permeate and concentrate conductivity; and permeate and concentrate flow were recorded every 12 hours.

Backwash Volume Reduction

A Memcor model 4M1 microfilter capable of producing up to 2 gpm of microfiltered water was set up to treat backwash water from a Memcor model 6M10C microfilter (Figure 1). The second microfilter used the backwash from the larger microfilter as a feedwater source. The second MF unit consists of four M1 modules (2.5 m^2/module) operated in dead-end mode. Permeate is collected on the lumen side of the fibers. Feed and filtrate pressure on the smaller unit were recorded every 12 hours. The unit was run at five separate fluxes: 13.4, 16.7, 20.1, 23.4, and 26.8 gfd. The unit was allowed to run under each flux condition until a trans-membrane pressure (TMP) of 20 was reached. When TMP reached 20 the amount of days the unit ran up to that point was recorded and the unit was taken out of service for a chemical cleaning. The smaller unit operated under a backwash interval of 20 minutes. The amount of backwash produced every 20 minutes was approximately 1 gallon. The theoretical volume of backwash produced by an

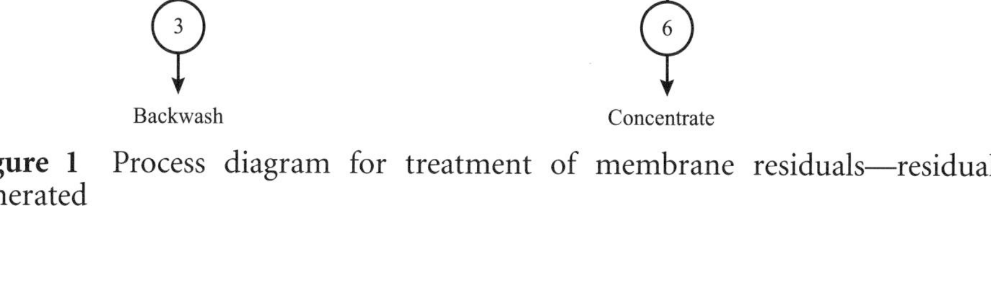

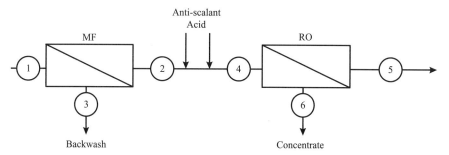

Figure 1 Process diagram for treatment of membrane residuals—residuals generated

MF/UF process in gallons per day (V) can be calculated by the following equation:

$$V = (T) \times \left(\frac{1}{BW_{int}}\right) \times (BW_V)$$

where T is equal to 1,440 which is the amount of minutes in one day, BW_{int} is the backwash interval in minutes, and BW_V is the volume of backwash water produced per backwash.

The theoretical concentration of solids in the backwash in tons per day (C_s) can be found by the following equation:

$$C_s = (S_c) \times \left(\frac{1}{(1 - R) \times (A) \times (V)}\right)$$

where S_c is the concentration of solids in the MF/UF feedwater in mg/L, R is the percent recovery the MF/UF process is operated at, and A is 4.2×10^9 (conversion factor).

Characteristics of Backwash Water

The Feed Fouling Index (FFI) and standard jar tests were used to characterize the filtration/settling properties of the MF/UF backwash water. The correlation between the FFI results obtained on both secondary effluent and MF backwash water will allow direct comparison of the filterability of these waters through a membrane process. The FFI number itself is generated by a computer program supplied by Memtec (MEMFFI Software). The FFI is a measure of the resistance of a membrane (and its filtercake) divided by the volume filtered through the membrane. A comprehensive explanation of Feed Fouling Index can be found elsewhere [6].

Settling studies were conducted on both the first- and second-stage backwash waters. The purpose of these studies was to investigate the feasibility of reducing the solids concentration in MF/UF backwash by conventional settling. The settling characteristics of the backwash water were measured using a jar test apparatus (Phipps & Bird Floc Illuminator model # 7790-500). Two different settling aids, ferric chloride and alum, were used in conjunction with an anionic polymer for the settling studies. Ferric chloride and alum were both tested at doses ranging from 5 to 15 mg/L. Both settling aids were combined with anionic polymer doses between 0.25 and 1 mL/L. Settling studies consisted of a rapid mix period followed by a slow mix period and finally a settling period with no agitation. Settling studies were conducted using 30- and 60-minute observation periods. Turbidity was measured at various time intervals during the settling tests. Also, floc size was noted at the time turbidity was measured.

In addition to the bench scale jar tests, a 65-gallon conical-shaped clarifier was set up downstream of the 6M10C Memcor MF. A portion of the backwash from the MF was pumped to the conical clarifier for solids settling (Figure 2). Backwash was fed to the conical clarifier at a rate of 208 mL/min. The clarifier was dosed with 5 mg/L of ferric chloride to increase settling. The turbidity of the backwash entering the clarifier and the backwash leaving the clarifier as effluent were measured. The conical clarifier operated

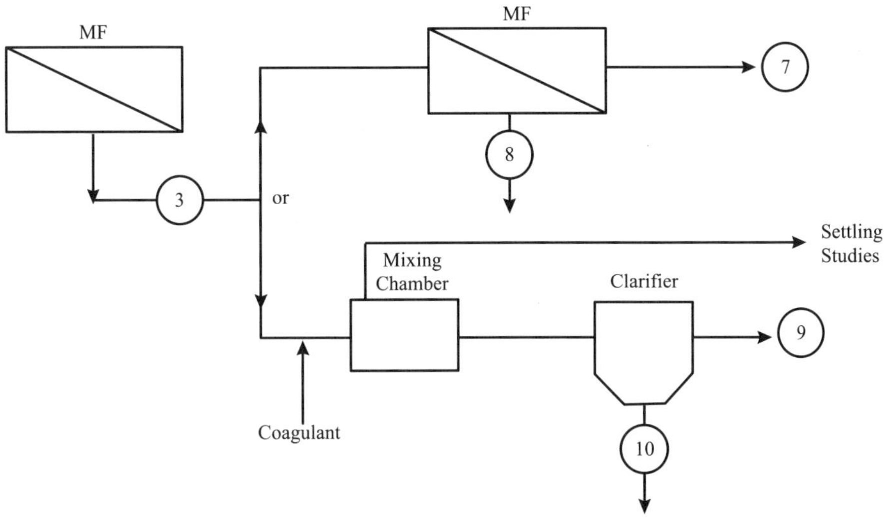

Figure 2 Process diagram for treatment of membrane residuals—MF backwash treatment

with a rotating arm that extended the width of the top circular section of the clarifier. The arm was attached to a central rotating axis that ran the length of the clarifier and was attached to a variable-speed motor. At both ends of the rotating arm were two square sheets of plastic that acted as flocculator paddles. A small peristaltic pump was operated at low speed to constantly draw settled sludge from the bottom of the conical section of the clarifier (Figure 2).

Reverse Osmosis Concentrate Toxicity Reduction

Reverse osmosis concentrate contained on average an ammonia concentration of 100 mg/L as nitrogen, a total dissolved solids concentration of 4,000 mg/L and a 2 mg/L chlorine residual (Table 2). The RO concentrate was routed to a 24 L holding tank and dosed with 100 mg/L of sodium thiosulfate in order to dechlorinate. Nitrification occurred in a series of glass column reactors (working volume of 2.4 L) filled with various media (granular activated carbon, anthracite, expanded clay, etc.) to form a bed approximately 6 inches in height (Figure 3). The RO

Table 2 Generation of reverse osmosis concentrate

	Process Stream		
Parameter	RO Feed (4)	RO Permeate (5)	RO Concentrate (6)
TDS (mgL^{-1})	1,000	24.8	4,000
TOC (mgL^{-1})	9.4	0.87	35
NH_3– N (mgL^{-1})	12.4	1.86	110
Org. Nitrogen (mgL^{-1})	1.28	0.47	3.5
NO_2/NO_3^-	3.1 / 1.2	< 0.4 / 0.33	< 0.4 / 0.33
coliform (cfu/100 ml)	< 1	< 1	–

NOTE: Numbers in parentheses correspond to labels on Figures 1, 2, and 3.

concentrate was fed into the top of these columns using a peristaltic pump. A portion of the flow was recirculated from the top of the reactors and fed under pressure at the bottom of the media using another peristaltic pump. The flow entering the bottom of the column allowed the media bed to become fluidized. The media were used to house various bacteria that would nitrify the RO concentrate. The columns could also be operated as a fixed film reactor by feeding the RO concentrate into the top of the reactor without recirculating any of the flow to the bottom of the media bed. An air supply was added to the columns to provide an oxygen source for the nitrifying bacteria. The reactors were operated with various empty bed contact times ranging between three and 12 hours (Table 3). For future studies a denitrification reactor will be set up downstream of the nitrification reactors (see Figure 3). The denitrification reactor will operate on effluent from the nitrification reactors. The denitrification reactor will be operated as a fluidized bed reactor under anoxic conditions with methanol being dosed into the reactor as a carbon source. The empty bed contact time for the denitrification reactor will vary between 30 minutes and one hour. TDS levels will not be artificially elevated with sodium chloride as was done with the nitrification reactors.

For direct comparison with bench-scale results, RO concentrate was fed to a pilot-scale nitrification reactor known as a Biofor

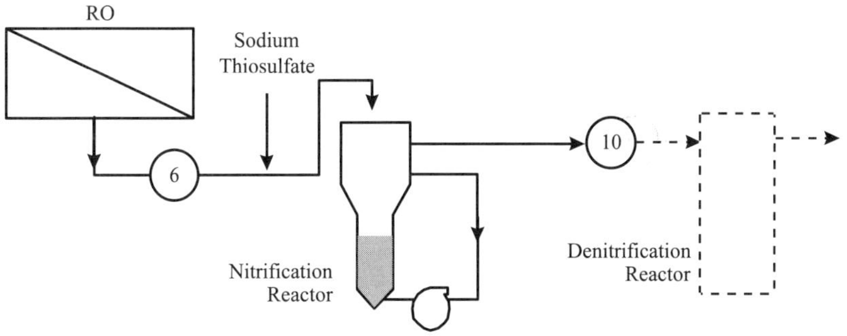

Figure 3 Process diagram for treatment of membrane residuals—RO concentrate treatment

Table 3 Loading conditions of bench-scale and pilot-scale nitrification reactors

Reactor Conditions	Bench–Scale Reactor	Bench–Scale Reactor	Pilot–Scale Reactor
Reactor Type	Fluidized Bed	Fixed Film	Fixed Film
Flow Rate	3–10 mL/min	3–10 mL/min	15 gpm
Air Flow	750 mL/min	750 mL/min	–
Packing Media	GAC	GAC	Expanded Clay
Empty Bed Contact Time	2–10 hours	2–10 hours	15 min

by Infilco Degremont Inc. of Richmond, Virginia. The pilot unit was capable of producing 15 gpm of effluent (Table 3). The pilot-scale nitrification reactor used expanded clay media as a habitat for biological growth. This reactor was operated in an upflow fluidized mode similar to the pilot-scale reactors. Air diffusers are located along the length of the reactor to provide air for the nitrifying bacteria. The pilot unit backflushed the media automatically on a timed basis. During testing the timer was set for twice per day. The backflush step is done in order to maintain the hydraulic flow through the reactor once excessive biological growth occurs.

RESULTS AND DISCUSSION

The first-stage MF reduced suspended solids to <1 mg/L, turbidity to 0.1 NTU, and total coliform to <1 CFU/100 mL (Table 1). This was expected due to the nominal pore size inherent in MF membranes. The MF does very little to reduce TOC or silicate due to their size in relation to the pore size of the MF membrane (Table 1). The MF backwash has a high concentration of suspended solids, turbidity, and bacteria because it concentrates all of these constituents as they are retained on the surface of the membrane (Table 1). An encouraging result found by the two-stage MF test is the water quality of the filtrate from the second-pass MF (see Table 4). The concentration of constituents in the second-pass filtrate is very similar to that of the first-pass filtrate. This is very encouraging given the fact that the feed to the second-pass MF was of significantly lesser quality than that fed to the first-pass MF (secondary effluent). To an extent the good water quality found in the second-pass MF filtrate was expected due to the nature of the MF process. MF is a separation process, so that in theory any particle larger than the pore size of the membrane will be retained no matter the concentration of the particles in the feedwater. The main area of concern was whether the high concentration of solids in the backwash would diminish the amount of time the second-pass MF could remain online without irreversible fouling (i.e., rise in TMP or trans-membrane pressure even after backwashing). The cleaning interval for the first- and second-stage MF system decreased from approximately 25 days to two days as the flux increased from 13.4 to 26.8 gfd (Figure 4). Figure 4 shows that cleaning interval decreases as the flux is increased, as expected. The significance of the graph is that it shows a very similar decline in MF performance with increased flux for both feedwater sources. This shows that the prospect of a two-stage MF process to treat

Table 4 Treatment of microfilter backwash residuals

	Process Stream		
Parameter	MF Backwash (3)	2nd-Pass Microfiltrate	2nd-Stage MF Backwash
TSS (mgL^{-1})	25	< 1	933
TOC (mgL^{-1})	22.4	9.84	139
Turbidity (NTU)	15	0.1	150
Coliform (cfu/100 ml)	4.9×10^6	< 1	4.2×10^7
Indig. Phage (PFU/mL)	1.2×10^2	< 1	1×10^3

NOTE: Numbers in parentheses correspond to labels on Figures 1, 2, and 3.

backwash is feasible as long as flux through the membrane is maintained at a reasonable level. The graph also shows that as long as the flux is kept low enough, the increased concentration of solids in the feedwater does not cause irreversible fouling to occur faster than it would with a feedwater containing a low solids concentration. Feed fouling index tests done on both secondary effluent and on the first-stage backwash indicate that the backwash water still has a much higher fouling potential for MF membranes. This must be taken into consideration when deciding upon fluxes and recoveries at which to operate the MF/UF units in a two-stage backwash treatment process.

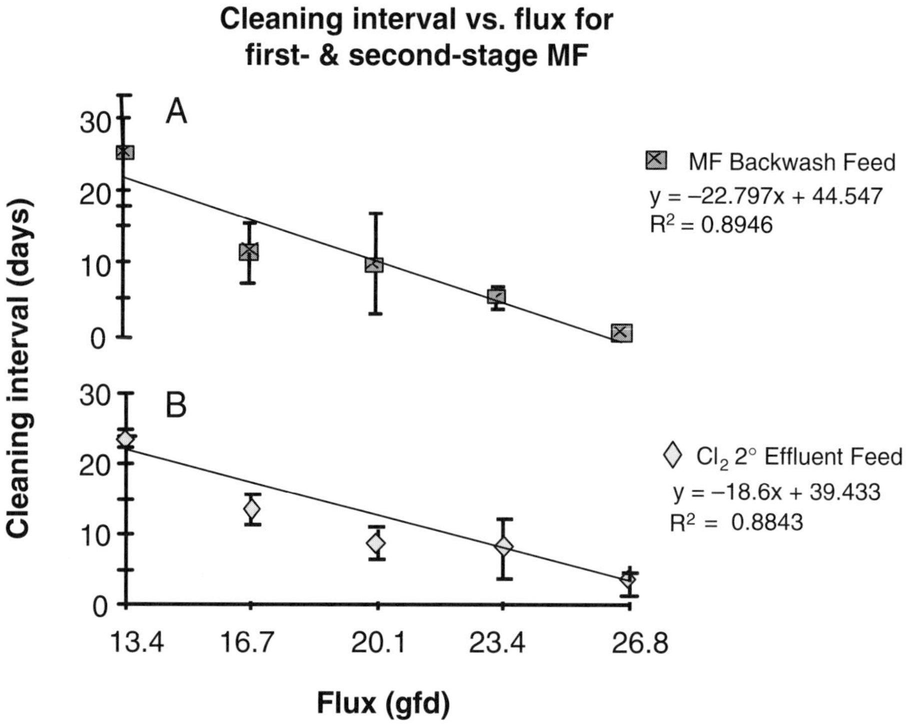

Figure 4 Cleaning interval (days) as a function of flux (gfd) for clarified secondary effluent (A) and first-stage MF backwash (B)

Now that it has been shown that a two-stage MF process can be operated without significantly increased membrane fouling, the reduction in backwash volume can be appreciated. The MF used in the first stage of the two-stage MF pilot test produced 80 gallons of backwash every 20 minutes. The smaller MF used for second-stage treatment was able to treat a backwash flow of 1 gpm. The amount of backwash produced by the second-stage MF was 1 gallon every 20 minutes. This corresponds to an 85% reduction in backwash volume through the concentration of the first-stage MF backwash by the second-stage MF. For a full-scale treatment plant using MF or UF this shows it may be possible to treat all backwash flows generated by routing all backwash flows to MF or UF units set aside exclusively to treat backwash flows from the main plant process. Also, the water coming out of the second-stage MF is treated and can be added to final product from the first-stage MF units or used around the plant. It may also be possible to reduce the amount of backwash further by using a third-stage MF to treat backwash from the second-stage MF units. Once the volume of backwash is reduced to a manageable level it can be disposed of into a sewer, routed to the head of the plant, sent to a DAF thickener, or sent to a solids digestor.

Limited results were found for settling studies conducted on both the first- and second-stage MF backwashes. An average turbidity reduction of 50% was demonstrated (Tables 5 and 6). It seems that ferric chloride in conjunction with an anionic polymer as a settling aid worked best to reduce turbidity. Unfortunately, without suspended solids concentration data before and after settling, the effectiveness of the process cannot be fully evaluated. Assuming that settling significantly reduces the solids concentration, it cannot be assumed that bacteria and virus are also reduced as is done in a two-stage MF process. Also, settling processes involve a significantly higher footprint area than membrane processes. More data are needed in order to judge the merits of using settling processes to treat membrane residuals.

Ammonia toxicity reduction of RO concentrate was done on a bench- and a pilot-scale basis. The bench-scale nitrification reactors

Table 5 Settling studies using ferric chloride and polymer

Ferric Chloride (*mg/L*)	Polymer (*mg/L*)	Time Floc Observed	Floc Size*	Initial Turbidity (*NTU*)	Final Turbidity (*NTU*)	Source Water Used
5	0.25	25 sec	B	6.0	2.80	1st-stage BW
10	0.5	17 sec	B - C	6.0	2.76	1st-stage BW
15	1	15 sec	C - D	6.0	3.50	1st-stage BW
5	0.25	60 sec	A	21.8	10.83	2nd-stage BW
10	0.5	40 sec	A - B	21.8	9.60	2nd-stage BW
15	1	20 sec	B - C	21.8	9.18	2nd-stage BW

NOTE: Rapid Mix: 100 rpm for 1 min.
Slow Mix: 70 rpm for 7.5 min., 40 rpm for 7.5 min., 20 rpm for 5 min.
Settling Time: 30 minutes

*Floc Size: A = 0.3–0.5 mm, B = 0.5–0.75 mm, C = 0.75–1.0 mm

Table 6 Settling studies using alum and polymer

Alum (*mg/L*)	Polymer (*mg/L*)	Time Floc Observed	Floc Size*	Initial Turbidity (*NTU*)	Final Turbidity (*NTU*)	Source Water Used
5	0.25	10 sec	A	5.23	3.11	1st-stage BW
10	0.5	8 sec	A - B	5.23	1.61	1st-stage BW
15	1	4 sec	B - C	5.23	3.96	1st-stage BW
5	0.25	12 sec	B - C	51.3	14.44	2nd-stage BW
10	0.5	10 sec	B	51.3	17.90	2nd-stage BW
15	1	5 sec	A	51.3	17.56	2nd-stage BW

NOTE: Rapid Mix: 120 rpm for 1 min.
Slow Mix: 70 rpm for 7.5 min., 40 rpm for 7.5 min., 20 rpm for 5 min.
Settling Time: 60 minutes

*Floc Size: A = 0.3–0.5 mm, B = 0.5–0.75 mm, C = 0.75–1.0 mm

were able to oxidize more than 95% of the ammonia and about 85% of the nitrite (Table 7). This was an encouraging result which showed that the toxicity of the concentrate can be reduced enough to make the RO concentrate suitable for disposal. The reactors were also resistant to shock loadings (i.e., unexpected increased ammonia levels). The limiting factor in this process is empty bed contact time of the reactors. The lowest contact time to produce significant ammonia reduction was three hours. This may not be efficient enough for a full-scale biological nitrification process, but it does show promise. Fortunately, the elevated TDS concentrations found in the RO concentrate did not affect the performance of the process. This was also the case for conditions of artificially elevated TDS by addition of sodium chloride. TDS concentrations as high as 20,000 mg/L did not affect the rate of nitrification as long as an empty bed contact time of three hours or more is maintained. Unfortunately the pilot-scale nitrification reactor only had a 50%

Table 7 Results from bench-scale and pilot-scale nitrification reactors

	Process Stream			
Parameters	RO Concentrate (6) Feed to Bench-Scale Reactor	Nitrified Concentrate From Bench-Scale	RO Concentrate (6) Feed to Pilot-Scale Reactor	Nitrified Concentrate From Pilot-Scale
TDS (mgL^{-1})	4,000	4,000	6,072	6,706
TOC (mgL^{-1})	35	–	59.1	56.9
NH_3–N (mgL^{-1})	110	< 5	74.5	42.6
Organic N (mgL^{-1})	3.5	–	3.5	–
NO_2/NO_3^-	< 0.4 / 0.33		5.5 / 4.7	40.5 / 15.6

NOTE: Numbers in parentheses correspond to labels on Figures 1, 2, and 3.

ammonia reduction, but this occurred with a 15-minute empty bed contact time (Table 7). The pilot-scale nitrification reactor was less efficient than the pilot-scale reactor in reducing the ammonia concentration. This may have been due to a lack of sufficient biological growth on the media in the reactor. More pilot testing will need to be done in order to fully evaluate the feasibility of reducing ammonia toxicity in RO concentrate on a pilot- or full-scale basis.

CONCLUSIONS

The management of membrane residuals is possible by reducing the volume of MF/UF backwash through two-stage MF treatment and by reducing the ammonia toxicity of RO concentrate by biological nitrification. Two-stage MF treatment was shown to reduce MF/UF backwash volumes by as much as 85% while still producing a usable effluent. This process is easily applicable to a full-scale membrane treatment process at the current time. The reduction of ammonia toxicity in RO concentrate is feasible through biological nitrification. Bench-scale testing showed a 95% removal of ammonia from RO concentrate even at TDS levels as high as 10,000 mg/L. The only drawback of the bench-scale results was a high contact time needed to achieve ammonia removal. Unfortunately, the pilot-scale results showed only a 50% removal of ammonia toxicity. With further pilot testing and optimization of the pilot-scale reactor it may be possible to improve this result. As membrane processes become increasingly prevalent it is apparent that more work will be required to develop effective residual management strategies.

ACKNOWLEDGMENTS

The authors would like to thank the California Energy Commission, Southern California Edison, and the Electrical Power Research Institute for their support. The authors also appreciate the assistance of the Operations, Maintenance, and Instrumentation and Electrical Departments of the Orange County Water District.

REFERENCES

1. Geselbracht, J., Microfiltration/Reverse Osmosis Pilot Trials for the Livermore, California Advanced Water Reclamation Project, Proc. Joint AWWA/WEF Water Reuse Conference, San Diego, CA Feb. (1996), 187.
2. Gagliardo, P., Indirect Potable Reuse in San Diego, Proc. Microfiltration II, NWRI, San Diego, CA Nov. (1998). 65.
3. Sudak, R.G., Jones, P.D. II, Municipal Reclamation with Microfiltration and Reverse Osmosis, Proc. Microfiltration II, NWRI, San Diego, CA, Nov. (1998), 51.
4. Metcalf & Eddy, Wastewater Engineering 3d Edition, McGraw Hill (1991).
5. Adham, S., et al., Water Repurification Project Final Draft Report on Membrane Prequalification Pilot Study (1995).
6. MEMFFI Operating Manual, Memtec Limited (1995).

CHAPTER 31

A Methodology for Calculating Actual Dilution of a Membrane Concentrate Discharge to Tidal Receiving Waters

Edward R. Weinberg, President
EW Consultants, Inc., Stuart, Florida

INTRODUCTION

In June 1996, Florida Department of Environmental Protection (FDEP) Permit No. FL0042358 and Administrative Order No. AO-005-SE were issued for the Town of Jupiter Reverse Osmosis facility concentrate discharge. The new permit and accompanying AO were issued to allow an increase of the concentrate discharge flow from two to four million gallons per day. Part of the requirements of the permit and AO were that a Dilution Study within the C-18 Canal be conducted to define mixing zone limits for Fluoride and Combined Radium 226 and 228 in the four million gallon per day (MGD) concentrate discharge.

Waters within the C-18 canal, where the Jupiter Reverse Osmosis (JRO) concentrate currently discharges, are designated as Class II waters (Chapter 62-301, F.A.C.) and are thus subject to Class II Water Quality Criteria. The state water quality standards require that even during "worst case" conditions the discharge cannot result in exceeding water quality criteria outside an identified mixing zone around the discharge location. The principle of a mixing zone is to provide sufficient dilution in a defined area such that beyond the mixing zone limits, all water quality criteria are met.

As part of the permit requirements to evaluate the proposed 4 MGD effluent discharge, field data collection and dilution calculations were undertaken to determine:

1. Actual dilution within the receiving waters.
2. Mixing zone boundaries for specific parameters.
3. The discharge to the C-18 canal will meet Class II water quality criteria outside of the delineated mixing zone.

Waters within the C-18 canal are controlled by a semidiurnal tide that influences flow regimes where the effluent discharges. Therefore, the initial mixing zone calculations were based on a theoretical tidal exchange calculation. The field data collection and analysis for the dilution study were conducted to develop "real world" limits of the mixing zone. The results of this dilution study defined the limits of the mixing zone for a known substance by utilizing a forty-eight-hour *in situ* sampling and data collection effort.

By analyzing the rate of effluent discharge, and the existing conditions within the C-18 canal, the theoretical borders of the mixing zone for each parameter of concern were previously projected. Field sampling of a known and controlled effluent was conducted for empirical testing of the projections. The projections were based on the following equations.

The change in concentration for a given parameter within the receiving water can be calculated utilizing:

$$\Delta C = \frac{M}{V}$$

where:

ΔC = Change in concentration from loading (mg/L)
M = Material loading (mg/day)
V = Dilution volume (L/day)

In order to accurately determine available dilution and compliance with water quality criteria, the background levels for each parameter of concern must be quantified within and around the proposed mixing zone. The background levels can then be combined with projected change in concentration due to the discharge, to calculate the total concentration.

$$C_T = \Delta C + B$$

where:

C_T = Total concentration
ΔC = Change in concentration due to discharge
B = Background concentration

METHODS AND MATERIALS

The methodology for developing the dilution calculation field data was through metered injection of a measured concentration of fluorometric dye into the RO plant discharge. Dye concentration was measured in the receiving water at discrete cross-section locations upstream and downstream of the discharge location. The data were collected at discrete time intervals over a 48-hour period of 4 MGD discharge. The specifications for sampling conditions, locations, equipment, and methodology are provided below.

The 48-hour study commenced on April 15, 1998 at 1000 hours and finished on April 17, 1998 at 1000 hours. Background sampling and equipment checkout using the flow-through fluorometer and other instruments were conducted, and the background levels were used to correct the fluorometry measurements collected during the 48-hour study. The sample stations were located and marked with anchored buoys.

The effluent discharge flow was initiated three hours before the first sample was collected and continued during the 48-hour study period at the rate of 4 MGD. The discharge was a combination of raw water and RO concentrate.

A pump connected to the RO plant and located prior to the effluent discharge pipe injected the dye. A dye/water solution was mixed and injected at a rate projected to achieve 100 parts per billion (ppb) dye in the discharge. The discharge water entering the outfall pipe was measured for actual dye concentration several times during the study. Dye injection began April 15, 1998 at 0900 hours and continued until the last sample was collected.

Samples (dye concentrations) were collected using a Turner Designs Model 10-AU Digital Fluorometer. According to literature from Turner Designs, this instrument has the capability to measure as little as one part per trillion of Rhodamine dye. Flow-through fluorometry samples were collected at depth with a sample pump and analyzed using the 25 mm one-piece continuous flow cell provided with the Turner Designs Model 10-AU Digital Fluorometer. Samples were automatically corrected for temperature by the digital fluorometer. Salinity, pH, DO, and temperature readings were collected at each station using a YSI Model 600 multiparameter environmental monitoring system. Finally, grab samples were collected at depth utilizing a 1.5 liter Nansen bottle. Grab samples were analyzed at each sample station during each sampling event to control for the effects, if any, of the sample pump.

Five sampling cross-section locations (sample stations) were established in the C-18 Canal, with each sample station including six sample points. The points were at 0.8 and 0.2 times recorded depth at locations in the center of the canal and points equidistant from the center to each shoreline. The sample stations were marked with anchored buoys so as to ensure consistent location of sampling. The sample stations were located as follows:

Station A - 500 ft. (150 meters) upstream of Discharge Pipe
Station B - 100 ft. (30 meters) upstream of Discharge Pipe
Station C - 50 ft. (15 meters) downstream of Discharge Pipe
Station D - 500 ft. (150 meters) downstream of Discharge Pipe
Station E - 1000 ft. (305 meters) downstream of Discharge Pipe

One additional sampling point was located at the Town of Jupiter Water Plant effluent discharge caisson. This point was used to collect flow-through and regular-interval grab samples of the discharge prior to entry in the C-18 canal. Caisson grab samples were stored in dark bottles for later analysis.

Stations were sampled with the flow-through fluorometer at approximately 1 hour, 6 hours, 12 hours, 18 hours, 24 hours, 36 hours, and 48 hours after study commencement, resulting in seven sample events per sample station during the 48-hour study.

From the field data, actual dilution of injected dye at five discrete locations was calculated. Based on known concentrations of Fluoride and Combined Radium 226 and 228 in the discharge, as well as background in the receiving water, a location where Class II Standards are met can be calculated based on the dilution of the Rhodamine WT dye that was measured at the cross sections. From this, mixing zone limits for defined parameters of concern were projected for the C-18 canal.

RESULTS

Sampling was conducted within the C-18 canal prior to dye release in order to determine the background fluorometry levels. From these results, a mean background fluorometry level was determined to be 1.47 ppb. This background level was used to correct the samples collected during the dye study by subtracting the mean background value from the measured sample concentrations.

In order to represent the composite dilution for each cross-section sampling event, the six individual fluorometry readings at each sample point were combined and a mean was calculated for each sampling event. This allowed for an accurate representation and analysis of the measured dilution at each individual sample station. These mean concentration data were the basis for analyzing the C-18 canal dilution capacity at discrete locations upstream and downstream of the outfall.

Station A was located at the far upstream point of the study. Figure 1 provides a graphic summary of the mean fluorometry readings collected at Station A. Due to its upstream location, this station was the least affected by the discharge. This is demonstrated by the lowest peak mean concentration of dye (6.9 ppb) of all stations during the study. The peak reading occurred approximately 36 hours after the dye injection commenced.

Station B is summarized in Figure 2. This station was located approximately 100 feet upstream of the discharge pipe. The peak mean dye concentration of 11.6 ppb was recorded during the 36th hour of the study.

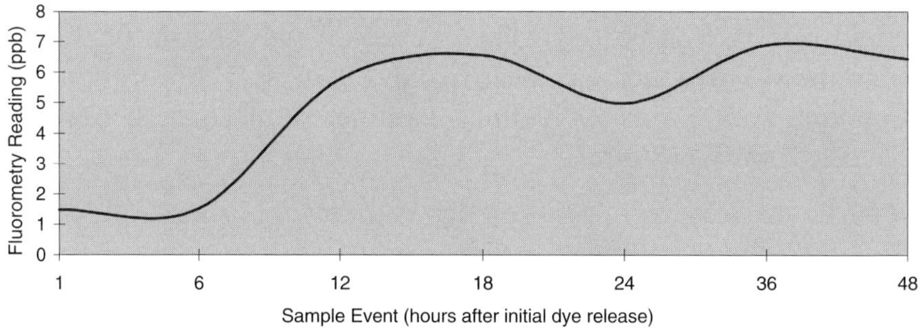

Figure 1 Station A mean fluorometry readings in parts per billion for each sample period during the 48-hour dilution study

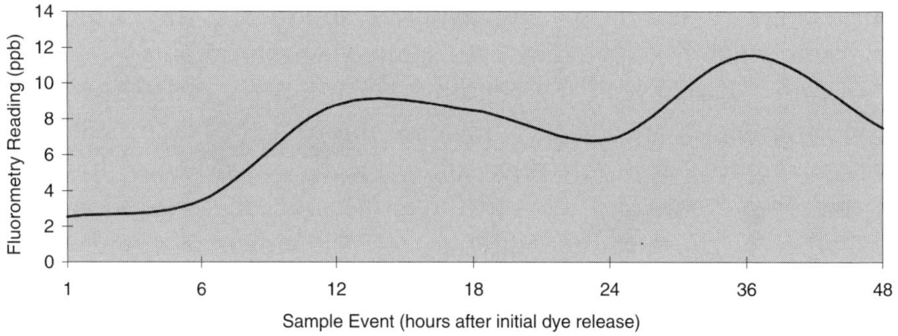

Figure 2 Station B mean fluorometry readings in parts per billion for each sample period during the 48-hour dilution study

Station C was located 50 feet downstream of the discharge pipe and the average concentrations of dye are summarized in Figure 3. The peak mean concentration of fluorometric dye at this station was recorded during the 36th hour of sampling at 11.1 ppb.

Station D was located 500 feet downstream of the discharge pipe and had a peak mean concentration of fluorometric dye during the 36th hour of the study at 8.8 ppb. Mean concentration data from this station are summarized in Figure 4.

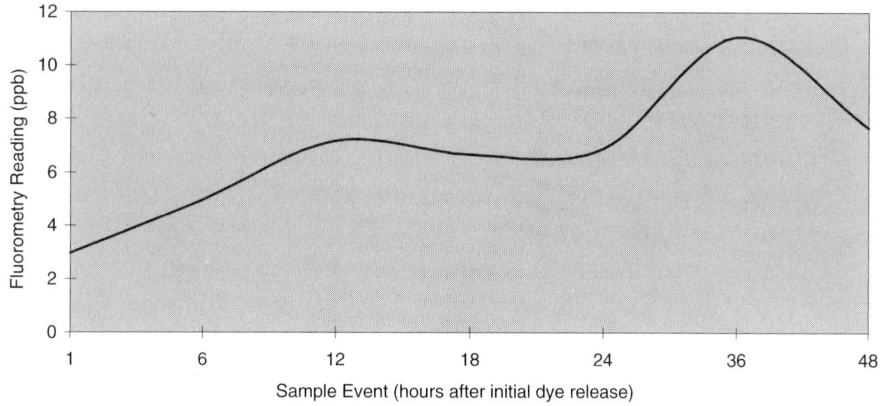

Figure 3 Station C mean fluorometry readings in parts per billion for each sample period during the 48-hour dilution study

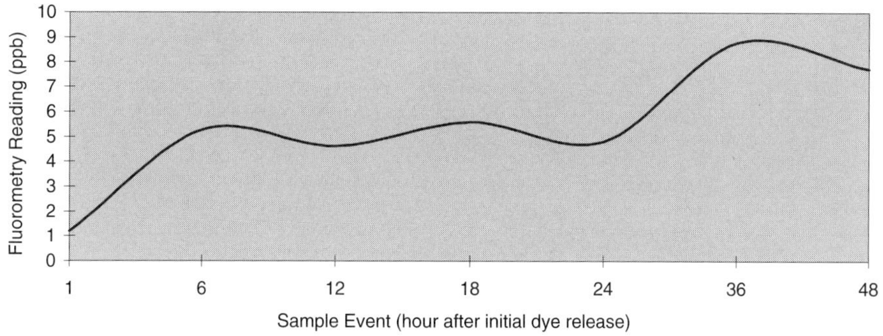

Figure 4 Station D mean fluorometry readings in parts per billion for each sample event during the 48-hour dilution study

Station E was located 1,000 feet downstream of the discharge pipe. The peak mean concentration of fluorometric dye was 8.0 ppb during the 36th hour of the study. Mean concentration data from this station are summarized in Figure 5.

Individual data from each sample station were combined in order to develop the worst-case scenario for dilution of the discharge in the C-18 canal. These data clearly demonstrated that during the 36th hour of the study, the worst-case conditions for in situ concentrations of the fluorometric dye occurred.

The caisson data were collected before or after C-18 canal sampling events with the flow-through fluorometer. Caisson fluorometric data were projected to be 100 ppb; however, actual readings ranged from 98 to 50 ppb. Tidal information from Reed's Nautical Almanac, 1998 East Coast Edition, was used to project tide change within the study area during sampling. Figure 6 provides a summary of the tide data during the data collection. The maximum mean dye concentrations (worst-case dilution) at each of the stations were measured during the 36th-hour sampling event. That event coincided with an incoming tide that followed the lowest high tide and second lowest low tide of the study period.

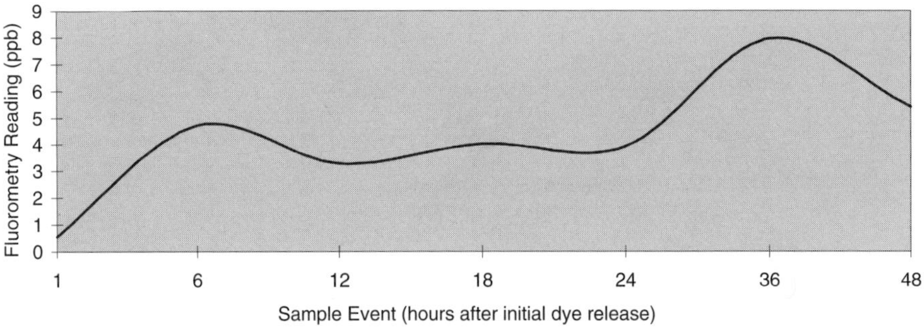

Figure 5 Station E mean fluorometry readings in parts per billion for each sample event during the 48-hour dilution study

DISCUSSION

Permit No. FL0042358 established an interim mixing zone of 1,000 feet (305 meters) to either side of the discharge pipe. This was established theoretically for two constituents in the concentrate, Combined Radium 226 and 228 and Fluoride, that exceeded Class II standards, and thus required a mixing zone. The receiving waters were analyzed using fluorometric dye to calculate actual required mixing zone limits for these constituents.

The "worst case" dye dilution during the study occurred during the sixth sampling event (36th hour). All sampling stations recorded the highest mean concentration of fluorometric dye during this sample event. Although the projected concentration of the dye in the discharge was 100 ppb, several samples collected from the caisson during dye injection were substantially lower. As such, a conservative adjustment was made by using the average measured dye concentration from the caisson (73 ppb) to calculate dilution factors. Measured dilution in the C-18 canal is summarized in Table 1.

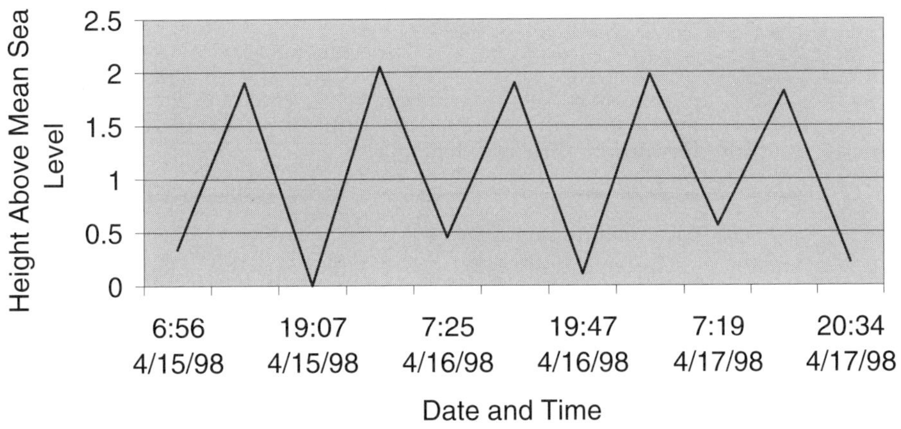

Figure 6 Southwest Fork spillway tidal fluctuation during 48-hour dilution study

The dilution factors were then used to calculate the projected concentration of Fluoride and Combined Radium 226 and 228 at each sample station. Fluoride and Combined Radium 226 and 228 data that were collected in the C-18 prior to permitted discharge (May 1990 to December 1990) were used to calculate a background concentration of 0.65 mg/l Fluoride and 1.3 pc/l Combined Radium 226 and 228. Fluoride and Combined Radium 226 and 228 concentrations from discharge monitoring reports for the period of October 31, 1996 to April 30, 1998 were used to determine worst-case concentrations of 9.5 mg/l and 22.8 pc/l, respectively, in the discharge.

The maximum level of Fluoride allowed by the Class II criteria is 1.5 mg/L, and the maximum level of Combined Radium 226 and 228 allowed by the Class II criteria is 5 pc/L. The worst-case background and discharge concentrations of these parameters were used to define the mixing zone required to meet Class II standards within the C-18 canal. The results of these calculations are shown in Tables 2 and 3.

The data summarized in Table 2 indicate that the measured dilution in the C-18 canal does not provide for the ability to meet Class II water quality criteria for fluoride at any of the dilution study stations. Therefore, the dilution factor at each station was plotted with the corresponding distance from the discharge pipe

Table 1 Worst-case fluorometric dye dilution in the C-18 canal and resulting dilution factors for each station

Station	Mean Fluorometric Dye Concentration (*ppb*)	Dilution Factor	Reduction Factor
A	6.9	0.09	0.91
B	11.6	0.16	0.84
C	11.1	0.15	0.85
D	8.8	0.12	0.88
E	8.0	0.11	0.89

Table 2 Projected Fluoride concentrations during a 4 MGD discharge and worst-case dilution conditions

Station	Fluoride Input (mg/l)	Dilution Factor	Projected Concentration (mg/l)	Background Concentration (mg/l)	Total Fluoride (mg/l)	Meets Class II Criteria (1.5 mg/l)
A	9.5	0.09	0.86	0.65	1.51	NO
B	9.5	0.16	1.52	0.65	2.17	NO
C	9.5	0.15	1.43	0.65	2.08	NO
D	9.5	0.12	1.14	0.65	2.79	NO
E	9.5	0.11	1.05	0.65	1.70	NO

Table 3 Projected levels of Combined Radium 226 and 228 during a 4 MGD discharge and worst-case dilution conditions

Station	Radium Input (*pc/l*)	Dilution Factor	Projected Concentration (*pc/l*)	Background Concentration (*pc/l*)	Total Combined Radium (*pc/l*)	Meets Class II Criteria (*5.0 pc/l*)
A	22.8	0.09	2.05	1.3	3.33	YES
B	22.8	0.16	3.65	1.3	4.93	YES
C	22.8	0.15	3.42	1.3	4.70	YES
D	22.8	0.12	2.74	1.3	4.01	YES
E	22.8	0.11	2.51	1.3	3.79	YES

(Figure 7). The slope from this plot was then used to project the dilution factors upstream and downstream of the discharge pipe beyond the sample station locations. The upstream dilution rate was calculated to be 0.01 units of dilution for every fifty feet and the downstream dilution rate was 0.004 units of dilution for every fifty feet.

Using the calculated dilution factors, Fluoride was projected to be 1.41 mg/L at 550 feet upstream and 1.41 mg/L at 1,325 feet downstream of the discharge pipe, which define the upstream and downstream mixing zone boundaries.

The data summarized in Table 3 indicate that the measured dilution within the C-18 canal allows for the discharge to meet Class II water quality criteria for Combined Radium 226 and 228 at all of the stations. Upstream of the discharge point, Station B (100' upstream) is 4.93 pc/L, which falls within Class II criteria for Combined Radium 226 and 228. As such, the upstream mixing zone boundary for Combined Radium 226 and 228 is at 100 feet and the downstream mixing zone boundary for Combined Radium 226 and 228 is at 50 feet from the outfall.

Closer examination of the dilution factors measured in the C-18 during worst-case conditions indicated an unexpected occurrence. The dilution at 100 feet upstream of the outfall was less than that measured 50 feet downstream of the outfall. At first glance, this would seem to be an unexpected result. However, considering

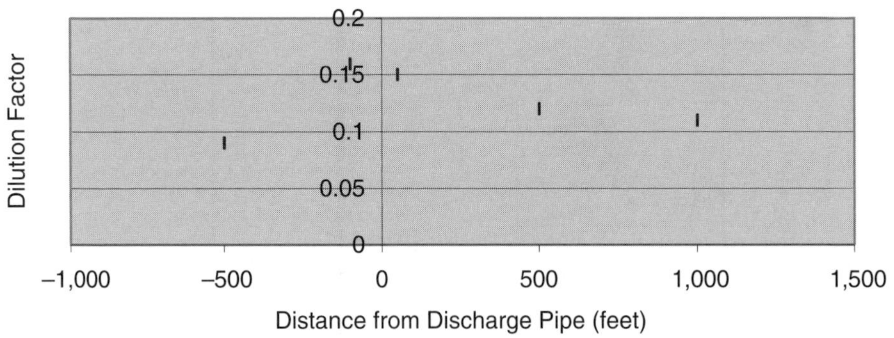

Figure 7 Dilution factor upstream and downstream of the discharge pipe

the unique characteristics of the receiving waters, this unexpected result can be explained.

The C-18 is an estuarine system that, under worst-case conditions, does not have any base flow of freshwater. As such, tidal exchange provides essentially all of the potential dilution volume. On an outgoing tide, the water column is well mixed, and the salinity data collected during the study did not indicate any apparent density stratification. As a result, the discharge is diluted throughout the entire cross section of the canal. However, on an incoming tide, the only source of volume input is higher-salinity water from the Jupiter Inlet downstream of the connection with the Southwest Fork of the Loxahatchee River. The salinity data collected during the study on incoming tides clearly indicate a density stratification in the lower portion of the water column. As a result, the discharge is carried upstream on an incoming tide, and is concentrated in a smaller portion of the water column, resulting in less dilution at the upstream stations close to the outfall. When the tide changes direction, the density stratification is broken, and the discharge is mixed into the entire water column and results in more efficient dilution of the concentrate at downstream locations.

CONCLUSIONS

The 48-hour dye release dilution study provided site-specific data on the dilution capacity of the C-18 canal receiving waters. These data, along with measured concentrations of Fluoride and Combined Radium 226 and 228 in the discharge (as well as background), were used to project concentrations of these constituents in the receiving waters under worst-case conditions of discharge (4 MGD) and available dilution. The results of this analysis indicated mixing zone lengths varied for each parameter tested.

The information developed in the dilution study, specifically the dilution factors and rates of dilution outside the sampling stations, can be applied throughout the receiving waters. It can be, and in fact has been, used to evaluate the mixing zone needs for other parameters that occur in the discharge. Additionally, the dilution factors and rates can be used to evaluate the effects of operational changes (i.e., recovery rate, discharge flow, etc.) as well as potential additional discharge locations.

ACKNOWLEDGEMENTS

The author would like to acknowledge the assistance of Toby Overdorf in preparing the manuscript. John Potts of Hutcheon Engineers provided the opportunity to work on this project and encouraged publication. Finally, sincerest thanks to David Brown, Director of the Jupiter Water Systems, for his patience and persistence in this challenging project.

Index

NOTE: *f.* indicates a figure; *t.* indicates a table.

A

B

G

H

I

J

N

O

P

U

V